Investigating Physical Geography

An Exercise Manual

INSTRUCTOR'S EDITION

Arthur N. Strahler
Alan H. Strahler

For use with *Elements of Physical Geography*, 4th Ed., 1989,
and *Modern Physical Geography*, 3rd Ed., 1987,
by the same authors, John Wiley & Sons, Inc.

John Wiley & Sons

New York • Chichester • Brisbane • Toronto • Singapore

ISBN 0-471-51681-3
Printed in the United States of America

10 9 8 7 6 5 4 3 2 1

PREFACE TO THE INSTRUCTOR'S

The Strahler exercise manual is largely a new work in terms of content, style, and format. About half of the material is entirely new with respect to previous Strahler exercise manuals. New exercises provide coverage of major text chapters and topics not formerly represented. Old exercises selected for inclusion have been revised, restructured, and reinforced. The manual contains over 100 exercises, for an average of about five per chapter of the textbook it serves.

Topics in the new manual conform in sequence with chapters in the Fourth Edition of ELEMENTS OF PHYSICAL GEOGRAPHY. In addition, each exercise is specifically keyed to the corresponding text and figures of Strahler and Strahler, MODERN PHYSICAL GEOGRAPHY, Third Edition, thus fully serving the needs of that textbook as well.

One aim of INVESTIGATING PHYSICAL GEOGRAPHY is to reach beyond the parent textbook in selected topics that lend themselves to problem-solving activities and illustrate the scientific method of research. Most exercises require the student to analyze and process data from nature to reveal processes of origin and change. Explanation and prediction are emphasized, calling for both inductive and deductive thinking. The manual is not a study guide to drill the student on all pertinent facts and processes covered in the text. The Strahler STUDY GUIDE fully covers this routine review function.

New introductory paragraphs relate the topic of each exercise to significant and familiar aspects of every-day life and include references to interesting places and to historical and current events. Environmental problems are referred to frequently and appear in the questions. Geographical relationships are stressed throughout the manual, requiring the student to locate on world maps in the textbook and manual many of the physical and cultural features referred to in the questions. Skills in cartographic analysis are continually reinforced.

The level of technical difficulty of the exercises shows a wide range and allows for selection to suit the individual student's capabilities. Most of the operations and questions are within the range of the average student, but special optional questions requiring basic mathematics, physics, or chemistry are offered as a challenge. Other optional questions call for ingenuity in problem analysis.

The exercise manual is ideally suited to home study because all information required to complete each exercise is found either in the manual or in the Strahler textbook; both being essential sources. Thus, all required tables, graphs, maps, and photographs are in these two sources. The manual provides all spaces and graphics needed to enter and calculate processed data and solutions. Equipment needs, described in the Introduction, are modest and minimal.

This Instructor's Edition of the exercise manual gives complete answers and solutions, including data entries on graphs, diagrams, and maps. Additional information offered with many of the answers adds background material the instructor will find useful.

Acknowledgments

Special thanks are due two reviewers who read the manuscript of the Instructor's Edition of this manual; both are on the faculty of Bellevue Community College, Bellevue, Washington: Professor Michael L. Talbott, Chairman, Social Science Division, and Professor Douglas Roselle, Chairman, Geography Department. The many suggestions they made for changes in the content and presentation of the exercises have materially improved the quality and usefulness of the manual.

ARTHUR N. STRAHLER
ALAN H. STRAHLER

Table of Contents

---

*Chapter title is that of the Strahler textbook, ELEMENTS OF PHYSICAL GEOGRAPHY, 4th Ed., 1989.

Chapter 3   Heat and Cold in the Life Layer

Chapter 4   Global Circulation of the Atmosphere and Oceans

Chapter 5   Atmospheric Moisture and Precipitation

Chapter 6   Air Masses and Cyclonic Storms

Chapter 7 The Global Scope of Climate

Chapter 8 Low-latitude Climates

Chapter 9 Midlatitude and High-latitude Climates

Chapter 10 Runoff and Water Resources

Chapter 19   The Soil Layer

Chapter 20   Environments of Natural Vegetation

INTRODUCTION

Investigating Physical Geography builds on your knowledge of physical geography developed through class lectures and study of your textbook. Not only will you put this knowledge to use in solving problems based on course topics, but also in reaching out in new directions to explore special topics not covered in the text. Some of these exercises are geographical journeys to study exemplary environmental features--landforms, climates, forest types, or soils, for example--in various parts of the globe. Other exercises are journeys of the mind into methods of doing scientific research; they require you to analyze sets of data in search of a better understanding of how the processes of nature act in varied global environments. In this way you will be taking part in a sampling of choice topics from several of the many fields of physical geography that are currently under intensive scientific investigation.

Each exercise is keyed to specific pages, illustrations, and tables in your textbook. Although many exercises contain all of the data required to solve the problems and answer the questions, for some you will need to refer to the textbook for working data in the form of photographs, tables, maps, graphs, and diagrams. In some cases, special explanations are provided in the manual for particular topics that extend beyond what your textbook covers.

Your exercise manual provides ample space in which to write out answers in full. Blank graphs for plotting data are provided, as well as diagrams and maps on which information is to be inserted. The perforated pages can be easily removed for convenience in doing the work and submitting it for evaluation.

You will need to have on hand a few essential tools and materials. Drawing equipment includes the following: Dividers with sharp points for measuring and laying off distances on maps and graphs. (A well-sharpened pencil compass will do.) Triangle of clear plastic, preferably 30°/60°/90°, at least six inches on the longest side. Protractor, spanning 180°. Color pens and/or pencils (red, blue, green, orange), needed for plotting point and line data and for coloring areas on maps and graphs. Tracing paper (8½" x 11" pad), required for several exercises.

A student atlas should be at hand as you work, and is essential for some of the exercises. (Recommended: a recent edition of Goode's World Atlas, Rand McNally, Publishers.) In addition, a good globe showing both physical and political features will be very helpful.

Exercise 1-A Latitude and Longitude
          [Text p. 11-12, Figure 1.5.]*

The geographic grid of parallels and meridians is an artifact, or cultural thing, created by humans as a concept originating in the brain. Does this mean that the geographic grid shouldn't be included in physical geography? Come to think about it, however, the earth's rotation actually describes the parallels and poles of our earth. Imagine a space creature who could hold its position in space just above the earth and direct a powerful laser beam directly down to the surface, searing a black line--a parallel--as the earth turned through 360 degrees. So all points on the earth's surface generate "natural" parallels as the earth turns. Not so for the meridians of longitude, for these are pure artifacts.

(1) Using a political globe, find each of the following cities and give as closely as possible the latitude and longitude of each. Estimate to the nearest degree or half-degree. (If a globe is not available, refer to any good atlas.)

New York, N.Y.	lat. __41°N__	long. __74°W__
Capetown, S. Africa	lat. __34°S__	long. __18½°E__
Shanghai, China	lat. __31°N__	long. __121½°E__
Honolulu, Hawaii	lat. __21°N__	long. __158°W__
London, England	lat. __51½°N__	long. __00°__
Rio de Janeiro, Brazil	lat. __23°S__	long. __43°W__

(2) What error has been made in each of the following notations of latitude and longitude? Encircle the error and explain below.

(a) Lat. 5°S, long. 191°W

Longitude exceeds maximum value of 180°. _____

_____

(b) Lat. 96°N, long. 88°E

Latitude exceeds maximum value of 90°. _____

_____

*Modern Physical Geography, 3rd Ed., p. 7-8, Figure 1.8.

(c) Lat. 21½°E, long. 177°E

Latitude cannot be "east." _____

_____

(d) Lat. 94°S, long. 103½°N

Latitude exceeds 90°. Longitude cannot be "north." _____

_____

(e) Lat. 48°N, long. 188°N

Longitude exceeds 180° and cannot be "north." _____

_____

(3) The total number of degrees of arc of latitude on the globe is

___180°__ .

(Note: The term <u>arc</u> refers to the angle in degrees of any part of a circle. There are 360 degrees of arc in a full circle.)

The total number of degrees of arc of longitude on the globe is

___360°__ .

Should your answer be the same for both latitude and longitude? Explain below:

No, because latitude treats each hemisphere (northern and southern) as a separate unit, each of which has only 90 degrees of arc of latitude. Doubling this value yields only 180 degrees of arc. Longitude treats the entire sphere as a unit; its circumference is a full 360 degrees of arc. _____

_____

_____

Minutes and Seconds

The use of latitude and longitude to designate position on the globe is attributed to Claudius Ptolemy (Second Century, A.D.), who probably followed a scheme invented earlier by Hipparchus. Ptolemy was the first to use the expressions "meridians of longitude" and "parallels of latitude." The ancient Chaldeans had long before divided the circle into 360 degrees, the degree into 60 minutes, and the minute into 60 seconds, but it from Ptolemy's adoption of these divisions, stated in Latin, that our words "minutes" and "seconds" come to us. Ptolemy called the 60 subdivisions of the degree "the first small parts," which in Latin is minutae primae, whence our word "minutes." The 60 subdivisions of the minute he called "the second small parts," which in Latin is minutae secondae, whence our word "seconds."

Using minutes and seconds, the full coordinates of latitude and longitude are stated as follows:

> lat. 34°12'31" N., long. 77°03'41" W.

This can be read as "latitude 34 degrees, 12 minutes, 31 seconds north, longitude 77 degrees, 3 minutes, 41 seconds west."

This rather archaic method, which involves awkward and time consuming calculations, has been relaced in modern science by a decimal system, using only the decimal parts of the degree. The above coordinates thus become:

> lat. 34.2086° N., long. 77.0614°W.

Warning: In the index of many modern atlases you will find what seems to be a decimal system. For example, Washington, DC, is listed as "38.54 N, 77.01 W." But when you scan down a long column of entries, the largest "decimal" part you can find is .59!

(4) One degree of latitude at the equator has a standard length value of 110.569 km (68.704 mi). What is the length of one minute of latitude? Of one second of latitude?

> Length of 1 min.: __1.8428__ km __1.1451__ mi
>
> Length of 1 sec.: __0.0307__ km __0.0191__ mi

(5) What is the length in meters and feet of one second of latitude?

> Length of 1 sec.: __30.7__ m __100.85__ ft

Exercise 1-B   How Long is the 60th Parallel of Latitude?
      [Text p. 10-12, Figures 1.4, 1.5.]*

Part A   Lengths of Parallels of Latitude

Thinking of the earth as a perfect sphere sets up an idealized geometrical model
for you to work with in this exercise. With text Figures 1.4 and 1.5 in view,
think about how meridians and parallels are geometrically alike and how they
differ. First, of course, the parallels are all true circles. (Disregard the two
poles, which are points.) Second, all meridians are half-circles and therefore of
equal length. While the parallels are full circles, they decrease in length
(circumference) from equator to either pole. In this exercise we pose the
following proposition: Somewhere in each hemisphere--northern and southern--is a
unique parallel whose length is exactly half the length of the equator.

The drawing below shows a quadrant of a circle representing the cross section of
one-half of the northern hemisphere. The quadrant is marked off into 10-degree
intervals of arc, corresponding to the positions of parallels ten degrees apart.

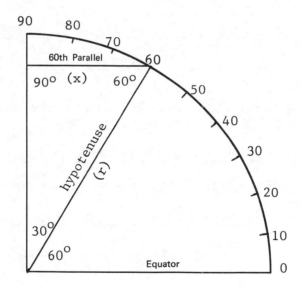

(1) Label the marked points from 0° through 90° to show latitude. Draw
horizontal, parallel lines through each of the numbered points on the quadrant.
Each is the radius of its respective parallel. Which radius appears to be half the
length of the radius of the equator?

The radius of the 60th parallel appears to be half that of the equator.

---

*Modern Physical Geography, 3rd Ed., p. 7-8, Figures 1.7, 1.8.

(2) How is the radius of a circle related mathematically to its circumference (length)? Give the equation.

Circumference, c, is equal to radius, r, times twice pi. (c = 2πr) (In other words, circumference is directly proportional to radius.)

(3) Prove the correctness of the answer you gave to Question 1. Make use of the diagram in any way you choose, drawing in whatever lines are needed and labeling distances or points. State your proof below.

[Two kinds of proof are acceptable here. For those with no knowledge of Euclidian geometry, direct measurements of the radii of the 60th parallel and equator can be made and compared. A compass is the easiest tool to use for this purpose. Set the compass points for the radius of the 60th parallel and apply them to the equatorial radius. A second and more rewarding proof is to create a 30°-60°-90° triangle, as shown, and cite the theorem that the hypotenuse is double the length of the shorter leg. A third step is contained in Question 4.]

(4) (Special credit question.) Can you use trigonometry to complete your proof? State the equation required.

The sine of an angle, α, is defined as the ratio of radius, r, to the shorter leg of a right triangle, y. Thus, sin α = y/r. The sine of 30° is 0.5. Hence, the radius of the 60th parallel is one-half that of the earth's radius, which is identical to the radius of the equator.

Part B Speed of the Earth's Rotation

We turn next to the speed of earth rotation, defined as the speed in miles per hour (or kilometers per hour) of a fixed point on the earth's surface in its circular path; i.e., the "linear velocity." That path is, of course, identical with a parallel of latitude. To make things as simple as possible, assume that the earth turns exactly 360° in 24 hours and that the length of the equator is exactly 25,000 mi. (For metric units, use 40,000 km.)

(5)   What is the speed of rotation of a point on the equator? at 60°? (Round off to nearest mi or km.)

At equator: <u>1041</u>   mi/hr,   <u>1667</u>   km/hr

At 60°:   <u>521</u>   mi/hr,   <u>833</u>   km/hr

A special mathematical problem: Keeping pace with the sun.

(6) Flying from Boston to San Francisco, you depart Boston when the sun is at its highest point in the sky; i.e., noon by local time. What average ground speed must you observe in order to touch down in San Francisco when it is precisely noon there? Assume for simplicity that your flight path is equivalent to following the 40th parallel. Use 71°W as the longitude for Boston; 122°W, for San Francisco. Show your calculations.

Longitude difference = 51°. Cosine of 40° = 0.766. Distance per degree of longitude at equator: 25,000 mi ÷ 360° = 69.4 mi. Distance per degree of longitude at 40°N: 69.4 X 0.766 = 53.2 mi. Total distance: 53.2 X 51° = 2,713 mi (round to 2,700). Sun's "travel time" at 4 minutes per degree: 4 X 51° = 204 min. = 3.4 hr. Divide sun's travel time by distance: 2,700 mi ÷ 3.4 hr = 794 m.p.h. Round to 800 m.p.h.

Name _____ Date _____

_____   _____

**Exercise 1-C How to Travel North, East, West, and South
        --and Be Back Where You Started**
        [Text p. 11-12, Figures 1.4, 1.5.]*

This is your question: From how many different starting points on the globe would it be possible to travel 100 km north, then 100 km east (or west), then 100 south, and be exactly at the starting point? The southern-hemisphere case is simple but can you solve this problem fully for the northern hemisphere?

Use the two circles below to construct your solution to the problem for each hemisphere.

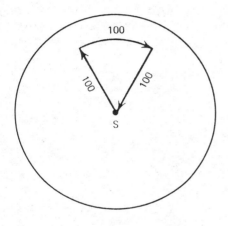

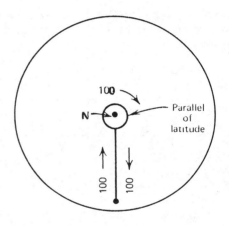

a. Southern hemisphere   b. Northern hemisphere

Use the space below to explain fully:

(a) Southern hemisphere: Our starting point is the south pole. Travel north on any meridian, then east or west on a parallel, returning on a meridian. The track will be an equilateral spherical triangle. (b) Northern hemisphere: There exists a parallel whose circumference is exactly 100 km (the radius of this parallel is $100/\pi$). The starting point will be located 100 km south of the 100-km parallel. Traveling north 100 km brings us to the 100-km parallel. After traveling either east or west 100 km we return to the same meridian. Tracking south 100 km brings us back to the starting point. There is an infinite number of points on

*Modern Physical Geography, 3rd Ed., p. 7-8, Figures 1.7, 1.8.

that parallel on which the starting point is located. Consider next that there exists a parallel exactly 50 km in circumference. This parallel can be circled twice to yield the required 100 km of eastward travel. A parallel that is 33 1/3 km in circumference can be circled three times, etc. The number of possible parallels form an infinitely long series. Therefore the requirements can be satisfied from an infinite number of starting points.

For the mathematical genius:

Derive a mathematical equation that will express fully and precisely the complete solution of the problem for the northern hemisphere. (Use a separate page, if necessary.)

Name _____   Date _____

_____   _____

Exercise 1-D Standard Time and World Time Zones
[Text p. 12-13, Figure 1.6, 1.7][*]

In this day of satellite television, when the Olympic Games can be viewed live from Korea, Japan, or Australia, nearly halfway around the world, we become very conscious of time differences from place to place, varying according to the differences in longitude. American presidential election polls, closing three hours sooner in the Northeast than in the Far West, pose a problem for voters and politicians alike. Advance word that a candidate is winning in the east can discourage California voters from even going to the polls. The artifact of global standard time is a purely cultural phenomenon, but it has long held a place in physical geography.

We investigate first the standard times of cities in the United States and Canada. Use the time-zone map, Figure 1.7, p. 13, in your textbook. If you don't know the location of the city, look it up in an atlas.

Use the following code designation of zones:

    AST   Atlantic Standard Time
    EST   Eastern Standard Time
    CST   Central Standard Time
    MST   Mountain Standard Time
    PST   Pacific Standard Time

(1) Give the time zone code and the standard meridian (in degrees of longitude) for each of the following places:

	Time Zone	Standard Meridian
Cleveland, OH	EST	75°W
Las Vegas, NV	PST	120°W
El Paso, TX	CST	90°W
Chicago, IL	CST	90°W
Denver, CO	MST	105°W
Montreal, PQ	EST	75°W
Halifax, NS	AST	60°W
Vancouver, BC	PST	120°W
Labrador	AST	60°W

[*]Modern Physical Geography, 3rd Ed., p. 35-38, Figures 2.17-2.19.

(2) Of the four time zones shown on the map, Figure 1.7, which one spans the greatest extent in latitude within the 48 contiguous United States ? ___EST___ . The estimated maximum width (extent) of this zone in degrees of longitude is ___23°___ .

(3) Find El Paso, Texas, and note its position in the CST zone. Find the location of Albuquerque, New Mexico, almost due north of El Paso. Compare clock times in these two cities:

Clocks in Albuquerque will read one hour earlier than those in El Paso. For example when it is 4 P.M. in El Paso, it is 3 P.M. in Albuquerque.

_____

(4) Clock differences between cities: When it is 11 AM in Los Angeles, the time in New York City is ___2 PM___ . When it is midnight in Pittsburgh, the time in San Francisco is ___9 PM___ . When it is 3 PM in New York City, it is ___10 AM___ in Honolulu, Hawaii.

Standard time throughout the world. Refer to the world map of time zones reproduced here on the oppoosite page. Designate the zones in terms of <u>hours fast</u> (minus sign) and <u>hours slow</u> (plus sign). Greenwich time (zero hours) is based on the Greenwich Meridian (zero longitude).

(5) Give the time zone for each for the following nations:

Japan	___-9___
Iceland	___0___
Malagasy Republic	___-3___
Iran	___$-3\frac{1}{2}$___
Chile	___+4___
India	___$-5\frac{1}{2}$___
Newfoundland	___$+3\frac{1}{2}$___
New Zealand	___-12___

Time zone map of the world. (U.S. Navy Oceanographic Office.)

(6) Which nation has the greatest number of time zones?   U.S.S.R.

How many zones are in that nation?   11

Special project: Reading simultaneous hour and day at different places in the world.

Text Figure 1.6 is reproduced below in two parts. Cut out both disks and center the smaller over the larger. Use a pin or thumbtack through the center points to fasten the disks to a base of cardboard or Styrofoam. Rotate the inner disk as needed to match the specified latitude zone with the specified hour. Count hours westward from the given city to the other city.

(7) When it is 10 AM Wednesday in Sydney, what time and day is it in London, England?   12 M Wed.   . In Anchorage, Alaska?   2 PM Tues.   .

(8) When it is 8 AM Friday in Los Angeles, what time and day is it in Tokyo, Japan?   12 N Sat.   .

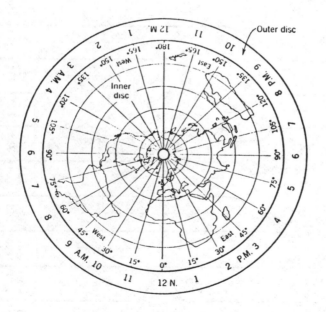
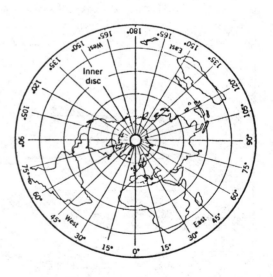

Exercise 1-E   Equinoxes and Solstices
[Text p. 14-17, Figures 1.10, 1.11, 1.12, 1.13]*

The diagram below represents a view of the earth from a point above the plane of the orbit (plane of the ecliptic). Label each equinox and solstice and give date. Add arrows to show direction of earth rotation and revolution. The dashed line halving each globe represents the circle of illumination. Shade the dark half of each globe. Mark the points of noon, midnight, sunrise, and sunset with the letters N, M, R, and S, respectively.

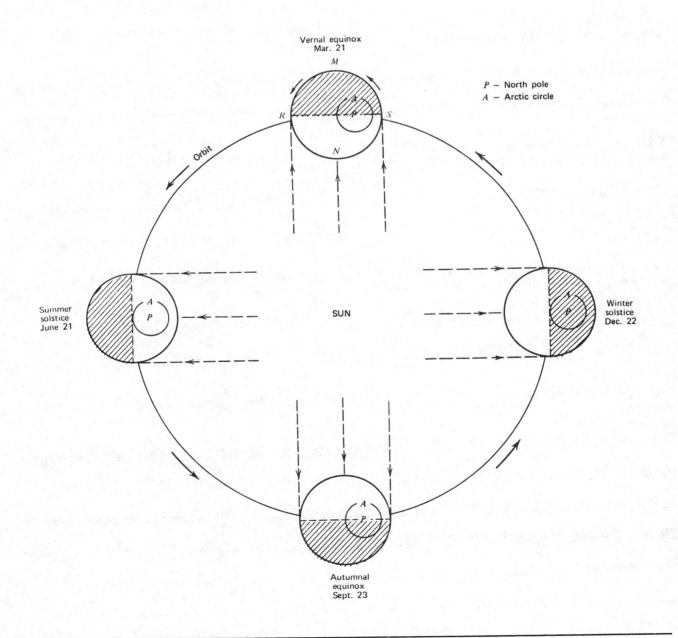

*Modern Physical Geography, 3rd Ed., p. 27-31, Figures 2.7-2.11.

(1) On what dates does the circle of illumination pass through the north pole? (Give name of solstice or equinox, and date)

        Vernal equinox   Date   March 21

        Autumnal equinox   Date   September 23

(2) On what dates is the circle of illumination exactly tangent to the Arctic Circle?

        Summer solstice   Date   June 21

        Winter solstice   Date   December 22

(3) At summer solstice, what portion of the earth's surface is exposed to the sun's rays for the entire day of 24 hours?

        All the area lying north of the Arctic Circle.

(4) At winter solstice, what portion of the earth's surface is exposed to the sun's rays for the entire 24 hours?

        All the area lying south of the Antarctic Circle.

(5) On what dates are the lengths of day and night exactly equal for all points on the globe? (Disregard effects of the earth's atmosphere.)

        Vernal equinox   Date   March 21

        Autumnal equinox   Date   September 23

(6) Justify your answer with a statement of geometrical proof, evident from the diagram.

On equinox dates, the circle of illumination passes through the two poles and must divide every parallel of latitude into two equal parts. Therefore, a given point on any given parallel spends exactly one-half of the 24-hour rotation period on the illuminated side and one-half on the dark (shadowed) side of the circle of illumination.

Exercise 1-F Comparing Map Projections
[Text p. 18-22, Figures 1.15, 1.16, 1.17.]*

When selecting a world map on which to present research data, a geographer must chose wisely, because each projection has its unique set of advantages and disadvantages. How much do you know about the three projections described in your textbook?

Listed below are statements describing a particular property of quality or quality of a map projection. For each of the three projections named, place a check mark to indicate that the property applies to that projection; leave blank if it does not apply.

	Polar Stereographic Projection	Mercator Projection	Goode's Homolosine Projection
(a) An equal-projection.	____	____	X
(b) A conformal projection.	X	X	____
(c) All the parallels are true circles.	X	____	____
(d) All parallels are straight, parallel lines.	____	X	X
(e) All the meridians are straight lines.	X	X	____
(f) Both meridians and parallels are straight lines.	____	X	____
(g) All meridians except the central meridians are curved lines.	____	____	X
(h) All the parallels and meridians intersect each other at true right angles.	X	X	____
(i) Neither pole can be shown.	____	X	____
(j) Only one pole can be shown on a single map.	X	____	____

---

*Modern Physical Geography, 3rd Ed., p. 12-17, Figures 1.17, 1.19, 1.21.

	Polar Stereographic Projection	Mercator Projection	Goode's Homolosine Projection
(k) Both poles can be shown on a single map.	____	____	__X__
(l) Map outline is circular when only one hemisphere is shown.	__X__	____	____
(m) Any straight line drawn on the map is a true rhumb line (loxodrome).	____	__X__	____

Exercise 1-G   Great-Circle Routes and Rhumb Lines
[Text p. 20-21, Figure 1.16.]*

Today, intercontinental flights of commercial aircraft are guided by automatic electronic systems, using satellites for navigation, and with radar and inertial systems automatically keeping the aircraft on the desired course. Not so in the earlier days of transoceanic flying in the period around World War II. Pilots of the pioneering Pan American Clipper aircraft, for example, carried out their navigation much as had the navigators of steam-powered oceangoing ships for decades before them. On the navigation chart, a great circle route was first laid out between origin and destination, then divided up into a series of straight legs, each one a rhumb line, or course of constant compass direction. At the end of each leg, an abrupt change of course was made. There was plenty of guesswork involved.

Figure 1.16 of your text is a Mercator map projection showing both great-circle routes and rhumb lines (loxodromes) between the same two points on the globe. A navigator following the great-circle route is taking the shortest surface distance between two points, but the compass direction in most cases is continually changing. Another navigator, following the rhumb line, is holding to a course of constant compass direction, but in most cases the distance will be longer than the great-circle route.

You are asked to plot and compare the two types of navigational routes. For this purpose you need both the Mercator chart and a special map projection called a Great-Circle Sailing Chart. (It is a gnomonic map projection.) Both are reproduced here. The great-circle sailing chart is in two sections, each designed to show the chosen route to best advantage. The following questions test your understanding of the concepts involved in the two kinds of map presentation.

(1) Examine the great-circle route for a commercial jet flight between Portland and Cairo. Name several regions, islands, or nations over which the route passes:

Alberta, northern Baffin Island, central Greenland, southern Norway and

Sweden, Poland, Romania, Bulgaria, western Turkey
_____

_____

(2) At what point in your flight would you be traveling due east? On both maps mark the point with a dot and label "E". Give the location of the point:

Lat.   75°N   Long.   44°W

*Modern Physical Geography, 3rd Ed., p. 12-17, Figure 1.19.

(3) When you depart Portland, in what compass direction are you headed?

___NNE___. As you pass over Sweden, what is your compass direction? ___SW___

(4) If a globe is available, estimate the distance between Portland and Cairo on the great circle route. Use a strip of paper tape to mark off the length of route; move the tape to a north-south orientation (along a meridian) and determine the arc in degrees of latitude; then multiply by 111 km.

___11,000___ km

(5) Examine the alternative routes from Rio de Janeiro to Capetown. Explain why the rhumb line is only slightly longer than the great circle, as compared with the two lines for Portland-Cairo.

The latitude is low; the direction is nearly west-east.

(6) In view of your explanation in Question 5, can you predict where on the globe the great-circle route and rhumb line are one and the same line (two answers)?

Along the equator and along any meridian.

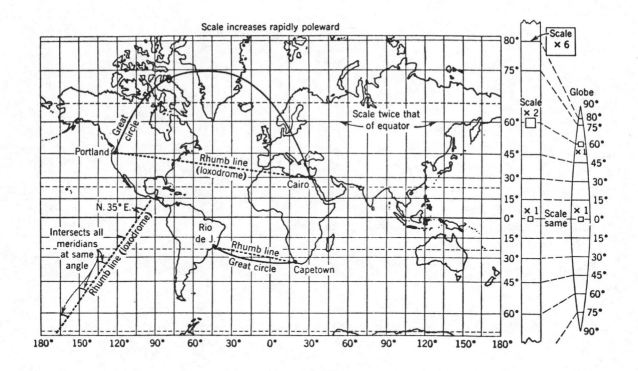

A. Mercator chart

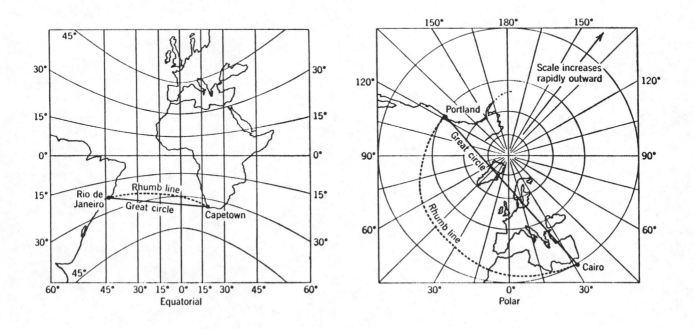

B. Great-circle sailing chart

Exercise 1-H   Using Graphic Scales to Measure Distances
[Text p. 22-23, Figure 1.18.]*

For this exercise, we reproduce text Figure 1.18. The three graphic scales are placed below the map; you may cut them apart into three separate strips for measuring distances on the map. To measure distance, place a division mark of the scale on one point, such that the zero point falls short of the second map point and the remaining distance falls within the subdivision scale at the left. The scale below shows how this is done:

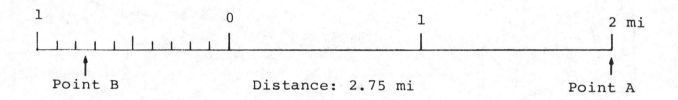

(1) What is the airline distance in kilometers between the Lighthouse and Santa Barbara Point?

___1.9___ km

(2) What is the length in feet of the pier named "Stearns Wharf"?

___1,850___ ft

(3) What is the distance in miles between the beacon on Point Castillo breakwater and Santa Barbara Point?

___0.95___ mi

(4) Calculate the surface area in square kilometers within the limits of the map. (Measure length and width and multiply the two numbers.)

Length ___3.96___ km;   Width ___2.50___ km;   Area ___9.90___ sq km

_____

*Modern Physical Geography, 3rd Ed., p. 482, Figure A.III.9.

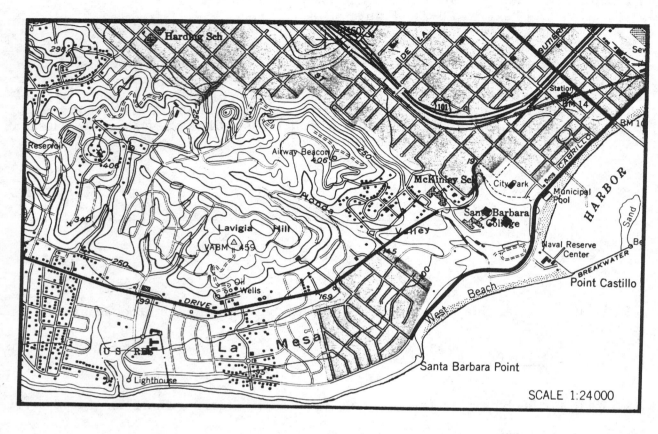

SCALE 1:24000

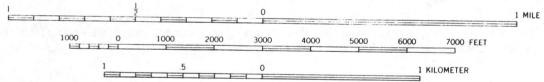

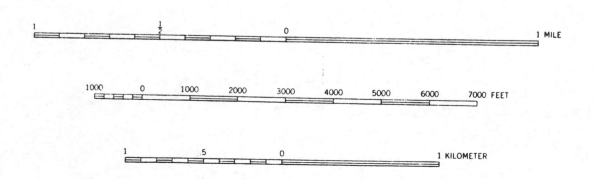

Exercise 1-I The United States Land Office Grid System*

Beginning in 1786, public lands lying west of the colonial states were surveyed and subdivided into units called <u>congressional</u> <u>townships</u>, which are squares 6 miles on a side. Thomas Jefferson is said to have proposed this plan to the Congress, and it was implemented under Thomas Hutchins, who had been appointed Geographer of the United States. At times, the surveyors' work had to be suspended because of danger of attack by hostile Indians. The Land Office grid system still dominates many cultural features of the landscape of the Middle West, the Great Plains, and valleys of the Far West. Patterns of highways, roads, city streets, farms, and counties follow this grid system.

Figures A, B, and C show details of the U.S. Land Office Grid System.

A. Townships are designated according to position north or south of a <u>base line</u> and east or west of a <u>principal meridian</u>. Note that the tiers of townships are numbered upward and downward from the base line (T.1 N., T.2 N.; T.1 S., T.2 S., etc.), while the vertical columns of townships are designated as <u>ranges</u>, numbered eastward and westward from the principal meridian (R.1 E., R.2 E.; R.1 W., R.2 W., etc.).

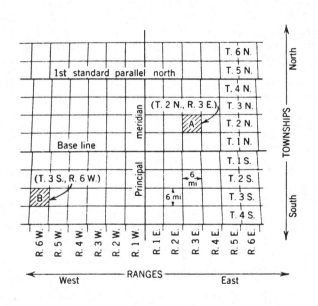

B. A township is divided into 36 <u>sections</u>, each one a square mile. Notice the unusual pattern of numbers, starting in the upper right corner.

R. 6 W.

6	5	4	3	2	1
7	8	9	10	11	12
18	17	16	15	14	13
19	20	21	22	23	24
30	29	28	27	26	25
31	32	33	34	35	36

T. 2 S.

*Modern Physical Geography, 3rd Ed., p. 485-7, Figures A.III.17-20.

C. A section may be subdivided into many units. This map shows the method of designating the various possible subdivisions, together with the area in acres of each part.

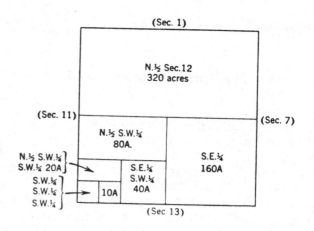

Printed below is a portion of the Redfield, South Dakota, Quadrangle surveyed nearly a century ago by the U.S. Geological Survey. It show sections of land and a standard parallel. Almost every section boundary is occupied by a road (double-line symbol) or a railroad. Convoluted lines are topographic contours and stream channels.

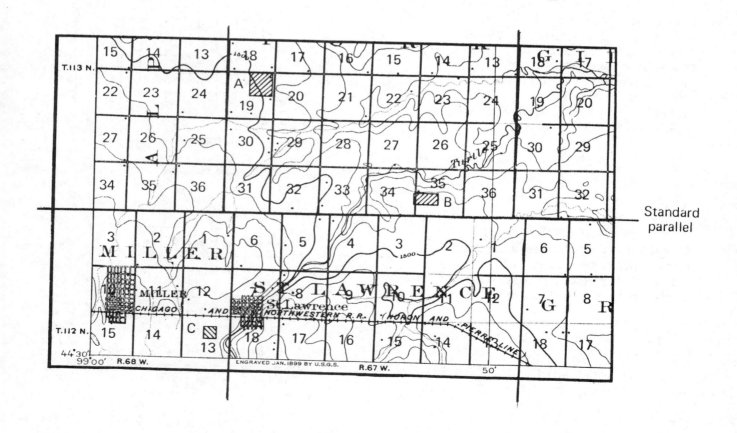

(1) Find the township boundaries and highlight them with horizontal and vertical color lines. (Boundaries are shown by long dashes printed over the symbol for roads.) Why do you suppose the two towns--Miller and St. Lawrence--happened to grow up so close to each other?

(a) They lie in different townships.(b) It was essential that both towns be located on the railroad. Otherwise, no satisfactory reason is obvious.(Today, Miller is much the larger of the two and is the county seat of Hand County)

(2) On the map, number all sections with their correct numbers according to the established system shown in Figure B on the previous page.

(3) Find and label the <u>standard parallel</u> on the map. Can you explain why north-south section and township boundaries are offset along this parallel? Hint: Are meridians of longitude truly parallel to each other on the globe?

Because the range lines on eastern and western boundaries of townships are meridians converging slightly as they are extended northward, the width of townships is progressively diminished in a northward direction. At each standard parallel the base line is resurveyed, and hence does not match the narrowed units south of it

(4) Speculate as to why the sections in the tier located beneath the word "MILLER" (large bold letters) are more than one mile in north-south extent.

The standard parallel is surveyed as an independent line from east to west, while the township grid is surveyed from south to north, using a different standard parallel. This discrepancy reflects errors in land surveying over long distances.

(5) State accurately and fully the location of the letter "T" in the word "Turtle" on the map. (Refer to Figure C on the previous page.)

S $\frac{1}{2}$ SE $\frac{1}{4}$ of Sec. 26, T. 113 N, R. 67 W

(6) Locate on the map each of the parcels of land described below. Carefully draw in the boundaries of each parcel and label with the corresponding letter.

Parcel A: NE $\frac{1}{4}$ of Sec. 19, T.113 N, R.67 W

Parcel B: N $\frac{1}{2}$ of SW $\frac{1}{4}$ of Sec. 35, T.113 N, R.67 W

Parcel C: SW $\frac{1}{4}$ of NE $\frac{1}{4}$ of Sec. 13, T.112 N, R.68 W

Name _____   Date _____

_____   _____

Exercise 2-A   Insolation and Latitude
              [Text p. 37, Figure 2.8.]*

In this exercise we investigate the way in which incoming solar energy, or insolation, diminishes from full strength at the equator to zero at the two poles. The conditions specified are that the earth is at either the vernal equinox or the autumnal equinox, when the sun's noon rays strike the equator at an angle of 90° (a right angle) with respect to a horizontal flat surface. We assume that no atmosphere exists to diminish the intensity of insolation. Examine Figure 1.11 in your textbook to get the picture.

The way in which insolation varies with angle made by the sun's rays is illustrated in Figure 2.8 of your textbook. Study the figure closely and read the text that goes with it (p. 37).

Figure A develops in greater depth the relationship between latitude and the intensity of insolation. Notice, first, that the angle of latitude plus the angle between the sun's rays and the horizontal surface ("angle of incidence") always equals 90°:

Angle of latitude	Angle of incidence	Right angle
0°	90°	90°
30°	60°	90°
60°	30°	90°
90°	0°	90°

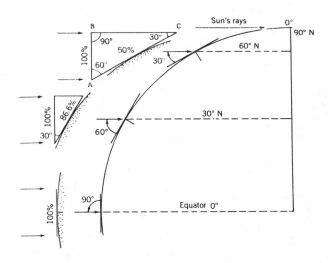

Figure A

Intensity is given as percentage of the maximum possible, which is 100%, and applies at the equator. At either pole, the sun's rays barely graze the surface, but do not strike it, so there the value is 0%. In Figure A, right triangles are drawn at 30°N and 60°N. Side AB cuts across the sun's rays at right angles and represents 100% intensity. The hypotenuse of the triangle, side AC, represents the horizontal ground surface. As latitude increases, AC becomes longer, indicating that the insolation is becoming spread out over a larger area, i.e., that it is becoming less intense. Notice that in these triangles, the angle A is the same as the latitude angle; the angle C is the same as the angle of incidence.

---

*Modern Physical Geography, 3rd Ed., p. 56-58, Figure 4.5.

EX. 2-A

The method we use in this exercise is to construct triangles like those in Figure A for the latitudes shown in Figure B, and on them make a direct measurement of the length of each hypotenuse (AC). Use Diagram C for this purpose. You will need a protractor to construct the latitude angles correctly. Diagram C provides a scale with which to measure length of each hypotenuse. One unit on this scale is equal to the length of AC, which is constant for all latitudes. Enter these measurements in the table provided below and calculate the percentage. The problem is solved for latitude 45°.

Lat.	AB	AC	AB/AC	x100
0°	1.00	1.00	1.00	100%
15°	1.00	1.03	0.97	97%
30°	1.00	1.16	0.86	86%
45°	1.00	1.41	0.71	71%
60°	1.00	2.0	0.50	50%
75°	1.00	3.9	0.26	26%
90°	1.00	(infinite)	0.0	0%

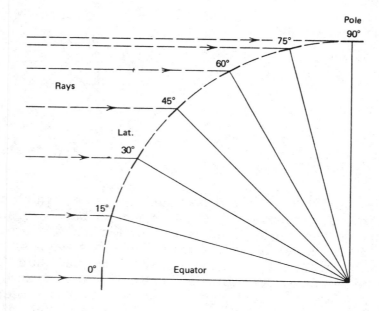

Figure B

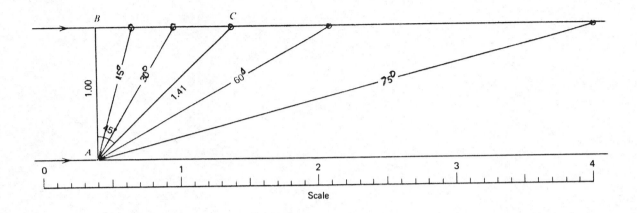

Figure C

Optional method using trigonometry. (Special credit.)

The values for each latitude can be found more easily and accurately by use of trigonometry, as shown in Figure D. Using a table of the values of the cosine of an angle, fill in the missing values in the table below.

Lat.	Cos lat.	x 100
0°	1.000	100.0%
15°	0.966	96.6%
30°	0.866	86.6%
45°	0.707	70.7%
60°	0.500	50.0%
75°	0.259	25.9%
90°	0.000	0.0%

By trigonometry:
At lat. 45° N

$$\frac{AB}{AC} = \cos\ 45° = 0.707$$

Figure D

_____ _____

Exercise 2-B The Annual Cycle of Insolation
[Text, p. 37-38, Figures 2.11, 2.12.]*

The annual cycle of insolation at a given location on the globe depends on two factors: (a) the latitude at which the observer is located and (b) the sun's changing angle above the horizon at noon.

You are asked to show the sun's changing angle throughout the year on Graph A on the following page. The angle used on the vertical scale of the graph is known as the sun's declination, which for our purposes can be thought of as that global parallel of latitude over which the sun at noon is in the zenith position (straight up in the sky). This condition occurs at the equator twice yearly on the dates of the two equinoxes (March 21 and September 23). The sun is directly overhead at noon over the tropic of cancer, lat. $23\frac{1}{2}$°N., at summer solstice, June 21; over the tropic of capricorn, lat. $23\frac{1}{2}$°S, at winter solstice, December 22.

Table A gives the sun's declination angle at ten-day intervals throughout the year. Plot these points on the blank graph (B) and connect them with a smooth curve. This is a kind of mathematical curve called a sine curve or sine wave. Label the horizontal line of 0° as "Equator." Draw horizontal lines at $23\frac{1}{2}$°N and $23\frac{1}{2}$°S; label them "Tropic of Cancer" and "Tropic of Capricorn."

Table A. Sun's Declination Throughout the Year

Date		Declination (degrees)	Date		Declination (degrees)
Jan	1	23 S	Jul	10	$22\frac{1}{2}$
	10	22		20	21
	20	20		30	$18\frac{1}{2}$
	30	$17\frac{1}{2}$	Aug	10	16
Feb	10	15		20	$12\frac{1}{2}$
	20	11		30	9
Mar	1	8	Sep	10	5
	10	$4\frac{1}{2}$		20	$1\frac{1}{2}$
	20	$\frac{1}{2}$		30	$2\frac{1}{2}$ S
	30	$3\frac{1}{2}$ N	Oct	10	$6\frac{1}{2}$
Apr	10	$7\frac{1}{2}$		20	10
	20	11		30	$13\frac{1}{2}$
	30	$14\frac{1}{2}$	Nov	10	17
May	10	17		20	$19\frac{1}{2}$
	20	20		30	$21\frac{1}{2}$
	30	22	Dec	10	23
Jun	10	23		20	$23\frac{1}{2}$
	20	$23\frac{1}{2}$			
	30	23			

*Modern Physical Geography, 3rd Ed., p. 56-58, Figures 4.7, 4.8.

The intensity of insolation throughout the year at various latitudes is given in Table B. The value of one unit used in this table is 889 gram calories per square centimeter, or 889 langleys. (See text p. 35 for explanation.) On the blank graph (B), plot these monthly values for the following latitudes: 0° (equator), 20°N, 40°S (south), 60°N, 90°N. Enter the data point in the middle of the month as shown on the partially drawn line for 60°N. Connect the points with a smooth curve and label the latitude.

Table B. Insolation Throughout the Year

Latitude		Jan.	Feb.	Mar.	Apr.	May	Jun.	Jul.	Aug.	Sep.	Oct.	Nov.	Dec.	Total for year
	90	...	...	1.9	17.5	31.5	36.4	32.9	21.1	4.6	...	...	...	145.9
°N	80	...	0.1	5.0	17.5	30.5	35.8	32.4	20.9	7.4	0.6	...	...	150.2
	60	3.0	7.4	14.8	23.2	30.2	33.2	31.1	24.9	16.7	9.0	3.8	1.9	199.2
	40	12.5	17.0	23.1	28.6	32.4	33.8	32.8	29.4	24.3	18.4	13.4	11.1	276.8
	20	22.0	25.1	28.6	30.9	31.8	32.0	31.8	30.9	28.9	25.8	22.5	20.9	331.2
Equator		29.4	30.4	30.6	29.6	28.0	27.1	27.6	28.6	30.1	30.2	29.5	28.9	350.0
	20	33.8	32.2	29.0	24.9	21.2	19.6	20.5	23.7	27.7	31.1	33.3	34.1	331.1
°S	40	34.8	30.4	23.9	17.4	12.5	10.4	11.6	15.8	21.9	28.5	33.6	36.0	276.8
	60	33.0	25.3	16.0	8.1	3.3	1.7	2.7	6.5	13.6	22.6	31.1	35.3	199.2
	80	34.2	20.5	6.3	0.3	...	...	...	...	3.8	16.0	31.0	38.1	150.2
	90	34.7	20.7	3.2	...	...	...	...	...	1.0	15.6	31.5	38.7	145.4

(1) Study the insolation curves in relation to the annual curve of the sun's declination. Compare times of maximum and minimum values at 60°N and 40°S. Explain how these two curves differ in timing.

The curve for 60°N shows a perfect correspondence in time of maximum and minimum with the insolation curve. For 40°S the curves are similar, but exactly out of phase. The curve for 90°N has no insolation from autumnal equinox to vernal equinox. At 20°N, the curve flattens near June solstice because the sun's declination exceeds that latitude slightly for several weeks.

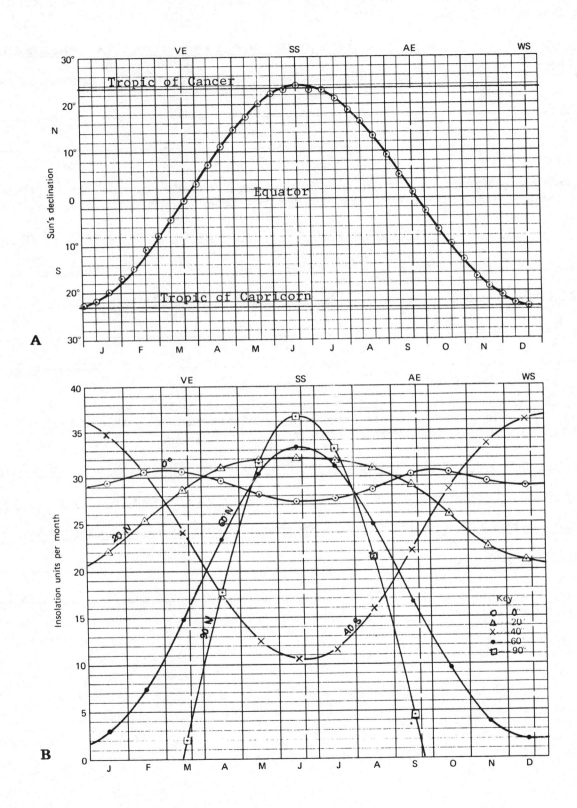

A

B

(2) Explain how the maxima and minima of insolation at the equator are related to the curve of the sun's declination.

The noon sun is at zenith over the equator at each equinox. The sun's noon altitude is less ($66\frac{1}{2}°$) at each solstice. Hence the curve has two maxima and two minima.

(3) What disadvantage lies in the use of calendar months for analysis of the yearly cycle of insolation? Cite specific figures from the data table..

Calendar months differ in number of days. A difference of more than 3% in insolation can be attributed to this cause. At the equator, March (31 days) has greater insolation than September (30 days); December (31 days) has a greater value than June (30 days).

(4) Special credit question. Why does the south pole receive more insolation in January then the north pole receives in July? (Information required is not in your textbook. Hint: The earth's orbit is an ellipse.)

Perihelion occurs early in January; aphelion early in July. Both months consist of 31 days each. Therefore, the difference is largely due to the fact that the earth is closer to the sun in January. (See Strahler: Modern Physical Geography, 3rd Ed., p. 27, Figure 2.5.)

Exercise 2-C   Duration of Sunlight at Different Latitudes
        [Text p. 38-39, Figure 2.10.]*

How long is the sun above the horizon at a given latitude and season? Is there romance in the "Land of the Midnight Sun"? Tourists by the thousands take the North Cape cruise to the northernmost tip of land in Norway, 71°N. There, if the weather cooperates, you can take a picture of the sun well above the horizon at midnight. The duration of daylight is an important factor in many aspects of everyday life, especially at high latitudes--the arctic and polar zones--where summer-to-winter inequalities of day length and night length are very great.

(1) Figure 2.10 of your textbook shows the path of the sun in the sky at 40°N latitude, which is about the latitude of Philadelphia, Indianapolis, Denver, and Salt Lake City. By simple addition, using the given times of sunrise and sunset, or by counting the hour intervals along each path, give the approximate number of hours of sunlight daily for each of the following points in the yearly cycle:

	Month and day	Number of hours
Equinox	March 21, Sept. 23	12
Summer solstice	June 21	15
Winter solstice	Dec. 21, 22	9

The graph on the next page shows the number of hours each day that the sun is above the horizon for selected north latitudes from the equator to the arctic circle. Check the number of hours you entered in the above blanks for 40°N by reading the hours from the graph for that latitude.

(2) On the graph, sketch in the curves for latitudes 70°, 80°, and 90°N.

(3) At the north pole, for how many consecutive months (or days) will the sun be above the horizon continuously for 24 hours of the day?

        6   months;   182   days

Same for the duration of darkness (defined as sun below horizon).

        6   months;   182   days

(Note: Twilight, which supplies important amounts of illumination when the sun is within a few degrees below the horizon, is not considered in making this answer.)

_____

*Modern Physical Geography, 3rd Ed., p. 29, 56-58, Figures 2.9, 24.4.

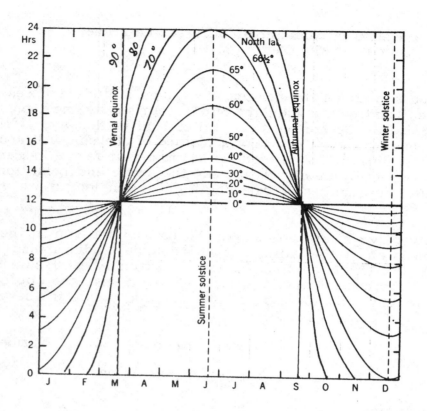

(4) If we define "day" as the period when the sun is continuously above the horizon and "night" as the period when it is below the horizon, how many days and nights per year are experienced at the north pole?

        <u> 1 </u> days, <u> 1 </u> nights

(5) What effect does length of daily sunlight period have on the summer growing season for food crops, such as grains and vegetables?

<u>Long days provide the added sunlight period needed for photosynthesis by green</u>

<u>plants. The long days in the arctic partly make up for the short summer growing</u>

<u>season.</u>

(6) What factors of climate would tend to modify the effects of increase in number of sunlight hours with increasing latitude?

Although at latitudes 60° to 65°N. the number of daylight hours is from 18 to 21, the sun's angle above the horizon is comparatively low throughout the day (as compared to low latitudes). In sum, the gain in insolation for the month of July (typically the warmest month), going poleward from 20°N to, say, 60°N is not significant, and actually decreases rapidly in August. Consult Table B in Exercise 2-B for figures.

Name _____ Date _____

_____   _____

Exercise 2-D   Remote Sensing--Interpreting Infrared Photographs
      [Text p. 49, Figures 2.21, 2.22, 2.23.]*

You will make use of text Figure 2.22, a color infrared photograph, in completing this exercise. Trim a sheet of tracing paper or acetate to a width of $7\frac{1}{2}$ in. and a height of 8 in. Place the sheet over Figure 2.22 and hold it in place by pieces of pressure tape at the top. Mark the corners of the photo.

Review on p. 49 of your textbook the significance of the colors seen on this kind of infrared photograph. Read carefully the caption of Figure 2.22.

(1) Red color on the photo is a response to reflected __infrared__ light; green

color is produced by ___red___ light; and blue color by ___green___ light.

(2) On the photograph, Figure 2.22, how is the red color in rectangular patches best interpreted?

Healthy growing vegetation, such as field crops. _____

_____

(3) What is the significance of the dark-brown to black rectangles and squares (labeled A), in the upper left part of the photograph?

Saturated soil is indicated. This is an effect of waterlogging, resulting from

prolonged irrigation and poor drainage. _____

(4) What is the significance of the white areas, labeled B, and other white squares and rectangles between the red areas?

Salts have been precipitated at the soil surface as a result of prolonged

evaporation of irrigation water. _____

_____

(5) On the tracing sheet, draw lines to delimit areas of (A) waterlogging, (B) salinization, (C) fields of growing crops mixed with salted fields, (D) uncultivated or abandoned land, (E) Hilly areas with numerous ravines. Label the major highway and irrigation canal crossing the area.

Remove the tracing sheet and fasten it over the blank rectangle on the next page.

_____

*Modern Physical Geography, 3rd Ed., p. 69, Plates A.2, A.3.

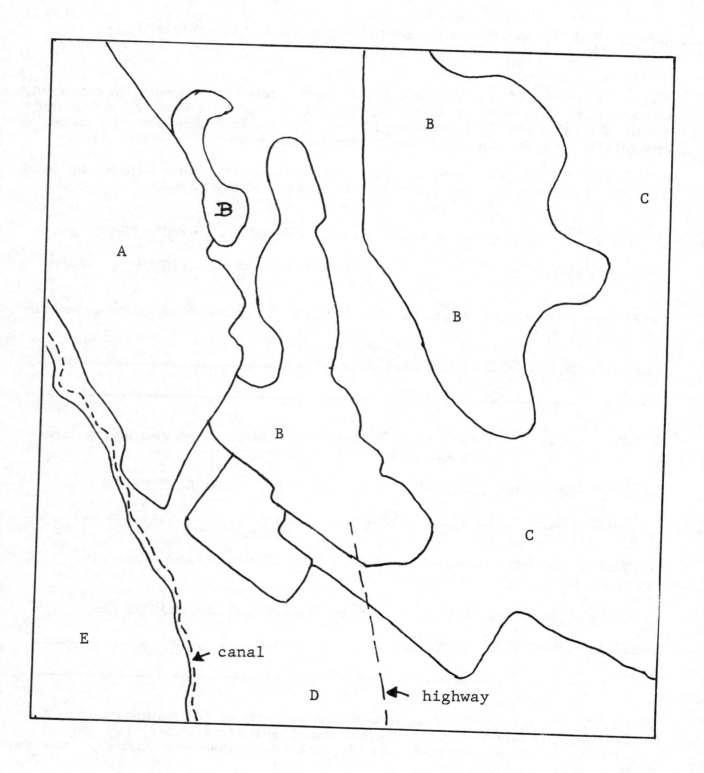

canal

highway

Name _____ Date _____

_____   _____

Exercise 2-E   Remote Sensing: Imagery of Emitted Infrared Radiation
[Text p. 55-56, Figure 2.27.]*

Seeing in the dark is one of the many specialties of remote sensors that extends the natural sensory apparatus of humans. Thermal infrared sensing is one of these. Rattlesnakes use it to find and strike their prey in the dark; sightless persons develop the natural ability of the human skin to sense the warmth of nearby surfaces.

Read over the description of thermal infrared sensing in your textbook, p. 55-56. Then refer to Figure 2.27, an infrared image of Brawley, California, taken between 2 and 4 A.M.

(1) How does the infrared radiation recorded on this image differ from that recorded in the infrared photo of Exercise 2-D?

The latter is an image of reflected infrared solar rays and is taken in the

daytime under solar illumination of the ground. The Brawley image records

radiation given off by the ground because it possesses sensible heat.

_____

(2) What is the physical significance of the range of tones from white, through gray, to black in terms of the intensity of infrared emission?

    White areas: warm surfaces, emitting intensely_____

    Black areas: cool surfaces, emitting weakly_____

    Gray areas: intermediate in temperature_____

(3) Classify the tone (white, gray, black) to be expected from each of the following kinds of surfaces on this image: (Place an "X" in the blank space.)

(Continued over.)

*Modern Physical Geography, 3rd Ed., p. 72, Figure 4.26.

(Scale gradation)

	White ---->	Gray ---->	Black
Irrigated fields of of growing crops	____	____	X
Dry fields, plowed	____	X	____
Paved highways and city streets	X	____	____
River channel	X	____	____
River floodplain	____	____	X
City park	____	____	X
Water treatment plant	X	____	____

Name _____   Date _____

_____   _____

Exercise 2-F   Remote Sensing: SPOT Image of New York Harbor
                [Text p. 60, Figure 2.30.]*

In this exercise you will be referring to text Figure 2.30, a digital image of the New York Harbor taken by the French SPOT satellite in May, 1986. Trim a sheet of tracing paper and mount it over the image in your textbook. Mark the corners carefully. With the aid of the outline map on the next page, answer the following questions, after first labeling features named at the start of each question. When complete, transfer your tracing sheet to the space provided.

(1) Greenwood Cemetary and Prospect Park, Brooklyn. Why are these areas red in color?

<u>Healthy, growing vegetation</u>

What are the black areas enclosed in the red areas?

<u>Water, in lakes or ponds.</u>

(2) Coney Island, Brighton Beach. What kind of material forms the white strip along the shore?

<u>Sandy beach</u>

(3) Staten Island, Staten Island Expressway. How does the overall color pattern of Staten Island differ from that of Brooklyn? Explain.

<u>Staten Island shows a high percentage of green vegetation, indicating large areas</u>

<u>of lawns, parks, golf courses, and undeveloped land.</u>

(4) Battery Park City, Battery Park. Explain the rectangular white areas bordering the west shore of Manhattan.

<u>These are bare surfaces of filled earth awaiting development of buildings.</u>

_____

(5) Liberty Island, Verrazano Narrows Bridge. What is the round-trip distance in kilometers by ferry from Battery Park to Liberty Island? <u>2.2</u> km How wide is the waterway (The Narrows) spanned by the Verrazano Bridge? <u>1,600</u> m

_____

*Modern Physical Geography, 3rd Ed., p. 75, Plate A.6.

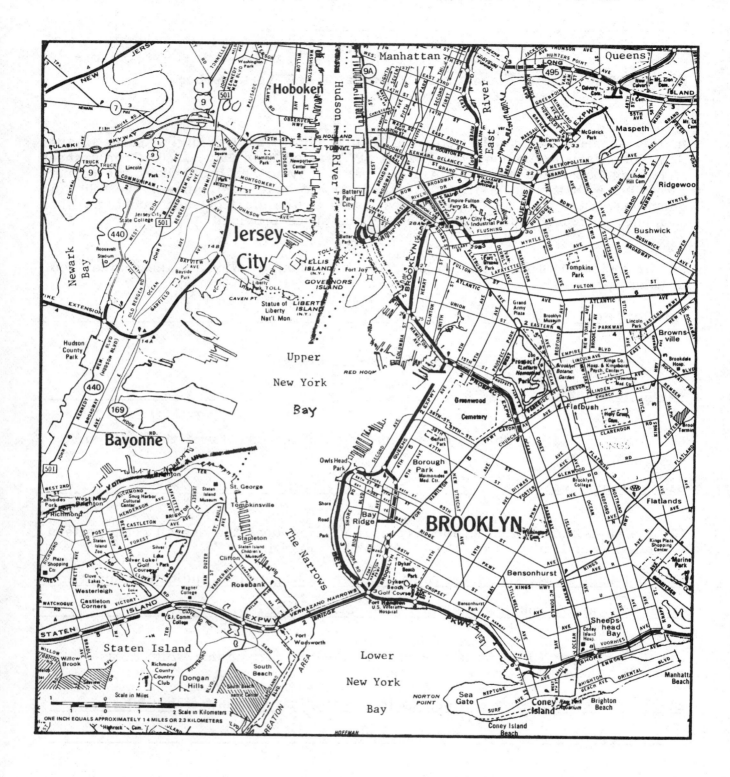

Map of the same area covered by the SPOT image. (Copyright © by the American Automobile Association. Used by permission.)

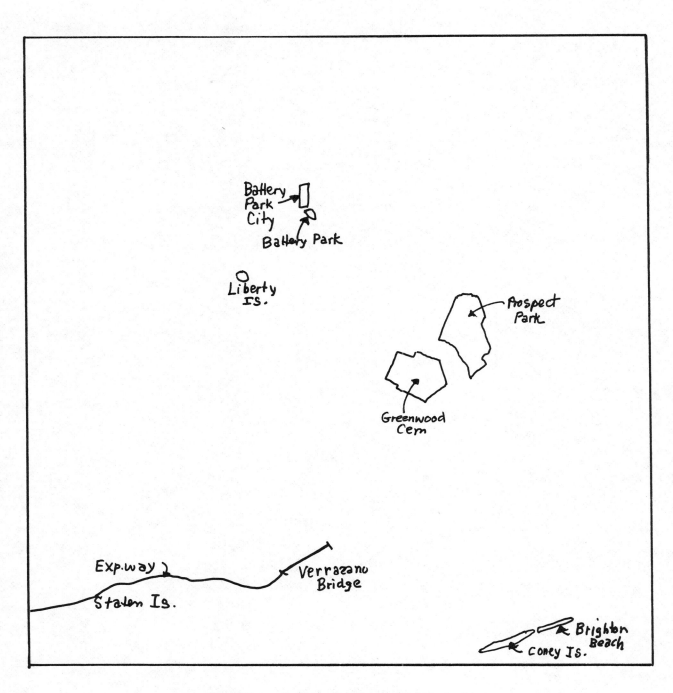

Tracing of features on SPOT image.

Name _____   Date _____

_____   _____

Exercise 3-A   Conversion of Temperature Scales
     [Text p. 62-63, Figures 3.1, 3.2.]*

Americans are chided not only from without by the world community but also from within by the local scientific community for not having converted to the metric system on a national scale. Despite some abortive attempts to conform, such as converting our gasoline pumps to liters, American commerce, industry, and news media continue to use gallons, miles, pounds, and the Fahrenheit temperature scale. In science education, on the other hand, not only are metric units used, but the awesome SI (Le Système International d'Unités) is de rigueur.

Formerly called the centigrade scale, the name switch to Celsius scale was made in 1948 to honor the Swedish physicist Anders Celsius (1701-1744), who invented the centigrade thermometer. The Celsius scale, as you know, runs from zero at the freezing point of water to 100°C at the boiling point. What few persons know, however, is that Anders Celsius' scale started with 100° at freezing and ran backwards to zero at the boiling point. So the physicists now pay homage to "Wrong-way Celsius."

(1) Using the formulas on p. 63 and in Figure 3.2, convert the following temperatures from °F to °C and from °C to °F. Show your calculations in the blank lines at left. The scales shown on the thermometers in Figure 3.2 should not be used to find the answers, because they can give only rough approximations.

Fahrenheit to Celsius:

C=5/9(32-32°) = 5/9(0) = 0.0	32°F	0.0 °C
C=5/9(0-32) = 5/9(-32) = -17.78	0°F	-17.78 °C
C=5/9(90-32) = 5/9(58) = 32.22	90°F	32.22 °C
C=5/9(-22-32) = 5/9(-54) = -30.00	-22°F	-30.00 °C

Celsius to Fahrenheit:

F=9/5(0)+32 = 0+32 = 32.00	0°C	32.00 °F
F=9/5(100)+32 = 180+32 = 212.00	100°C	212.00 °F
F=9/5(11)+32 = 19.8+32 = 51.80	11°C	51.80 °F
F=9/5(-40)+32 = -72+32 = -40.00	-40°C	-40.00 °F

*Modern Physical Geography, 3rd Ed., p. 78-79, Figures 5.1, 5.3.

(2) Can you think of a practical reason why the U.S. National Weather Service continues to use the Fahrenheit scale in recording and reporting air temperatures on its daily weather maps?

The costs of replacing thermometers in its numerous instrument shelters,

including the maximum/minimum thermometers, and of altering its entire data

collection system have deterred the change-over to Celsius.

(3) Examine Figure 3.1, the photograph of a standard instrument shelter. In which compass direction does the opening of the shelter appear to be facing? How can you tell? Is this the best orientation for an instrument shelter? Explain.

Shadows point in a northerly direction during the middle of the day. The open

door probably faces due north. This is desirable in middle and high latitudes in

the northern hemisphere, because facing north prevents the sun's rays from

falling directly on the instruments while the door is opened during observation

periods.

Name _____   Date _____

_____   _____

Exercise 3-B Lapse Rate and Tropopause in Middle and Low Latitudes
[Text p. 65-66, Figure 3.6.]*

The environmental temperature lapse rate is of great importance in understanding how weather works in the troposphere. Generally, the air gets colder as you go higher. Why should that be so? Think of being close beside a roaring bonfire; you feel warm, of course. Slowly, you back away from the fire; the warmth diminshes with distance. In this analogy, the earth's surface is the bonfire, heated by absorption of the sun's radiant energy. As you ascend, you move away from the radiating heat source and the air becomes cooler.

An example at middle latitude:

At Omaha, Nebraska, on a mild afternoon in January, the surface air temperature measures 50°F. A vertical sounding of the atmosphere by balloon reveals a nearly constant temperature lapse rate of 3.6F° per 1000 ft. At the tropopause, a temperature of -77°F is reported. (Assume the elevation of Omaha to be sea level.)

(1) Calculate the altitude of the tropopause (show your computations at left).

_____+50°- (-77)° = 127F° difference;_____

_____127 ÷ 3.6 = 35.3;   35.3 x 1000 = 35,300 ft_____   Ans. __35,300__ ft

(2) Convert your answer for (1) into kilometers.

_____1000 ft = 304.8 m = 0.3048 km_____

_____35.3 x 0.3048 = 10.76 km_____   Ans. __10.8__ km

(3) On the blank graph provided, plot the lapse-rate line and draw a horizontal line at the tropopause. Label fully.

The specified value of the environmental temperature lapse rate we use here-- 3.6F°/1000 ft--is only a long-term average value for the global troposphere. Our examples display that average. Actually, the lapse rate at Omaha might be quite different on the day you actually measure it. On a warm summer afternoon, the rate might be greater (the line slants less steeply); on a winter day it might be much less (the line slants more steeply). In fact, on a cold winter morning the air temperature might be lowest at the ground and actually increase upward (a low-level inversion, shown in Figure 3.8).

*Modern Physical Geography, 3rd Ed., p. 42, Figure 3.6.

EX. 3-B

An example at low latitude:

Nauru Island is located in mid-Pacific, close to the Equator. On the same day that observations are being made at Omaha, a similar sounding is being made at Nauru. Here the sea-level air temperature is 30°C and the lapse rate is constant at 6.6C°/km. The tropopause is encountered at 16.8 km.

(4) Calculate the air temperature (°C) at the tropopause (show computations).

____16.8 km x 6.6 = 110.88C° (round to 111C°)____

____111C° - 30°C = -81°C_____   Ans. ___-81___ °C

(5) Convert your answer to degrees Fahrenheit.

_____F = 9/5C + 32; 9/5C = -145.8;_____

____-145.8 + 32 = -113.8°F (round to -114°)____   Ans. ___-114___ °F

(6) Plot the data for Nauru on the same graph with Omaha, using the metric scales in kilometers and degrees Celsius along the top and right side of the graph. Draw a horizontal line at the tropopause. Label lapse rate line and tropopause.

The difference in altitude of the troposphere at these two stations is quite large. The explanation is not given in your textbook. Altitude of the troposphere is lowest at the poles and highest over the equator, as shown in the accompanying figure. Moreover, at high latitudes there are seasonal differences in altitude, as the diagram shows. You might guess that centrifugal force of the earth's rotation causes the thickening of the troposphere over the equator, but physicists would reject that idea. More likely, the vertical mixing of the atmosphere, which is much more intense and extends to a greater height at low latitudes, causes the thickening of the troposphere. (Take a look at text Figure 5.22, which shows deep vertical mixing by convectional storms in the Hadley cell circulation.)

(7) Do your findings as to the altitude of the troposphere at Omaha and Nauru agree in a general way with the data shown on the diagram?

Yes, the agreement is quite good, especially for Nauru, since seasonal variations

are small near the equator. That for Omaha is a bit on the low side.

(Considerable variation in altitude in winter accompanies the wandering path of

the polar jet stream.)

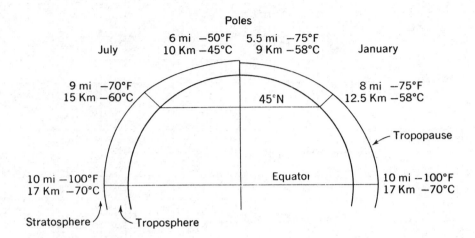

Average altitude and temperature of the troposphere in January and July. (Data from <u>Handbook</u> <u>of</u> <u>Geophysics</u>, U.S. Air Force.)

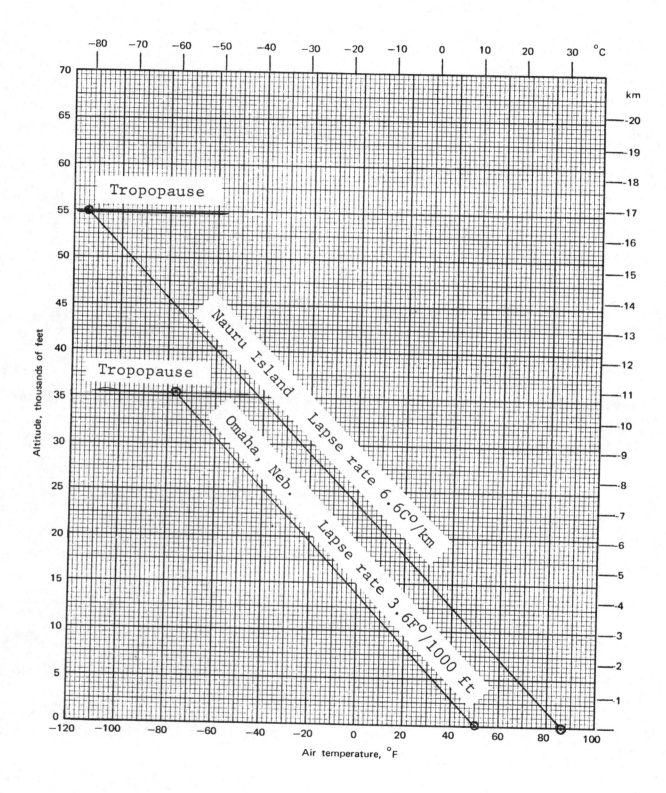

Name _____   Date _____

_____   _____

Exercise 3-C   The Daily Cycles of Net Radiation and Air Temperature
         [Text p. 64-65, Figure 3.4.]*

The relationship between air temperature and net radiation is important, but also difficult to understand. From experience, we know about the daily rise and fall of air temperature because we feel its effects and can easily read temperatures from a thermometer. Net radiation is something we can't feel directly, because it is the numerical difference between two sets of numbers. We suggest you go back to Chapter 2 and reread the section on p. 44-45 titled "Latitude and the Radiation Balance." Be sure you have a clear idea of the difference between an <u>energy</u> <u>surplus</u> and an <u>energy</u> <u>deficit</u>.

Given in Table A are data for net all-wave radiation (langleys per hour) at the ground and air temperature (°C) for a station in the midwestern United States at about latitude 40°N on a day in mid-August. Sunrise occurred at 5:15 A.M.; sunset about 6:50 P.M., local time. Plot the data on the blank graphs provided.

(1) Using color pencils, color red the areas of radiation deficit on the upper graph. Color blue the area of surplus. Label these areas.

Estimate the time at which deficit changed to surplus: <u>About 6:10 A.M.</u>

Estimate the time at which surplus changed to deficit: <u>About 5:50 P.M.</u>

(2) Estimate the time of occurrence of minimum temperature on the lower graph:

<u>About 6:10 P.M.</u>

Explain the time of occurrence of minimum temperature in terms of the net all-wave radiation curve.

<u>So long as a radiation deficit continues, air temperatures continue to fall slowly.</u>

<u>As soon as a radiation surplus sets in, air temperature begins to rise.</u>

_____

_____

*Modern Physical Geography, 3rd Ed., p. 79-80, Figure 5.4.

(3) Estimate the time of occurrence of maximum temperature: About 2:30 P.M.

Explain the time of occurrence of maximum temperature in terms of the radiation curve and other factors.

On theoretical grounds, air temperature should continue to rise as long as a radiation surplus exists, i.e., until about 6 P.M. It is therefore necessary to find another explanation for the falling off of temperature after about 2 to 4 P.M.

(4) Why does the air temperature decline rapidly in late afternoon, despite the fact that a radiation surplus exists?

Air turbulence increases throughout the day. Mixing of warm air close to the ground with cooler air aloft becomes important in mid-afternoon, causing a lowering of air temperature.

(5) On the upper graph sketch hypothetical radiation curves to represent typical conditions at June solstice and at December solstice. Label these curves.

(6) Construct a temperature scale in degrees Fahrenheit along the right-hand side of the lower graph. Label the scale.

Table A

	A.M.						P.M.					
	12M	2	4	6	8	10	12N	2	4	6	8	10	12M
Net All-Wave Radiation, ly/hr	-5	-5	-5	-2	+25	+41	+52	+45	+24	-1	-5	-5	-5
Temperature, °C	14	13	12.5	12	15	22	25.5	27	26.5	23	19	16.5	14

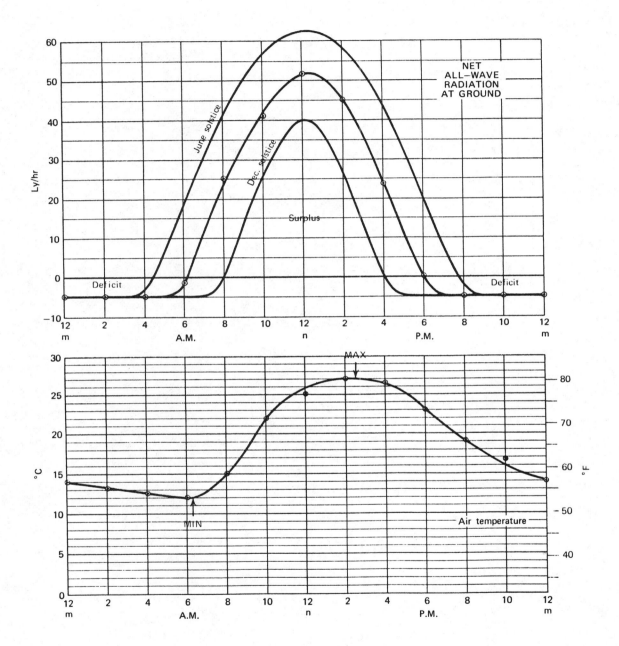

Exercise 3-D The Daily Temperature Range
[Text p. 66-67, 69-71, Figures 3.7, 3.12.]*

The daily temperature range figures importantly in personal comfort. We think of the ideal as "warm days and cool nights," but would not be happy with alternation of extreme heat and extreme cold. On December 25, 1878, at Bir Milrha in the Sahara Desert, the thermometer recorded 31°F (-0.6°C) just before dawn and 99°F (37°C) in late afternoon. That was a world's record daily range, holding for over a century. Here, we look at two basic causes of variation in the daily temperature range: (a) maritime vs. continental location and (b) increasing altitude in a mountainous region.

(1) Refer to Figure 3.12 in your textbook. You are asked to measure as accurately as possible from the graphs the daily temperature ranges at El Paso, Texas. To do this, use your straightedge (side of plastic triangle) and a very sharp soft pencil. Draw horizontal lines across each graph to pass through the high point (max) and low point (min) of each of the four temperature curves. Next, mark off the C° scale on the edge of a piece of paper (small pad or card) and use it as a ruler to measure the vertical distance between the horizontal lines. Enter the measurements in the table below. (Estimate to the nearest whole degree.)

	Jul. °C (°F)	Jan. °C (°F)	Apr. °C (°F)	Oct. °C (°F)
	10 (18)	12 (21)	12 (21)	15 (27)

(2) Range appears to be lowest in July and highest in October. Speculate on the reason or reasons why this difference should be present in the data. Keep in mind that the graphs show averages of hourly values over an entire month and for a period of several years of observation.

Lower daily range may be associated with greater average cloud cover during the day in July; high range in October with clear skies both night and day. Relative humidity may also be a factor, with high humidity in July reducing outgoing longwave radiation; lower humidity in October would intensify daytime heating and nightly cooling.

_____

_____

*Modern Physical Geography, 3rd Ed., p. 80-81, 84-85, Figures 5.6, 4.15.

(3) Using the same method as in Question 1, measure the daily range at North Head, Washington, also shown in text Figure 3.12. Estimate to nearest one-half degree C. Record below:

Jul. __2.5__ C°   Jan. __1.0__ C°   Apr. __2.0__ C°   Oct. __2.5__ C°

Compute the average (arithmetic mean) range of the year at both El Paso and North head. How much greater is the range at El Paso?

El Paso: __12__ C°   North Head: __2__ C°   Difference factor: __X 6__

(4) Refer to textbook Figure 3.7. Reread the explanation of this graph given on p. 66-67. Although, as stated, the mean daily temperature range increases generally with increasing altitude, the range for Cuzco (15C°) is less than for Arequipa and equal to that for La Joya, which is much lower in altitude. Can you think of one or more reasons why the range at Cuzco is not consistent with the overall trend of increase in range?

Cloud cover or fog, especially at night, may be more persistent at the level of Cuzco. Note that minimum values are about the same as at Arequipa. The observing station at Cuzco may be located in a more sheltered location that reduces the intensity of nightly outgoing radiation. Urban air pollution at Cuzco may also be a factor; perhaps a heavy smoke pall lies over the city at night.

(5) In Figure 3.7, the daily range at Mollendo, on the coastline of the Pacific Ocean, is a very small 6C°. At La Joya, which lies only 40 km inland at an altitude of 1261 m, the range has jumped to 15C°, about $2\frac{1}{2}$ greater than at Mollendo. Can this increase in daily range be attributed entirely to altitude difference? Is another factor involved? Support your answer with evidence based on data of all four stations.

No, altitude by itself is not a satisfactory explanation. Higher up and farther inland, the increase with elevation is in a much lower ratio. Mollendo's small range must be due in large part to its marine location (maritime effect). (This is a cool desert littoral with much fog. See text p. 184 and Figure 8.20.)

_____   _____

Exercise 3-E   The Annual Cycle of Air Temperature
[Text, p. 67-68, Figure 3.9.]*

If you had to make the choice, in which of these two places would you rather spend a full year for independent study? Auckland is the leading city on the North Island of New Zealand. It is a major port, corresponding in location (37°S) about to San Francisco and Lisbon in the northern hemisphere. Once the capital city of New Zealand, it has a major university and is surrounded by volcanoes. Churchill, in northern Manitoba at lat. 59°N., is a town on the western shore of Hudson Bay, lying at the northern terminus of the Hudson Bay Railroad that connects it with Regina, far to the south. Churchill has had a long, rich history as a fur-trading center, for it was the site of a Hudson's Bay Company trading post, established in 1717. One factor of the many influencing your decision would surely be climate, and in particular the annual temperature cycle.

The figures given in Table A are monthly mean air temperatures (°F) for Auckland and Churchill. On the blank graph provided, plot each monthly mean value as a point. Connect the points with a smooth curve. Label each curve. Because monthly values are rounded to the nearest one-half degree, points may lie slightly to one side or the other of your fitted curve.

(1) Compare the temperature curve of Auckland with the insolation graph for 40°S, which you drew in Exercise 2-B. Describe and explain the relationship between months of maximum and minimum temperatures and months of maximum and minimum insolation.

The general form of the temperature curve is the same as that of insolation at

40°S, but with important differences in the times of maximum and minimum

values. Minimum temperature lags more than a month past time of minimum

insolation (June solstice). Maximum temperature occurs more than a month after

December solstice.

_____

(2) Make a similar comparison between the temperature curve for Churchill and the insolation graph for 60°N.

The two curves are very similar in form and amplitude, but the times of maximum

and minimum values are not the same. Maximum temperature is one month after

June solstice; minimum temperature one month after December solstice.

_____

*Modern Physical Geography, 3rd Ed., p. 82-83, Figure 5.10.

(3) Calculate the annual temperature range for each station.

Auckland: ___14.5___ F°   Churchill: ___72.5___ F°

(4) Calculate the mean annual temperature for each station. On the graph, draw a horizontal line at the mean value and label it appropriately.

Auckland: ___59___ °F   Churchill: ___18___ °F

(5) Offer two good reasons why the annual range of temperature at Churchill should be so vastly greater than at Auckland.

(a) The latitude of Churchill is greater than for Auckland, hence the annual cycle of solar radiation has a greater annual range. (b) Churchill has a continental location, favoring severe radiational heat loss in winter. Auckland is a coastal station with strong marine influence from prevailing westerly winds off the adjacent ocean.

(6) Compare these two annual temperature curves with those in text Figure 3.13, p. 70. (a) Compare Churchill with Winnipeg; (b) Auckland with Scilly Islands.

(a) Churchill's maximum and minimum are about 10F° lower than Winnipeg's, as expected by the former's higher latitude, but the annual ranges are almost the same. (b) Auckland's maximum and minimum are about 5F° higher than Scilly's, because of Auckland's lower latitude. The ranges are nearly the same, despite the higher latitude of Scilly. The seasons are reversed on the calendar in the southern hemisphere, but this is not an essential difference.

Table A

	J	F	M	A	M	J	J	A	S	O	N	D
Auckland	66.5	66.5	63.0	61.5	56.5	53.0	53.0	52.0	54.5	57.5	60.0	63.5
Churchill	-19.0	-16.5	-6.0	14.0	30.0	43.0	53.5	52.5	41.5	27.0	5.5	-11.0

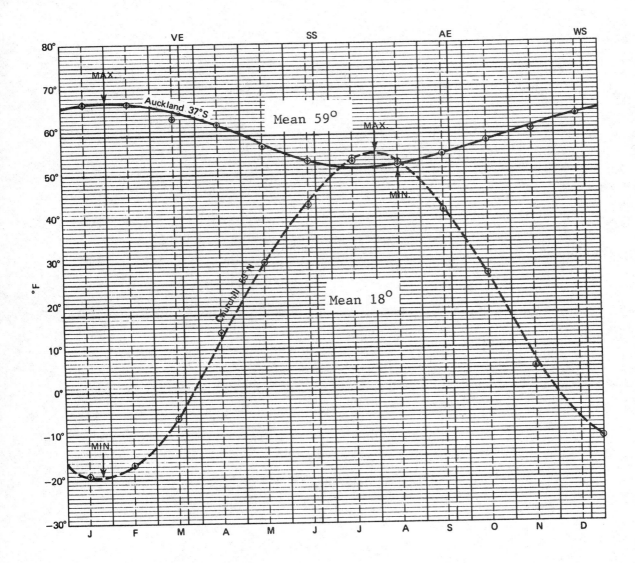

Exercise 3-F   Drawing Isotherms
          [Text p. 71, Figure 3.14.]*

In Chapter 1 of your textbook, p. 26-28, the subject of isopleths was introduced and illustrated by Figure 1.23. In Table 1.1, the isotherm was listed as an example of one kind of isopleth. We now offer you a chance to practice the drawing of isopleths, given a map on which the data have been entered as irregularly spaced numbers.

On the reverse side of this page is a map of a portion of the United States on which air temperatures in degrees Fahrenheit have been plotted. Readings were taken at 1:30 A.M., E.S.T. at U.S. National Weather Service observing stations. The date is early in May. The 32° (freezing) isotherm is shown. You are asked to draw isotherms for every 5 degrees, i.e., 25°, 30°, 35°, 40°, 45°, etc.

The method of drawing isotherms is illustrated by a portion of the 40° isotherm. Begin with a station numbered "40." Draw the line so as to pass between pairs of adjacent numbers, one of which is greater than 40, the other less than 40. Space the distance between the two numbers in proportion to the value of the number 40 with respect to the higher and lower numbers. If the numbers on either side are 35 and 45, the 40° isotherm will pass midway between them. Label each isotherm.

Use a soft pencil and draw the isotherms very lightly at first. When the map is completed, adjust the lines to form smooth sweeping curves. Note that isotherms may not intersect one another. They may, however, form closed loops surrounding centers of high or low temperatures.

---

*Modern Physical Geography, 3rd Ed., p. 85-86, Figure 5.16.

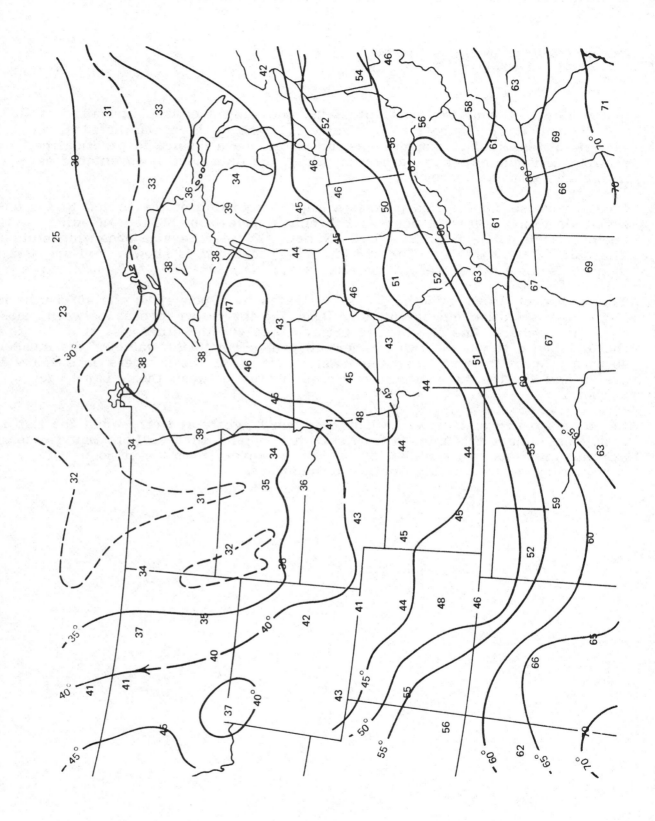

_____   _____

Exercise 3-G Continentality and the Air Temperature Range
[Text p. 74-75, Figure 3.17.]*

To investigate the concept of continentality of air temperatures, we now take a great sweep from west to east near the southern border of Canada, just below the 50th parallel of latitude. We start at Victoria, British Columbia, on the North Pacific coast, pause at Winnipeg, Manitoba, in the deep continental interior, and end at St. Johns, Newfoundland, a port on the North Atlantic Ocean.

Reproduced below is a composite graph of the daily maximum and minimum air temperatures during a single month at these three North American stations. We make this sweep first in mid-summer, then in mid-winter. Summer temperatures are for July; winter temperatures are for February or January, whichever is the month of lowest mean monthly temperature at the station. The monthly means are indicated by a labeled dashed horizontal line on each graph.

Look up the locations of these stations on a map of Canada. It will be helpful to know that in this latitude zone, the prevailing winds are westerlies, sweeping generally from west to east.

(1) Calculate the annual temperature range for each station and enter the values in F° and C° in the blank spaces below. (Calulate first from F°, then convert to C°.)

	F°	C°
Victoria	21.0	11.7
Winnipeg	59.0	32.8
St.Johns	33.5	18.6

(2) Why is the annual range least at Victoria?

Victoria lies on an estuary not far from the open Pacific Ocean and receives much of its weather from masses of air moving from ocean to land with the prevailing westerly winds. This air tends to be mild in its winter temperatures because of passage over a large ocean body that is warmer in winter than continental interiors at the same latitude. In summer, this same effect keeps the air cooler than in the continental interior.

_____

*Modern Physical Geography, 3rd Ed., p. 86-87, Figure 5.18.

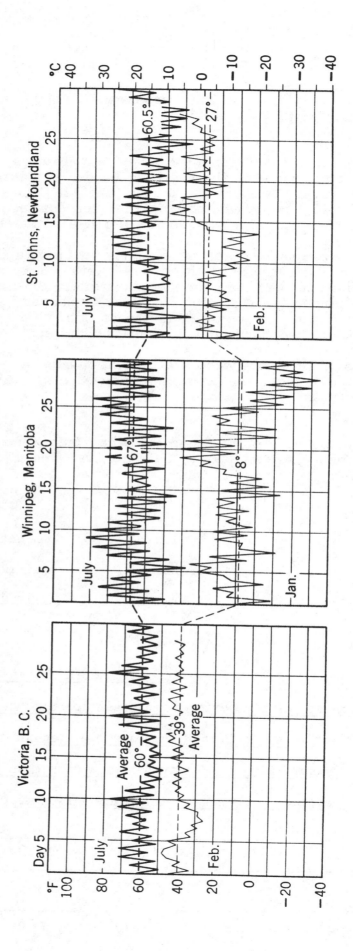

(3) Why is the annual range greatest at Winnipeg?

Winnipeg lies in the core of the continent, where surface temperatures become extremely cold in winter because of the severe deficit in the net radiation budget. Another factor is the inflow of extremely cold air from the arctic zone to the north. In summer, heating of the land surface brings a strong net radiation surplus. (See the net radiation curve for Yakutsk in Figure 3.9.)

---

(4) Explain why the annual range for St. Johns is intermediate between that for Victoria and Winnipeg.

St.Johns, located to the lee side of the continent with respect to the prevailing westerlies, frequently receives cold masses of air in winter from the continental interior, but also receives milder easterly winds from the North Atlantic. Much the same situation prevails during the summer. These alternations in air flow lead to a mixing of the continental and marine influences at St. Johns.

---

(5) Look next at Figure 3.17, the world map of annual temperature ranges. Locate as closely as you can the three stations on the map, marking each with a dot. Estimate from the map the annual range for each station and compare these values with those which you recorded in Question (1). Keep in mind that the graphs are for a single month of observation, whereas those of the world map are calculated from July and January means of a long period of record.

Summarize your data in the following table, using Celsius degrees:

	Victoria	Winnipeg	St.Johns
Annual range from graphs:	11.7C°	32.8C°	18.6C°
Annual range from world map:	15C°	35C°	18C°

Do you judge the degree of agreement to be good or poor? Think of a reason, or reasons, why the two sets of data should not agree closely.

The level of agreement is quite good, considering that the graph gives data for only one particular year, whereas the other is a long term average. The monthly means can be expected to vary from year to year by perhaps one to three degrees. Another factor is that the graph gives data for February--the coldest month--in the data of Victoria and St. Johns, whereas the world map is for January.

Name _____   Date _____

_____   _____

Exercise 4-A   Converting Barometric Pressures
      [Text p. 80-81, Figure 4.2.]*

Barometric pressure is mentioned in most news media weather reports, if only to refer to "highs" and "lows." When the pressure value is stated, the unit is usually inches of mercury. The aneroid barometer, which is the kind most amateur weather buffs use, usually features the inch units, but some have a millibar scale as well. If you tune in on the National Weather Service continuous weather information broadcast on the VHF band, as do farmers, aircraft pilots, and yachters, you will hear pressure reported in millibars. Pressure in millimeters (or centimeters) of mercury is rarely used in weather information, but you're bound to encounter that scale when your physician reports your blood pressure as "130 over 70."

(1) As review, from p. 80-81 and Figure 4.2, the value of standard sea-level

barometric pressure is: 29.92   in.;   760   mm;   1013.2   mb

(2) You are asked to make conversions of pressure readings from one scale to another, filling in the blanks of the table below. You should calculate the conversions using the formulas shown. The graphic conversion scale on the following page can serve only as a rough check on the inches/millibars conversions.

Inches	Millimeters	Millibars
30.12	765	1020.2
27.95	710	946.4
29.73	754.5	1006
29.60	751.8	1002.6
29.84	758	1010.6
14.75	375	500

Conversion formulas:

1.0 in.(mercury) = 33.87 mb = 25.40 mm (mercury)

1.0 mb = 0.0295 in. = 0.75 mm

1.0 mm = 0.03937 in. = 1.3333.. mb

---

*Modern Physical Geography, 3rd Ed., p. 40-42, Figure 3.4.

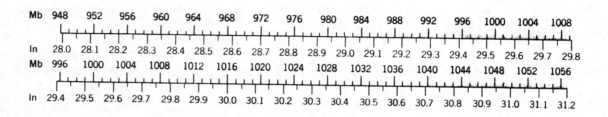

_____   _____

Exercise 4-B   Pressure Versus Altitude
              [Text p. 80-81, Figure 4.3.]*

The curve of decreasing atmospheric pressure with increasing altitude, shown in Figure 4.3, represents the average condition of the atmosphere. In technical language, it is the U.S. Standard Atmosphere. At any given time and place, the actual pressure curve would differ somewhat from the standard, and it is those small differences that are most important in weather phenomena, such as the interpretation and forecasting of cyclonic storms and weather fronts.

The table below gives data for the standard atmosphere in units of millibars and meters.

Pressure mb	Altitude m
400	7,425
500	5,643
600	4,186
700	2,955
800	1,889
900	984
1000	106
1013	0

(1) Plot the data from the table on the blank graph provided. Connect the points as a smooth curve.

(2) Using the graph, estimate the standard pressure in mb for each of the following places: (First, convert altitude from feet into meters. See p. 29 of your textbook for conversion data.)

Canton, OH	1,030 ft	314	m	970	mb
Las Vegas, NV	2,030 ft	619	m	935	mb
Cheyenne, WY	6,100 ft	1,860	m	805	mb
Mt. Hood, OR	11,235 ft	3,430	m	655	mb
Mt. Whitney CA	14,494 ft	4,420	m	580	mb

---

*Modern Physical Geography, 3rd Ed., p. 41-42, Figure 3.5.

(3)   Using the curve on the graph, estimate the amount of decrease in pressure in mb for 1 km of altitude increase at the following levels. Then calculate the percentage of decrease.

      Between   1 km and 2 km: __121__ mb/km   __11.9__ %

      Between   2 km and 3 km: __90__ mb/km   __11.5__ %

      Between   6 km and 7 km: __57__ mb/km   __11.9__ %

(4) Based on the percentages you have estimated, make a general statement about the rate of change of pressure with altitude:

The rate of pressure decrease with altitude is a constant quantity. This can be stated as: "the pressure falls by approximately 1/12 of its initial amount for each 1 km of rise."

(5) Refer to Figure 4.3 and read off the approximate values in answer to the following questions:

Barometric pressure in inches of mercury on Mt. Everest: __9__ in.

Express this pressure as a percentage of sea-level pressure: __33__ %

Barometric pressure in cm at 18 km, the cruising altitude shown for the Concorde aircraft: __6__ cm

Express this pressure as a percentage of sea-level pressure: __8__ %

For the physicists and mathematicians only:

Investigate the <u>barometric law</u> governing the decrease of pressure with altitude. Arrive at the statement of the <u>hydrostatic equation</u> that includes the variables of pressure, altitude, air density, and the acceleration of gravity. (Consult any standard textbook of meteorology.)

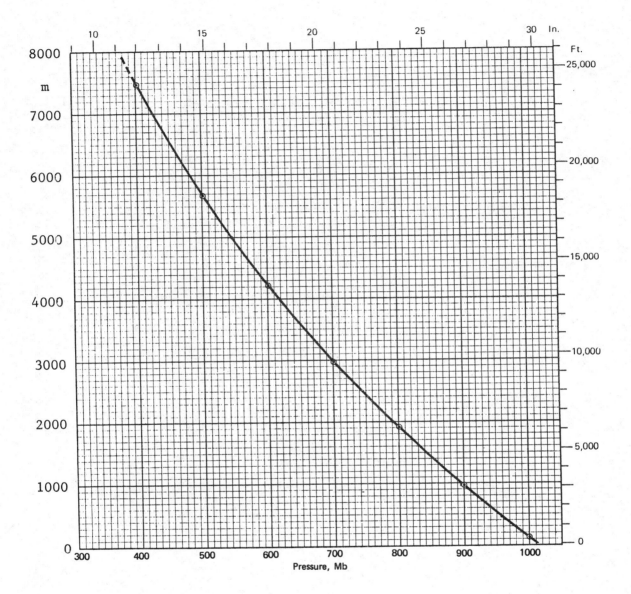

Name _____   Date _____

_____   _____

Exercise 4-C   Isobars on the Surface Weather Map
          [Text p. 81-82, Figure 4.4; p. 84-85. Figure 4.9.]*

Practice in drawing isobars on a weather map and figuring out how the surface winds move diagonally across them will increase your ability to read and interpret not only daily weather maps, such as those shown in Chapter 6, Figure 6.8, but also the global maps of pressure and winds, Figure 4.10.

How can you check on whether you have the wind arrows right on your map? Many decades ago, when weather forecasting was in a primitive state, a Dutch meteorologist, Buys Ballot, formulated a rule-of-thumb (actually "rule-of-hand" would be more accurate) to keep things straight. He advised: "Stand with the wind at your back with you arms outstretched to either side, and the lower pressure will be toward your left." Called Ballot's Law, and since then almost forgotten, it works quite well, but with two provisions. First, you must be in the northern hemisphere; second, your left arm should be aimed midway between straight left and straight to the front. Try this out when your exercise is done by imagining yourself standing on a wind arrow and looking in the direction in which the arrow is pointing. How would you amend Ballot's Law to apply in the southern hemisphere?

The weather map on the reverse side of this page shows barometric pressures observed simultaneously at many National Weather Service stations. Pressures are in millibars, but only the last two digits are given. Thus, "16" designates 1016 mb; "96" designates 996. The station is located at the dot beside the number. Draw isobars for the entire map, using an interval of 4 mb. Isobars should run thus: 992, 996, 1000, 1004, 1008, 1012, etc. Label the isobars. Label lows and highs. In drawing the isobars, use a soft pencil lightly at first, allowing for many corrections. Then draw the final isobars as smooth, flowing curves. Finally, draw many short straight arrows across the isobars to show the directions of surface winds.

---

*Modern Physical Geography, 3rd Ed., p. 91-95, Figures 6.2, 7.6-7.9.

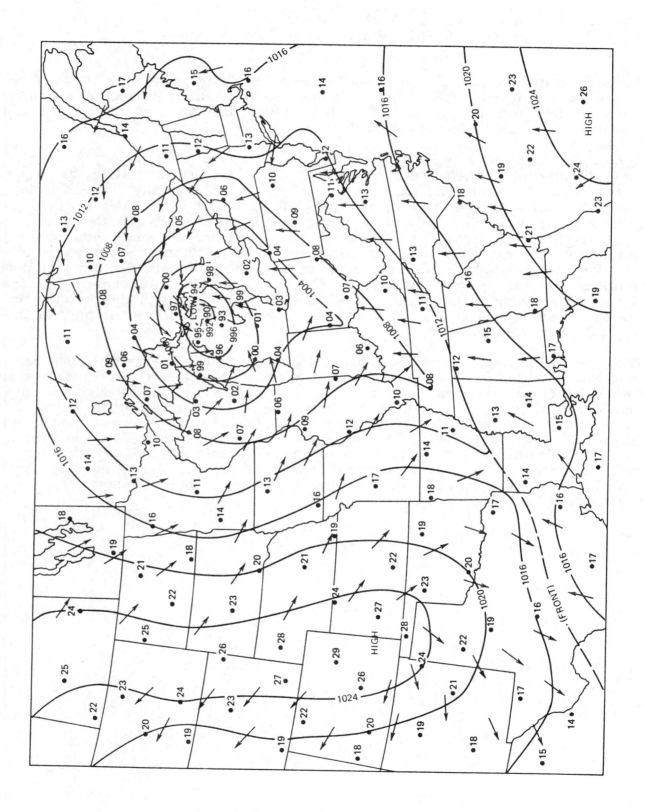

Name _____   Date _____

_____   _____

Exercise 4-D   The Wind Rose and Global Wind Belts
        [Text p. 83, 89-90, Figures 4.6, 4.10, 4.11, 4.12.]*

For this exercise we go back to the era of the great sailing ships—the clippers—that made use of the belts of prevailing winds to make their great transoceanic and circum-global cruises. Long before anemometers were available to measure wind strengths at sea, Admiral Sir Francis Beaufort of the British Navy in 1806 devised a system to standardize the estimation of wind force through its effects. It became known as the Beaufort Scale of Winds. It is reproduced on a following page. The knot, equal to one nautical mile per hour, is used in the Beaufort scale for marine applications. (One knot is equal to 1.15 miles per hour.) Under this sytem, wind speeds were given by numbers designating forces from 0 to 12, 0 being calm and 12 being hurricane force (above 65 knots). For example, a force of 5 was termed a "Fresh Breeze" and included velocities of 17-21 knots; a force of 9 was a "Strong Gale" (41-47 knots). Although of practical value when applied to the rigging of sailing vessels, the Beaufort scale was not adequate for modern scientific studies and has been abandoned. It is, however, used extensively on U.S. Navy Oceanographic Office pilot charts still available for study in most college libraries.

Maps A and B are portions of two U.S. Navy pilot charts on which average wind strengths and directions for a particular month are shown for each five-degree square of latitude and longitude by means of the <u>wind rose</u>. Our exercise makes use of this graphic device to gain a closer understanding of the belts of prevailing winds.

(1) Refer to the right-hand wind rose below. Using the scale under it, estimate the percentages of time during which the wind blows from each of the sectors of the rose. Measure shaft length from the center of the circle. Total the percentages and subtract from 100 in order to determine the percentage of light air and calm. Enter your data in the following table.

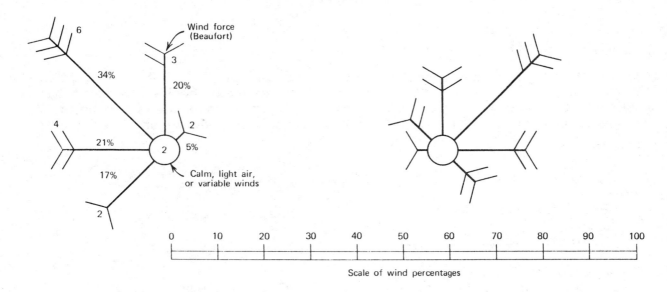

Scale of wind percentages

---

*Modern Physical Geography, 3rd Ed., p. 95-97, Figures 6.11, 6.12, Pl. B-3.

Sector	Percent of Time	Force
N	16	4
NE	30	5
E	18	4
SE	10	4
S	0	--
SW	0	--
W	6	2
NW	10	3
Total %	90	
% of calms	10	

(2) Referring to the two pilot charts, identify the wind belt from which the above data are taken. Estimate the latitude.

      Name of belt: <u>northeast trades</u>

      Latitude: <u>15°-20°N</u>

(3) Examine the left-hand rose (previous page), on which percentages and wind force are labeled. From what wind belt is this rose taken? Estimate the latitude.

      Name of belt: <u>northwesterlies</u>

      Latitude: <u>35°-45°S</u>

(4) Identifying the major belts of prevailing winds. On the pilot charts, label the following belts of winds or calms. (Place labels vertically beside the right-hand margin of each map.) In bold lines, draw the boundaries between the belts. In the spaces below, enter the latitude range of each belt shown on these charts:

Doldrums	Lat. <u>5°N</u>	to	<u>10°N</u>
Prevailing westerlies	Lat. <u>28°S</u>	to	<u>40°S</u>
Southeast trades	Lat. <u>5°N</u>	to	<u>20°S</u>
Subtropical belt of variable winds and calms	Lat. <u>20°S</u>	to	<u>28°S</u>
Northeast trades	Lat. <u>10°N</u>	to	<u>20°N</u>

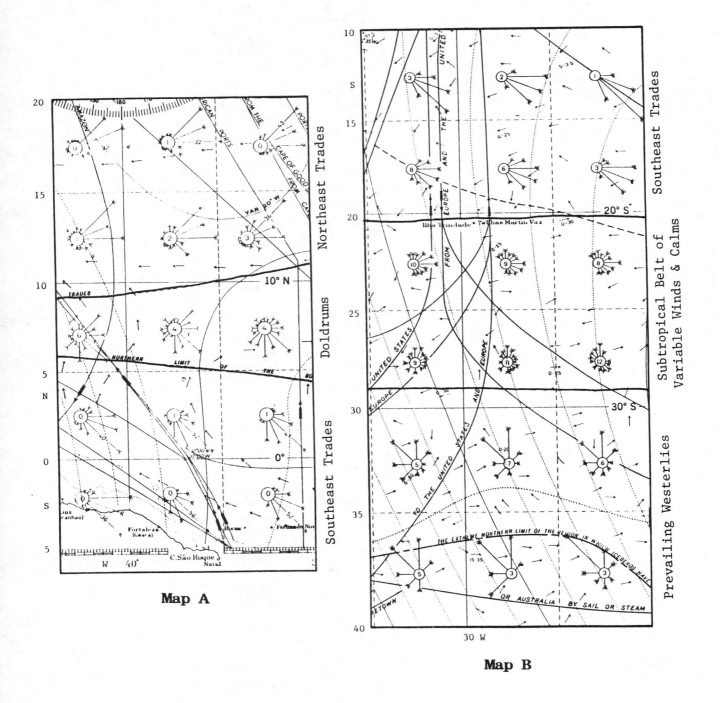

Map A

Map B

THE BEAUFORT SCALE OF WIND (NAUTICAL)

Beaufort No.	Name of Wind	Wind Speed		Description of Sea Surface	Sea Disturbance Number	Average Wave Height	
		knots	km/hr			ft	m
0	Calm	<1	<1	Sea like a mirror.	0	0	0
1	Light air	1–3	1–5	Ripples with appearance of scales are formed, without foam crests.	0	0	0
2	Light breeze	4–6	6–11	Small wavelets still short but more pronounced; crests have a glassy appearance but do not break.	1	0–1	0–0.3
3	Gentle breeze	7–10	12–19	Large wavelets; crests begin to break; foam of glassy appearance. Perhaps scattered white horses.	2	1–2	0.3–0.6
4	Moderate breeze	11–16	20–28	Small waves becoming longer; fairly frequent white horses.	3	2–4	0.6–1.2
5	Fresh breeze	17–21	29–38	Moderate waves taking a more pronounced long form; many white horses are formed; chance of some spray.	4	4–8	1.2–2.4
6	Strong breeze	22–27	39–49	Large waves begin to form; the white foam crests are more extensive everywhere. Probably some spray.	5	8–13	2.4–4
7	Moderate gale	28–33	50–61	Sea heaps up and white foam from breaking waves begins to be blown in streaks along the direction of the wind. Spindrift begins to be seen.	6	13–20	4–6
8	Fresh gale	34–40	62–74	Moderately high waves of greater length; edges of crests break into spindrift. The foam is blown in well-marked streaks along the direction of the wind.	6	13–20	4–6
9	Strong gale	41–47	75–88	High waves. Dense streaks of foam along the direction of the wind. Sea begins to roll. Spray affects visibility.	6	13–20	4–6
10	Whole gale	48–55	89–102	Very high waves with long overhanging crests. The resulting foam in great patches is blown in dense white streaks along the direction of the wind. On the whole the surface of the sea takes on a white appearance. The rolling of the sea becomes heavy. Visibility is affected.	7	20–30	6–9
11	Storm	56–65	103–117	Exceptionally high waves. Small- and medium-sized ships might be for a long time lost to view behind the waves. The sea is covered with long white patches of foam. Everywhere the edges of the wave crests are blown into foam. Visibility is affected.	8	30–45	9–14
12–17	Hurricane	above 65	above 117	The air is filled with foam and spray. Sea is completely filled with driving spray. Visibility very seriously affected.	9	over 45	over 14

Source: After R. C. H. Russell and D. H. Macmillan (1954), *Waves and Tides*, London, Hutchinson's Sci. and Tech. Publ., p. 54, Table 7; and N. Bowditch (1958), U.S. Navy Oceanographic Office Publ. No. 9.

Name _____ Date _____

_____ _____

Exercise 4-E Wind Roses for Four Wind Belts
[Text p. 89-90, Figure 4.10.]*

This exercise is an extension of Exercise 4-D, in which you will find all information you need about the Beaufort force scale and the construction of the wind rose used on the traditional U.S. Navy Oceanographic Office pilot charts.

On the blank wind roses on following pages, construct wind roses for the four localities for which data are given in the table below. The scale of percentages printed below the table may be cut out and used directly in plotting. After each wind rose is completed, identify the wind belt from which it comes and label the wind rose accordingly. U.S. Navy pilot charts A and B of Exercise 4-D will be most helpful in making these identifications.

Wind Rose Data

	A %	A Force	B %	B Force	C %	C Force	D %	D Force
N	6	2	7	3	15	2	1	4
NE	12	3	4	4	14	3	64	4
E	11	2	4	4	9	3	33	4
SE	9	3	5	4	13	3	2	3
S	17	3	16	4	19	3	0	0
SW	17	3	24	4	11	3	0	0
W	11	3	24	5	4	2	0	0
NW	9	3	15	4	4	2	0	0
Calms	8	--	1	--	11	--	0	--

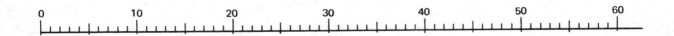

*Modern Physical Geography, 3rd Ed., p. 95-97, Pl. B-3.

A Horse latitudes (lat. 35°N, long. 45°W)

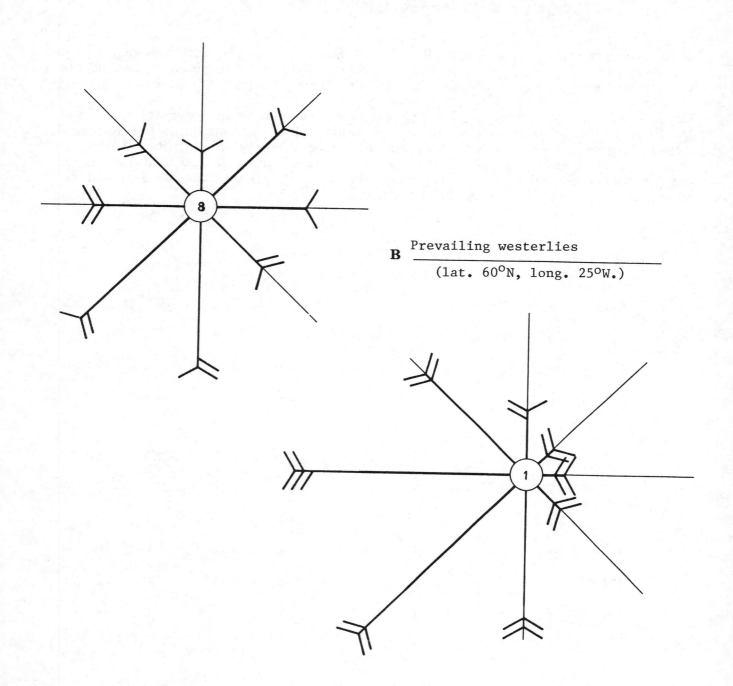

B Prevailing westerlies

(lat. 60°N, long. 25°W.)

C Doldrums (lat. 10°N., long. 20°W.)

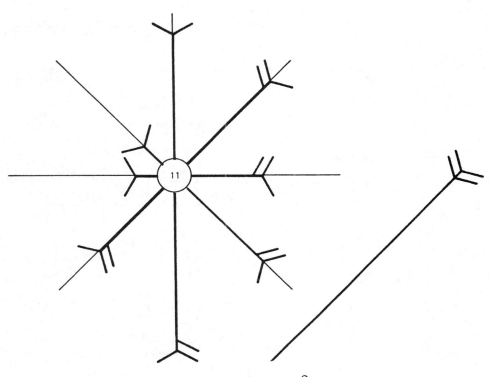

D Northeast trades (lat. 20°N., 40°W.)

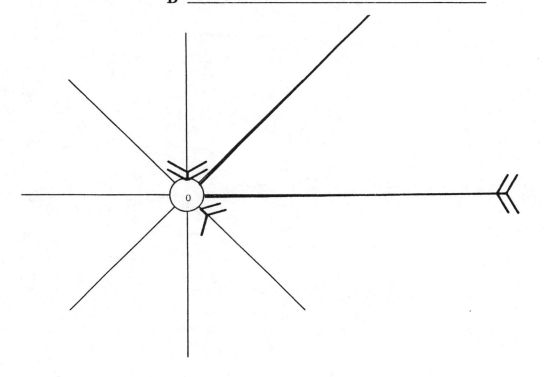

Exercise 4-F   Upper-Air Winds
    [Text p. 91-92, Figures 4.14, 4.15.]*

Knowledge of air flow at upper levels is crucial for weather forecasting. The patterns of lows and highs are relatively simple and can be shown in smooth circular or oval isobar patterns. Sharply defined changes in the direction of isobars are largely absent. The jet stream closely follows the pressure pattern, and it is the jet streams that "steer" the low-level cyclones.

Depiction of upper-level pressure patterns is done somewhat differently than on the surface map. The imaginary flat base, or datum, of the map is a surface of equal standard pressure. What look like isobars are actually elevation contours. An example is the U.S. National Weather Service map reproduced below for conditions at 7:00 P.M., E.S.T. on a day in early April. The lines are contours drawn upon the 500-millibar pressure surface. In keeping with U.S. practice, contours are labeled in thousands of feet. You notice, however, that low elevation numbers correspond with low pressure of a conventional map; high numbers with high pressure.

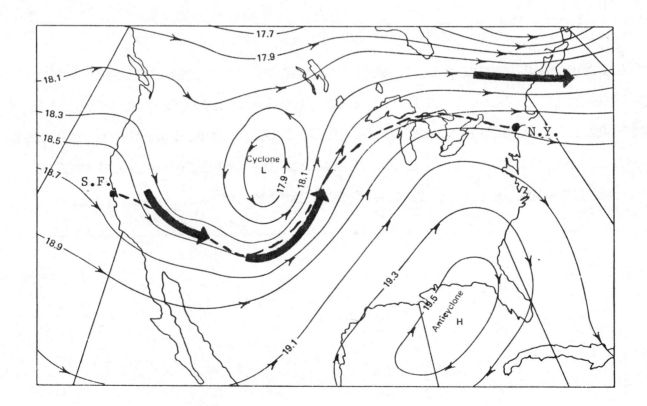

---

*Modern Physical Geography, 3rd Ed., p. 98-99, Figure 6.14.

EX. 4-F

(1) Treating the contours as if they were isobars, draw numerous arrow points on the contours to show the direction of air motion. Before starting, examine Figures 4.14 and 4.15. Explain the relationship between wind direction and isobars at this level. Does Ballot's Law apply here?

Air flow follows the isobars, rather than cutting diagonally across them, because

frictional resistance of air with ground surface has no effect at this level.

Because of the Coriolis effect, the motion of the air is turned until it runs

parallel to the isobar. Yes, Ballot's Law applies perfectly in this case, with left

arm outstretched directly to the side.

(2) Draw broad, sweeping arrows to show the probable position of the jet stream in two places on the map. (The jet stream would actually be at a much higher level of perhaps 35,000 ft.) Label a cyclone and an anticyclone.

(3) Imagine yourself piloting a private plane from San Francisco to New York at 18,500 ft altitude. You decide to try pressure-pattern flying, visibility conditions permitting, in order to keep the maximum tailwind in your favor. Describe your course and draw the path on the pressure map.

Leaving San Francisco, you head due east, getting into a position below the

overlying jet stream, crossing Arizona and New Mexico and bearing eastward to

cross the Texas Panhandle. Then you turn northeast, crossing Kansas, and veer

eastward over the Great Lakes. You would then turn southeastward, departing

somewhat from the optimum course to arrive at New York.

Name _____   Date _____

_____   _____

Exercise 5-A   Relative Humidity and the Dew-Point Temperature
[Text p. 101-3, Figures 5.1, 5.2, 5.3.]*

How humid is it? Measuring the level of relative humidity is actually a simple matter of reading the temperature simultaneously shown by two ordinary mercury-in-glass thermometers mounted side-by-side. One is just a regular thermometer from which we read the temperature of the air. The other is its twin, but with something added: a cloth sleeve over the bulb. In one typical arrangement, the cloth sleeve is a long tube that extends down into a small container of pure water. Capillary rise of the water keeps the sleeve thoroughly wet at all times. On a warm, dry day you will find that the thermometer with the wet sleeve reads a lower temperature than the "dry bulb" thermometer. The reason is obvious; evaporation of moisture from the cloth sleeve cools the bulb. The lower the relative humidity, the more rapid the evaporation, the greater the cooling, and the greater is the difference in temperatures. So we have a <u>dry</u>-<u>bulb</u> <u>temperature</u> and a <u>wet</u>-<u>bulb</u> <u>temperature</u> to compare.

Because the device described above--a hygrometer--works well only if air is moving quite rapidly past the wet bulb, a fan can be installed to create a strong air flow. Even better is the <u>sling</u> <u>psychrometer</u>, shown below. It has the two thermometers mounted side by side and linked to a handle. After wetting the cloth sleeve, you swing the psychrometer rapidly in a circle to get the maximum evaporation rate, then stop it and quickly record the two temperatures.

Our exercise shows you how to read the relative humidity from a prepared table. Obtaining the dew-point temperature follows next, using the same pair of observed temperatures, but referring to a second table.

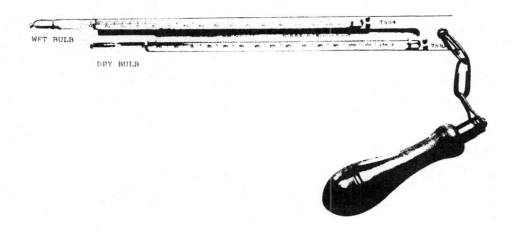

WET BULB

DRY BULB

A sling psychrometer. (National Weather Service.)

_____

*Modern Physical Geography, 3rd Ed., p. 107-8, Figures 7.2-7.5.

PSYCHROMETRIC TABLES

Table A. Relative Humidity, per cent. Temperatures in °F

Air Temperature	1	2	3	4	5	6	7	8	9	10	11	12	13	14	15	16	17	18	19	20	21	22	23	24	25	26	27	28	29	30
0	67	33	1																											
5	73	46	20																											
10	78	56	34	13																										
15	82	64	46	29	11																									
20	85	70	55	40	26	12																								
25	87	74	62	49	37	25	13	1																						
30	89	78	67	56	46	36	26	16	6																					
35	91	81	72	63	54	45	36	27	19	10	2																			
40	92	83	75	68	60	52	45	37	29	22	15	7																		
45	93	86	78	71	64	57	51	44	38	31	25	18	12	6																
50	93	87	80	74	67	61	55	49	43	38	32	27	21	16	10	5														
55	94	88	82	76	70	65	59	54	49	43	38	33	28	23	19	14	9	5												
60	94	89	83	78	73	68	63	53	48	43	39	34	30	25	21	17	13	9	5	1										
65	95	90	85	80	75	70	66	61	56	52	48	44	39	35	31	27	24	20	16	12	9	5	2							
70	95	90	86	81	77	72	68	64	59	55	51	48	44	40	36	33	29	25	22	19	15	12	9	6	3					
75	96	91	86	82	78	74	70	66	62	58	54	51	47	44	40	37	34	30	27	24	21	18	15	12	9	7	4	1		
80	96	91	87	83	79	75	72	68	64	61	57	54	50	47	44	41	38	35	32	29	26	23	20	18	15	12	10	7	5	3
85	96	92	88	84	81	77	73	70	66	63	59	57	53	50	47	44	41	38	36	33	30	27	25	22	20	17	15	13	10	8
90	96	92	89	85	81	78	74	71	68	65	61	58	55	52	49	47	44	41	39	36	34	31	29	26	24	22	19	17	15	13
95	96	93	89	86	82	79	76	73	69	66	63	61	58	55	52	50	47	44	42	39	37	34	32	30	28	25	23	21	17	15
100	96	93	89	86	83	80	77	73	70	68	65	62	59	56	54	51	49	46	44	41	39	37	35	33	30	28	26	24	22	21
105	97	93	90	87	84	81	78	75	72	69	66	64	61	58	56	53	51	49	46	44	42	40	38	36	34	32	30	28	26	24
110	97	93	90	87	84	81	78	75	73	70	67	65	62	60	57	55	52	50	48	46	44	42	40	38	36	34	32	30	28	26
115	97	94	91	88	85	82	79	76	74	71	69	66	64	61	59	57	54	52	50	48	46	44	42	40	38	36	34	32	30	28
120	97	94	91	88	85	82	80	77	74	72	69	67	65	62	60	58	55	53	51	49	47	45	43	41	40	38	36	34	33	31

Table B. Dew-Point Temperature, °F

Air Temperature	Vapor Pressure	1	2	3	4	5	6	7	8	9	10	11	12	13	14	15	16	17	18	19	20	21	22	23	24	25	26	27	28	29	30
0	0.0383	-7	-20																												
5	0.0491	-1	-9	-24																											
10	0.0631	5	-2	-10	-27																										
15	0.0810	11	6	0	-9	-26																									
20	0.103	16	12	8	2	-7	-21																								
25	0.130	22	19	15	10	5	-3	-15	-51																						
30	0.164	27	25	21	18	14	8	2	-7	-25																					
35	0.203	33	30	28	25	21	17	13	7	0	-11	-41																			
40	0.247	38	35	33	30	28	25	21	18	13	7	-1	-14																		
45	0.298	43	41	38	36	34	31	28	25	22	18	13	7	-1	-14																
50	0.360	48	46	44	42	40	37	34	32	29	26	22	18	13	8	0	-13														
55	0.432	53	51	50	48	45	43	41	38	36	33	30	27	24	20	15	9	1	-12	-59											
60	0.517	58	57	55	53	51	49	47	45	43	40	38	35	32	29	25	21	17	11	4	-8	-36									
65	0.616	63	62	60	59	57	55	53	51	49	47	45	42	40	37	34	31	27	24	19	14	7	-3	-22							
70	0.732	69	67	65	64	62	61	59	57	55	53	51	49	47	44	42	39	36	33	30	26	22	17	11	2	-11					
75	0.866	74	72	71	69	68	66	64	63	61	59	57	55	53	51	49	47	44	42	39	36	33	30	26	22	17	11	2	-11		
80	1.022	79	77	76	74	73	72	70	68	67	65	63	62	60	58	56	54	52	50	47	44	42	39	36	33	29	25	21	15	8	-2
85	1.201	84	82	81	80	78	77	75	74	72	71	69	68	66	64	62	61	59	57	54	52	50	47	44	42	39	36	32	28	24	19
90	1.408	89	87	86	85	83	82	81	79	78	76	75	73	72	70	69	67	65	63	61	59	57	55	53	51	48	45	42	39	36	32
95	1.645	94	93	91	90	89	87	86	85	83	82	80	79	78	76	74	73	71	70	68	66	64	62	60	58	55	53	51	48	43	39
100	1.916	99	98	96	95	94	93	91	90	89	87	86	85	83	82	80	79	77	76	74	72	71	69	67	65	63	61	59	57	55	52
105	2.225	104	103	101	100	99	98	96	94	93	91	90	88	87	86	83	82	80	78	77	75	74	72	71	69	67	65	63	61	59	57
110	2.576	109	108	106	105	104	103	102	100	99	98	97	95	94	93	91	90	89	87	86	84	83	82	80	78	77	75	74	72	70	68
115	2.975	114	113	112	110	109	108	107	106	104	103	102	101	99	98	97	96	94	93	92	90	89	87	86	84	83	81	80	78	76	75
120	3.425	119	118	117	115	114	113	112	111	110	108	107	106	105	104	102	101	100	98	97	96	94	93	92	90	89	87	86	80	78	81

(1) Using the tables on the opposite page, determine relative humidity (RH, %) and dew-point (°F) for each of the following pairs of temperature observations taken with the sling psychrometer.

			RH, %	Dew-point, °F
(a)	Dry bulb	70°		
	Wet bulb	50°		
	Difference:	20	19%	26°
(b)	Dry bulb	95°		
	Wet bulb	90°		
	Difference:	5	82%	89°
(c)	Dry bulb	30°		
	Wet bulb	22°		
	Difference:	8	16%	-7°
(d)	Dry bulb	105°		
		77°		
	Difference:	28	28%	65°

Extra-credit problems:

(2) Air having a temperature of 60°F and a relative humidity of 53% is heated to 90° without water vapor being added or removed. What is the RH at 90°? (Use both tables.)

Step 1:	Difference in wet-bulb and dry-bulb temperatures for 60° and 53% (Table A)	9	°F
Step 2:	Corresponding dew-point temperature as shown in Table B.	43	°F
Step 3:	In Table B, follow 90° line right to find same dew-point temperature value of Step 2. Read up to "dry bulb minus wet bulb" and enter here.	27°	F°
Step 4:	Take value of Step 3 to Table A; read down to value in body of table on 90° line. Ans.	19	%

(3) The same air described in (2) is cooled to 45°F without gain or loss of water vapor. What is the RH at 45°?

Step 1:	9	°F
Step 2:	22	°F
Step 3:	1°	F°
Step 4:	93	%

Name _____ Date _____

_____ _____

Exercise 5-B   Adiabatic Cooling of Rising Air
        [Text p. 103-4, Figure 5.6, 5.7.]*

Few words in weather science are more awkward than "adiabatic." It comes from the Greek word <u>adiabatos</u>, meaning "impassable." What is impassable, and what is trying to get past it? A certain volume of air is imagined to have impassable boundaries on all sides. Physics tells us that there are no boundaries impassable to heat--no perfect Thermos jug in nature. Rising and descending bodies of air can and do gain or lose heat to the surrounding air or ground by longwave radiation. Loss or gain by simple conduction or mixing also occurs, but is too slow to take into account. When the rise and descent of the air is very rapid, as it would be in the core of a dense cumulus cloud or in the powerful downdraft of a chinook descending the lee side of the Rockies, the dry adiabatic lapse rate holds true for all practical purposes.

Your textbook diagram, Figure 5.6 offers a descriptive explanation of adiabatic temperature change. In this exercise we treat the process quantitatively, using the dry adiabatic lapse rate given on p. 103 of your textbook. We shall use units of feet and °F. Use the blank graph provided to plot your data.

(1) A mass of rising air has a temperature of 65°F at 1000 ft altitude. If the air temperature is decreasing at the dry adiabatic rate of 5.5F° per 1000 ft, what will the temperature be at 4000 ft?

<u>   48.5°F   </u>

(2) Assuming that this air initially has a dew-point temperature of 50°F, at what altitude will condensation set in? (Dew-point lapse rate is 1 F° per 1000 ft.)

<u>   4300 ft   </u>

(3) Plot the data of this problem on the blank graph. Label fully.

(4) Under what natural conditions might a column of air rise spontaneously from 1000 ft to reach 4000+ ft? (See text Figure 5.7.)

<u>On a clear day, solar warming of an area of barren ground surface would heat</u>

<u>the air above it by longwave radiation and conduction. Warmer than the air</u>

<u>adjacent to it at the same level, buoyancy would cause the air to rise in a</u>

<u>convection column.   </u>

_____

*Modern Physical Geography, 3rd Ed., p. 108-9, Figure 5.8.

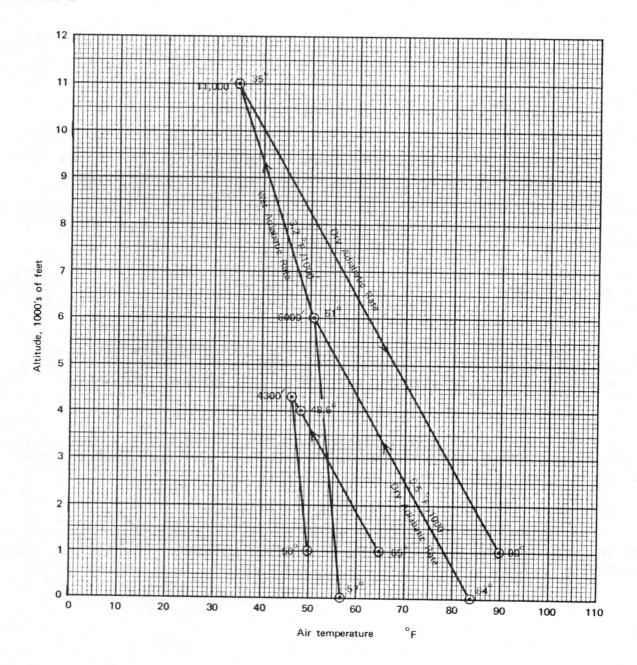

(5) Will the air cease to rise when it reaches the level where its temperature equals the dew-point temperature? What will happen? Explain. (See text Figure 5.7.)

No. Assuming that the surrouding air is cooler than the rising air, it will continue to rise, because of the condensation of water vapor, liberating latent heat and compensating in part for the adiabatic cooling. The wet adiabatic rate would take over.

Name _____ Date _____

_____   _____

Exercise 5-C Adiabatic Warming and the Rain Shadow
        [Text p. 104-5, 110-11, Figure 5.15, 5.16.]*

For emigrants from Europe, the Americas--both north and south--proved to be a "New World" in terms of abrupt climate zonation. Nowhere else on the globe are great marginal mountain chains such as the Sierra Nevada, Cascades, and Andes, to be found in the zone of the prevailing westerly winds, adjacent to low interior plains or basins to the east. Intense rain-shadow deserts in middle latitudes are thus uniquely American. For many emigrants seeking the paradise of California, Oregon, and Washington by the overland route, these rainshadow deserts were deadly traps for human and beast alike. The cause of such rainshadows lies in the adiabatic process, working first in rising air and then in sinking air. It's nature's process of wringing out the available moisture by forced ascent of air over a mountain barrier.

You are asked to solve a problem based on the steps pictured in Figure 5.15 of your textbook. Air at temperature of 84°F at sea level rises up over the slopes of a coastal mountain range. By adiabatic cooling, the air at the dry adiabatic rate (5.5F°/1000 ft) reaches the dew-point temperature of 51°F. The air continues to rise, but at the wet adiabatic rate of 3.2F°/1000 ft until, at the mountain summit, temperature has dropped to 35°F. The air then descends the lee side of the range, reaching the floor of an interior basin at 1000 ft altitude.

(1) Plot the above data on the same graph used in Exercise 5-B. Label fully. Using the graph as a direct source of information, answer the following questions:

(a) At what altitude was the dew-point temperature reached?

_____6000_____ ft

(b) Determine the initial dew-point temperature at sea level.

_____57_____ °F

(c) Give the altitude of the summit of the mountain range.

_____11,000_____ ft

(d) What was the air temperature on arrival at the floor of the interior basin?

_____90_____ °F

---

*Modern Physical Geography, 3rd Ed., p. 116-17, Figure 7.20.

(2) Compare the relative humidity of the air at the start and finish of this series of changes. Has relative humidity been increased or decreased? Explain.

The relative humidity is much lower at the finish than at the start. Water vapor was lost through condensation during the wet stage of ascent and was not replaced during descent. Air temperature at the finish is also much higher than at the start.

_____

_____

(3) Suppose the same air was next forced to pass over another mountain range of the same summit altitude farther inland. Would the same series of changes take place again? Explain.

No. There would be no essential change in water vapor content. Adiabatic warming would balance adiabatic cooling.

_____

_____

Name _____   Date _____

_____   _____

Exercise 5-D   Thunderstorm Rainfall
        [Text p. 112-14, Figure 5.18, 5.19.]*

The "cloudburst" seen in text Figure 5.18(a) can produce rainfall at a rate exceeding 5 inches per hour, but usually that level of intensity lasts only a quarter of an hour or less from a single thunderstorm cell. A rapid succession of cells can, of course, ring up a total of several inches in only a few hours.

In this exercise, we study the rainfall intensity of a brief thunderstorm at Hays, Kansas, in early August. Rainfall was recorded for each five-minute period. The accompanying table gives the rainfall intensity record of ten 5-minute periods. Plot these data as a bar-graph on the blank graph provided, using the left-hand scale.

Next, calculate the amount (depth, inches) that fell in each 5-minute period. (Divide intensity by 12.) Enter in the second column. In the third column, cumulate the depths and plot them to make a step graph, using the depth scale on the right-hand side of the graph. ("Cumulate" means to add each new value to the sum of the previous ones.) The final entry will be the total depth of rainfall during the storm.

(1) What was the total duration of the storm?   ___40___   min.

(2) What was the greatest depth of fall in any one 10-minute period?

                                ___0.783___   in.

(3) From both the table and graph, it is obvious that intensity values rise sharply from zero in the first three periods, then taper off gradually through the remaining periods. (a) How does this kind of rising and falling trend show itself in the cumulative rainfall curve? On the graph, draw a smooth curve through the steps to show this curve. (b) Can you offer an explanation of the trend in terms of the way a thunderstorm works?

(a) The cumulative curve starts steeply and lessens in steepness to the end,

where it is essentially flat. (b) The onset of the downpour is sudden because the

front edge of the downdraft of the moving storm is an abrupt phenomenon, with

practically no rain ahead of it. After that, the supply of falling rain is gradually

diminished, being continued in the after part of the cloud formation where the

updrafts are rapidly weakening.

_____

*Modern Physical Geography, 3rd Ed., p. 114-15, Figure 7.16.

Table

Period	Intensity (in./hr)	Depth (in.)	Cumulated depth (in.)
5:05 - 5:10	0.0	0.000	0.000
5:10 - 5:15	2.8	0.233	0.233
5:15 - 5:20	4.8	0.400	0.633
5:20 - 5:25	4.6	0.383	1.016
5:25 - 5:30	2.3	0.192	1.208
5:30 - 5:35	2.8	0.233	1.441
5:35 - 5:40	1.1	0.092	1.533
5:40 - 5:45	0.2	0.017	1.550
5:45 - 5:50	0.2	0.017	1.567
5:50 - 5:55	0.0	0.000	1.567

A special thought question:

(4) Suppose that in another example, storm rainfall intensity increases more gradually from zero and, after peaking, falls in a manner just the reverse of the rising curve before the peak. In other words, we have a bell-shaped curve that is symmetrical on both sides of the peak. How would the cumulative rainfall curve look on the graph? Sketch this hypothetical curve on your graph.

The curve would be S-shaped and symmetrical, with the inflection of the curve

occurring at the midpoint in time. (The two periods spanning 5:25 to 5:35 would

have equal values, as would corresponding periods before and after those

times.

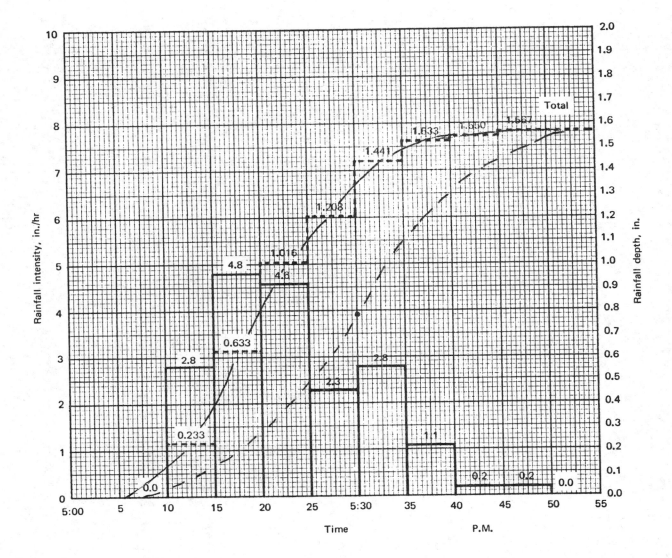

Exercise 5-E   The Global Transport of Water Vapor
  [Text p. 115, Figure 5.22.]*

It is often said that the hallmark of science and the scientific method is that it is quantitative. This simply means that the qualities of the objects of scientific investigation must be measured to give us numerical data. Quantification is not always practical in all branches of physical geography, but in the fields of meteorology and climatology, which are mostly applications of physics, nearly everything gets measured by means of instruments, yielding tables of numbers as our working data. What we do in this exercise is to take a particular part of a flow system of matter that is described in your textbook in words only (a descriptive analysis) and put it into quantitative form. The text description uses such expressions as "much" and "enormous amounts." We aim to answer the question: How much is "much"?

Reread the textbook section on latent heat and the global balances of energy and water (p. 115). Study Figure 5.22 carefully, because we will be reconstituting that figure as a graph. The figure tells us that water vapor is being transported either poleward or equatorward by the atmospheric circulation system. We now measure the rate of water transport (as vapor) in units of kilograms per year (kg/yr). The motion is described as <u>meridional</u>, because it deals only with motion northward or southward along the meridians of longitude. Part of the actual motion is eastward or westward, but we disregard that component in our analysis. We also disregard vertical motions, such as those displayed in Figure 5.22.

Table A gives the quantity of water vapor transported across each 10-degree latitude belt in one year. For plus values the transport is northward; for minus values, southward. The unit used is 10^{15}kg/yr.

(1) Plot the data as points on the blank graph provided. Place each point in the middle of the latitude zone. Then sketch a smooth curve that passsses through all the points. Parts of the curve arch above the zero line, while other parts dip below that line.

(2) Describe the curve you have drawn in relation to the zero line. What is the meaning of this relationship?

<u>Two parts of the curve lie above the zero line; two lie below. Where the curve is</u>

<u>above the zero line, water vapor is being transported northward; where below</u>

<u>the zero line, it is being transported southward.</u>

_____

*Modern Physical Geography, 3rd Ed., p. 117, Figures 7.22, 7.23.

Table A

Latitude belt (degrees)	Water vapor transport (10^{15} kg/yr)
80-90 N	0.0
70-80	+0.3
60-70	+1.2
50-60	+6.0
40-50	+13.5
30-40	+18.5
20-30	+7.0
10-20	-14.0
0-10	-1.0
0-10 S	+23.0
10-20	+15.5
20-30	-5.0
30-40	-26.0
40-50	-25.5
50-60	-10.5
60-70	-1.0
70-80	-0.1
80-90	0.0

(3) Draw a horizontal arrow inside each area bounded by the curve and the zero line. Point the arrow in the direction of movement of water vapor. What happens to water vapor that moves into higher latitude from about lat. 30° to the pole? Does it just "disappear"? Explain.

The rate at which water vapor moves poleward decreases in each latitude zone it crosses. This decrease is accounted for by condensation, in which the vapor passes into the liquid or solid state and falls out of the atmosphere. (Remember, the quantity of water vapor moving across each latitude zone is only the average net quantity.)

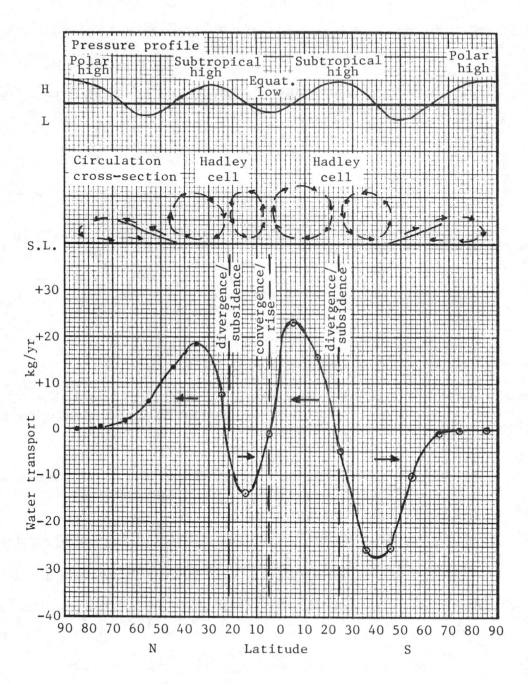

(4) Draw a dashed vertical line through each point where the curve crosses the zero line. What is the significance of this crossing point in terms of meridional atmospheric motion? Explain. (Take your cue from the labels on Figure 5.22.) Label each vertical line appropriately.

The crossing of the curve at the zero line marks the position of either (a) divergence accompanying subsidence of air or (b) convergence accompanying rise (lift) of air. (Use labels of two forms: (a) divergence/subsidence and (b) convergence/rise). There is only one line of convergence/rise; it is close to the equator. There are two lines of convergence/subsidence, both are between 20° and 30°.

(5) In the space on the graph labeled Circulation Cross-Section, sketch the atmospheric circulation by means of small, short arrows. Use Figure 5.22 as a guide. Label the Hadley cells. Show the fronts where cyclonic-frontal precipitation occurs in middle latitudes. Make sure the parts of your cross-section are correctly centered over the three zones you marked by vertical lines in Question 4.

(6) In the space at the top of the graph labeled Pressure Profile, sketch a curve from pole to pole generalizing the rise and fall of average barometric pressure along the meridian. Label polar and subtropical highs and the equatorial low. Get your information from the world map of pressure and winds, text Figure 4.10. (Use the southern hemisphere as the basis of your profile. Seasonal pressure centers of the northern hemisphere can be ignored.)

(7) What does your textbook mean (p. 115) by the term "fire box" applied to the equatorial zone of convergence/rise? Trace the water vapor from its source to its exit from the system in this zone.

A "fire box," such as that in a steam locomotive, generates sensible heat by burning fuel. In the equatorial zone, evaporation from the ocean surface first transforms sensible heat from insolation into latent heat (the fuel) present in the water vapor. The vapor, in turn, condenses to liquid water, thereby liberating the latent heat and transforming it back into sensible heat. This heat causes the air to expand and rise (transformation to kinetic energy) and drives the convection of the adjacent Hadley cells. Eventually, this energy is dissipated by friction and long-wave radiation.

Name _____   Date _____

_____   _____

Exercise 6-A Qualities of Various Air Masses
[Text p. 124-126, Table 6.1.]*

It can be said with a high degree of truth that "the air mass makes the climate."
An air mass affects the human senses simultaneously through both its
temperature and its water vapor content. The climate classification system
described in Chapter 7 is based in part on the kinds of air masses that are
present in each season and how they determine whether there will be
precipitation, and whether it comes as snow or rain. Know your air masses and
your knowledge of weather and climate will be both profound and far-reaching in
its applications.

Refer to text Table 6.1 for information on typical air temperatures and specific
humidities for six major varieties of air masses. You are asked to plot these data
on the graph provided here.

Compare this graph with Figure 5.4 on p. 102. We have copied a portion of this
graph and extended the graph toward the left to include temperatures as low as
-50°C. The curve on the graph shows the maximum specific humidity (SH) on the
vertical scale for any given air temperature (T) on the horizontal scale.

From text Table 6.1, read the values of T and SH given for each air mass and
locate the appropriate point on the field of the graph. Make a dot, surrounded
by a tiny circle. Be as precise as you can in locating the point. Beside the point
write in the symbol for the air mass: cA, cP, mP, etc.

(1) Recall that relative humidity, RH, expresses the ratio of the actual specific
humidity of an air mass to the maximum specific humidity associated with the
temperature of the air mass.

Which of the three low-latitude air masses (mE, mT, cT) probably has the

highest relative humidity? ___mE___

Which has the lowest? ___cT___

How did you arrive at these answers? Explain.

By estimating the relative vertical distance that the point lies below

the curve. Relative closeness to the curve must mean high relative humidity.

_____

(2) Which three of the air masses would you expect to yield heavy showers or

thunderstorms along a front? (give symbols) ___mE, mT___

_____

*Modern Physical Geography, 3rd Ed., p. 126-28, Table 8.1.

(3) In winter on Cape Cod, Massachussets, a powerful "nor'easter" (deep low off coast) brings ashore an air mass with T = 1°C and SH = 3 g/kg. Plot this point on the graph and label it "A".

Assign an air mass symbol to this point: mP

Describe its position relative to the mP station you have already plotted, which is from the Pacific coast of the U.S. Northwest. Is this relationship to be expected? What does it mean?

The points lie very close together, indicating that the two air masses are nearly

alike for both North Pacific and North Atlantic ocean source regions.

_____

_____

(4) At 2 P.M. in midsummer in northern Ontario, Canada, a clear air mass registers T = 25°C and SH = 5 g/kg. Plot this point and label it "B".

Assign an air mass symbol to this point: cP

Compare the position of this point to that of the winter cP airmass previously plotted. Explain your findings.

The main difference lies in the much warmer temperature of the Ontario air mass,

reflecting the continentality of that inland location.

_____

_____

(5) In summer in Seattle, Washington, an air mass registers T = 15°C, SH = 8 g/kg. Plot this point and label it "C".

Assign an air mass symbol to this point: mP

Compare this air mass with the two mP air masses previously plotted, both of them being winter air masses. Explain any differences you observe.

Seattle is much warmer because of season, and has a much greater water vapor

content. The relative humidity is much higher, perhaps because of longer

residence time of the air mass over the ocean.

_____

_____

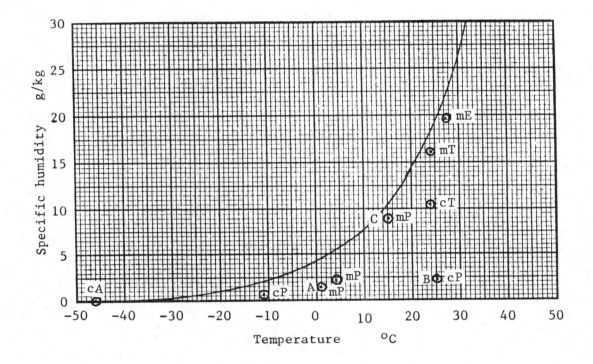

Name _____   Date _____

_____   _____

Exercise 6-B   World Weather on a Day in July
[Text, p. 131-2, Figure 6.11.]*

If you moved to Australia or New Zealand, your daily weather map might seem a bit crazy. Latitude gets higher as you move toward the bottom of the map, but the prevailing westerlies still move from left to right across the map and so do the cyclonic storms. The map, as compared with one in the northern hemisphere, is flipped top-to-bottom, but not left-to-right. This takes some getting used to, but will be easier after you have completed this exercise.

Reproduced below is text Figure 6.11, p. 131. This is a composite daily weather map for a typical day in July or August, when it is summer in the northern hemisphere and winter in the southern.

(1) How many low-pressure centers are shown on this map?   30

(2) How many high-presssure centers are shown?   19

---

*Modern Physical Geography, 3rd Ed., p. 130-31, Figure 8.13.

(**3**) How many lows fall into each of the following classes?

      (a) Middle-latitude cyclones: 19

      (b) Tropical cyclones: 1

      (c) Weak lows of the equatorial trough: 10

(**4**) At what approximate latitude, or latitude belt, is each of the following located? (Give location of center of pressure cell.)

      (a) Equatorial trough (ITC):   5° to 30°N

      (b) Subtropical cells of high pressure, northern hemisphere: 23° to 37°N

      (c) Subtropical cells of high pressure, southern hemisphere: 20° to 23°S

      (d) Middle-latitude cyclone centers, northern hemisphere: 35° to 70°N

      (e) Middle-latitude cyclone centers, southern hemisphere: 30° to 65°S

(**5**) Of the middle latitude cyclones shown on this map, how many are occluded, how many are open?

      Occluded: 9   Open: 9 or 10   (one indeterminate)

Name _____   Date _____

_____   _____

Exercise 6-C   Air Masses Around the World on a Day in July
[Text p. 131-32, Figure 6.11.]*

This exercise extends the study of the synthetic world weather map to the identification of air masses in both hemispheres.

Place a sheet of tracing paper over the world weather map in Exercise 6-B. Trace the rectangular border of the map. Label each of the major highs and lows with the symbol for the air mass most likely to be associated with the pressure system. Use the standard air mass symbols, as in Exercise 6-A. Label the airmass found along the equatorial trough (ITC). Transfer the finished overlay to the rectangle below.

*Modern Physical Geography, 3rd Ed., p. 130-31, Figure 8.13.

Name _____ Date _____

_____   _____

Exercise 6-D   Interpreting the Daily Weather Map
[Text p. 129-31, Figure 6.8.]*

The spread of the telegraph made possible the daily weather map. It was introduced to Americans by Cleveland Abbe, our first official government weather forecaster. His weather service started out in 1870 under the aegis of the Army Signal Corps, then turned civilian in 1891 under the Department of Agriculture as the Weather Bureau. From then on through the first half of this century, the daily weather map had enormous popularity. Many families in remote rural areas subscribed simply to be assured that the mail carrier would never fail to deliver daily at least that one message from the outside world. And it was cheap. Today, good daily weather maps are hard to find. You'll find some of the best ones in the better newspapers. Most we see on television deserve the rating of "disgraceful," but those on National Educational Television's daily A.M. Weather, put on by the National Weather Service of NOAA, are a delight.

Two weather maps on following pages show changes in position and form of a cyclonic storm on successive days in early April. Notice that whereas temperatures are given in C°, speeds of winds and cyclone centers are given in knots (nautical miles per hour). We changed F° on the maps to conform with the metric system, but retained the National Weather Service preference for knots as the units of wind speed used on the wind arrows. Refer to Exercise 6-E for a diagram explaining the reading of the wind arrows.

(1) Map areas experiencing precipitation are enclosed in a line of bold dots. Color these areas lightly in blue or green pencil or crayon.

(2) Construct pressure profiles across the maps on lines designated along the map margins as 1-A, 1-B, 2-A, and 2-B. Profiles 1-A and 1-B (Map A) are to be drawn on Graph A). Profiles 2-A and 2-B (Map B) are to be drawn on Graph B. Instructions are as follows: Cut out each profile graph, trimmed to the top dashed line. Attach the graph to the map with pressure tape so that the dashed line falls on the profile line. Where each isobar interects the profile line, follow that point down on the graph to the correct pressure and place a dot. Connect the dots with a smoothly curving line. To get you started, several points have been plotted and a partial profile drawn on Graph A. Label each profile. Mark high pressure crests with the letter "H", low troughs with "L". Fasten the completed profiles over the rectangular blank spaces provided.

(3) What distance has the center of low pressure traveled in the 24-hour period? Use the distance scale on Map B. Give speed in both knots (nautical miles per hour) and kilometers per hour.

Distance: ____1320____ km; ____830____ nautical mi

Speed: ____55____ km/hr; ____$34\frac{1}{2}$____ knots

*Modern Physical Geography, 3rd Ed., p. 130, Figures 8.9, 8.10.

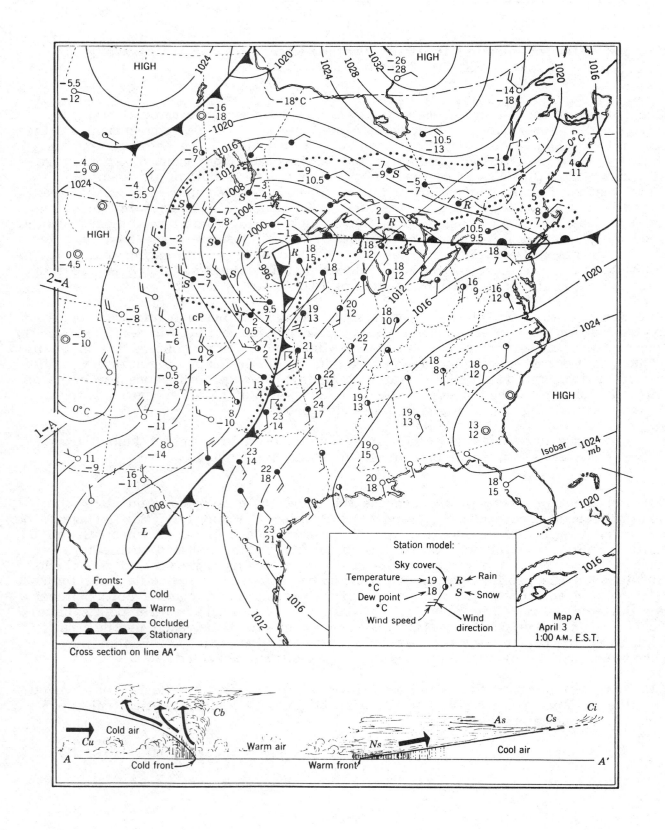

Fronts:
- Cold
- Warm
- Occluded
- Stationary

Station model:
Sky cover
Temperature °C → 19   R ← Rain
Dew point °C → 18   S ← Snow
Wind speed   Wind direction

Map A
April 3
1:00 A.M., E.S.T.

Isobar 1024 mb

Cross section on line AA'

Cold air   Cb   Warm air   Ns   Cool air   Ci   As   Cs
Cu
A   Cold front   Warm front   A'

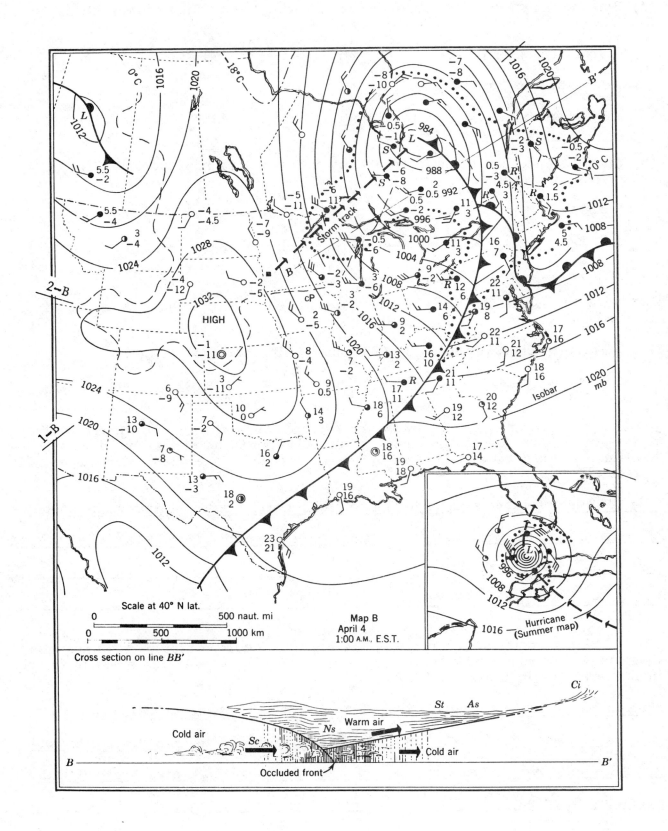

Scale at 40° N lat.

0   500 naut. mi

0   500   1000 km

Map B
April 4
1:00 A.M., E.S.T.

Hurricane
(Summer map)

Cross section on line BB'

Cold air   Sc   Ns   Warm air   St   As   Ci

B   Cold air   B'

Occluded front

(4) By how many millibars has pressure fallen at the cyclone center during the 24-hour period?

Pressure on Map A: 993 mb

Pressure on Map B: 981 mb

Difference: 12 mb

(5) What distance has the cold front traveled in the 24-hour period? Give the speed in both km/hr and knots.

Distance: 980   km;   610   naut. mi

Speed: 41   km/hr; $25\frac{1}{2}$ knots

(6) By what amount has pressure in the cold front trough changed in the 24-hour period?

Pressure on Map A: 1006   mb   Direction of change: +

Pressure on Map B: 1016   mb   Difference: 16   mb

(7) Temperature changes. On each weather map draw the 16C° isotherm of air temperature. Compare the positions of these isotherms on successive days. Explain how the form and location of the 16° isotherm is related to fronts and air masses.

The 16C° isotherm runs close to the cold and warm fronts, outlining the sector of warm air in the southeastern quadrant of the cyclone. The isotherm has moved east along with the cold front.

(8) On Map A, note temperatures in the high presssure center over Hudson Bay. Compare with temperatures in the high over Georgia. Explain the great difference in temperatures in these two highs.

Hudson Bay, -18°C; Georgia, 13°C. The high over Hudson Bay represents a mass of extremely cold Arctic air. High pressure is due to high density of cold air. The high over Georgia is a cell of the subtropical high-pressure belt with subsiding warm air.

(9) Winds. Compare wind directions in southwestern Missouri on Maps A and B. What change of direction has taken place? What is the cause of the change?

On map A the wind is SSE at 15 knots. On map B it has shifted to NW at 15 knots. The change was first to S, then to SW, to W, and finally NW. This kind of wind change is known as a "veering windshift" and is caused by the passage of a cold front. A change of air mass has occurred.

(10) Compare wind directions along the western shore of Lake Superior on Maps A and B. Answer the same questions as above.

On map A the wind is ENE at 10 knots. On map B it has changed to NW at 30 knots. The change was first from ENE to NE, then to N, and finally to NW. This change is known as a "backing windshift" and results from the passage of a cyclone center to the south of the observing station.

(11) Dew-point temperatures. Study the dew-point temperatures in relation to corresponding air temperatures at various stations on Map A, particularly in the vicinity of the cyclone and on either side of the cold front and warm front. Where is the dew-point temperature equal to the air temperature? What is the significance of this equality?

At a station in Minnesota, both air temperature and dew point are -1°C. This equality signifies that the air at ground level is saturated, with relative humidity 100%. Fog, drizzle, or sleet can be expected.

(12) What is the significance of the relatively high dew-point temperatures in the area east of the cold front and south of the warm front?

Here the air temperatures are abnormally high for early April. This condition indicates a warm, moist tropical (mT) air mass in this area. Thunderstorm activity on the cold fronts indicates air mass instablility.

(13) What is the significance of the large difference between dew-point temperature and air temperature in southeastern New Mexico?

The very large dew-point spread here indicates that the air mass has very low humidity. This condition results from the warming of continental polar (cP) air mass traveling south across the Great Plains, while at the same time little water vapor has been added to the air mass.

(14) Air masses. Based on your study of dew-point temperatures and air temperatures, identify the various air masses present. Label these air masses by means of conventional symbols on both maps.

(15) Clouds and precipitation. Examine the cross-section below each map, showing clouds and upper-air fronts. What type of cloud and precipitation are indicated along the cold front in Kansas and Oklahoma on Map A?

Cumulonimbus clouds with thunderstorm precipitation (convectional).

(16) What type of cloud and precipitation are found in northern Michigan on Map A?

Nimbostratus clouds. Warm-front precipitation from warm air aloft.

(17) On Map A, where are stations showing calm with clear sky located? Explain this distribution.

One location is in the ridge of high pressure over Idaho and Utah; here the cold dry air is subsiding. A second location is in the subtropical high over coastal Georgia; here air is subsiding in a cell.

Profile Graph Sheets   (See completed profiles on p. 123.)

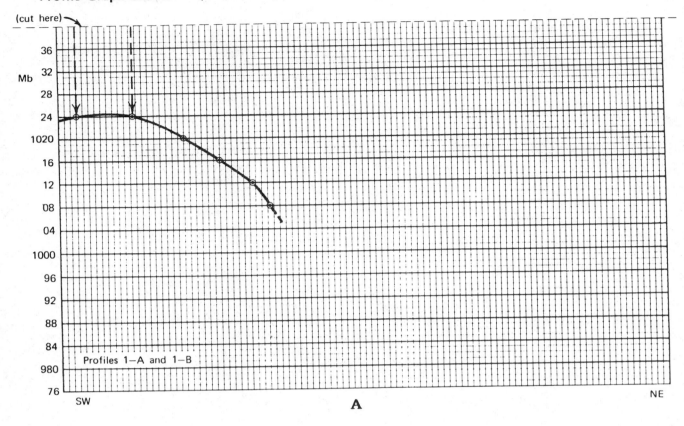

Profiles 1—A and 1—B

A

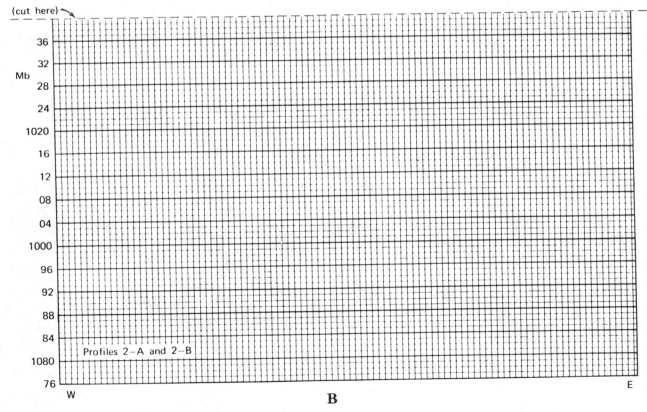

Profiles 2—A and 2—B

B

Space for Completed Profiles

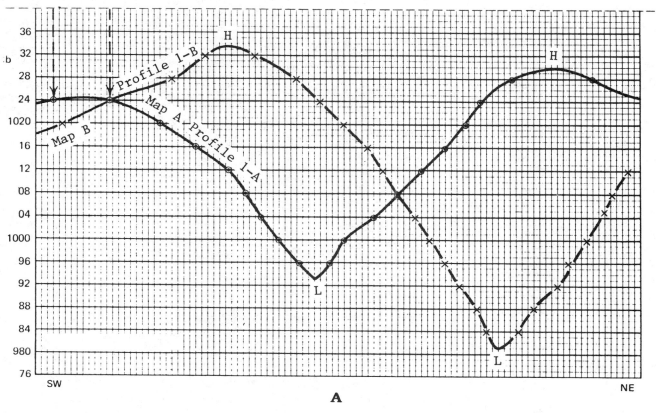

A

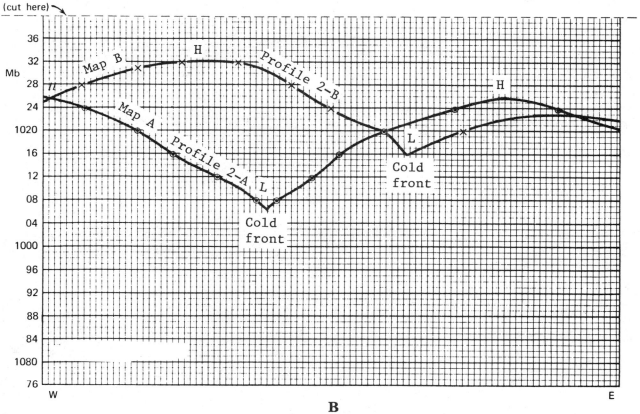

B

Name _____   Date _____

_____   _____

Exercise 6-E   Anatomy and Geography of Tropical Cyclones
          [Text p. 135-37, Figures 6.17-6.20.]*

The tropical cyclone goes by a variety of regional names. The hurricane of the Caribbean region is said to have been named by an ancient tribe of Central American aborigines, the Tainos, after Huracan, their God of Evil; but Webster's big dictionary cites the Spanish word huracan, meaning "wind." Perhaps word of Huracan's favorite toy was leaked back to Spain by the conquistadores. The typhoon of the China Sea is our anglicized version of the Cantonese term taai fung, meaning "great wind." Strangely, Webster's also links typhoon with the Greek typhon, a whirlwind. In the Philippines the storms are called baguios from the city of Baguio, where a cyclone in 1911 brought 46 inches of rain in a 24-hour period. For Australians, the tropical storm that visits their northeastern coast has long been advertised in other lands as the willy-willy, but the aborigines actually used that term only for small dust storms.

First, examine the hurricane shown on the inset area on Map B of Exercise 6-D. Note its small diameter in comparison with the cyclonic storm shown on the main part of the map. Note the steep barometric gradient and strong winds, forming a counterclockwise in-spiral.

For this exercise we use the weather map of Figure 6.17 in your textbook, but have reproduced it here somewhat enlarged for easier use. It shows a mature tropical storm of large size and severe intensity, centered over western Cuba at 1:00 A.M. on September 6. The storm track shows positions of the center on preceding and following dates. The precipitation pattern, interpreted from radar observations, is outlined by rows of dots. You may wish to color the precipitation areas for easier viewing.

Graph of a Hurricane Passage

The accompanying graph shows weather elements at the Tampa, Florida, weather station in the path of the storm. Let us assume that the storm follows the track as shown by the dashed line on the weather map, passing over Tampa, Florida. The ground speed of the storm center is assumed constant, requiring exactly two days to bring the eye over Tampa at 1:00 A.M. September 8. Observations show meteorological conditions into September 9.

The horizontal scale on the graph storm track is divided into segments representing 4-hour time intervals. The circled dots on the pressure profile are 8 hours apart. Wind arrows above each circled dot show direction and speed in knots, following the system shown on the accompanying diagram.

(1) What was the barometric pressure in the eye of the hurricane ?   __953__ mb

(2) At Tampa, as the storm approached, was the rate of pressure fall more rapid

or less rapid than the pressure rise as the storm moved away?   __less rapid__

_____

*Modern Physical Geography, 3rd Ed., p. 137-39, Figure 8.20.

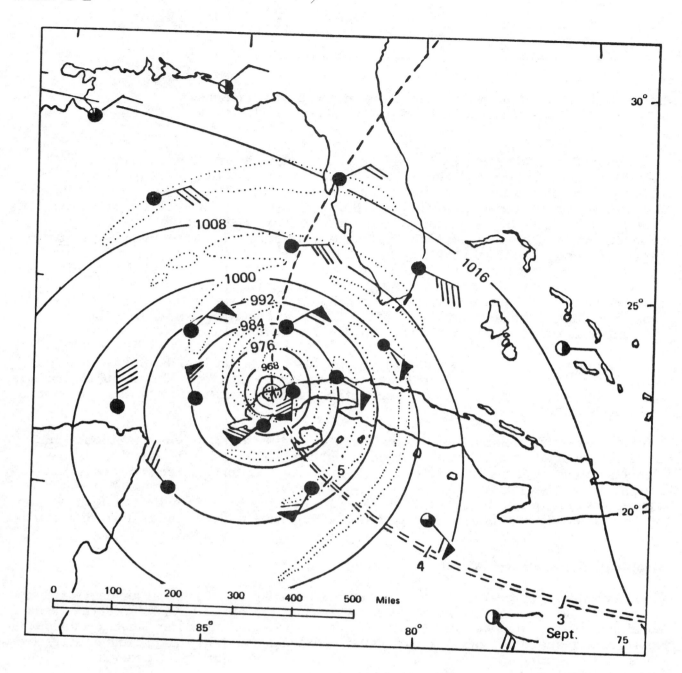

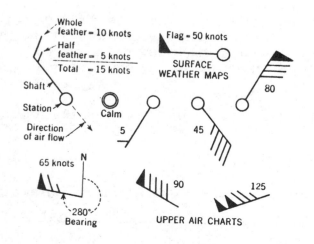

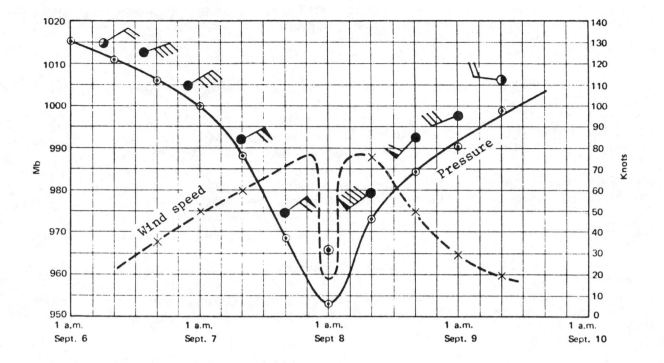

(3) Estimate the highest wind speed observed. ___90___ knots

(4) What conditions of wind speed existed in the eye of the storm? ___calm___

World Distribution of Tropical Cyclones

The following table gives data on the annual average number of tropical cyclones of hurricane intensity in each of five important regions of occurrence for which adequate data are available. Also given for each region are those four consecutive months in which hurricane-intensity storms were most frequent, i.e., the largest total of any four consecutive months.

	Annual mean no.	Months of most storms
1. Western N. Pacific and N. China Sea (1886-1958)	15.3	Jul-Oct
2. N. Atlantic (1886-1958)	4.6	Jul-Oct
3. Eastern N. Pacific (1965-1969)	4.8	Jul-Oct
4. Bay of Bengal (1890-1950)	4.7	Aug-Nov
5. S. Pacific Ocean, 150°E to 150°W (1940-1951)	4.6	Dec-Mar

(Data adapted from J-H. Chang, <u>Atmospheric Circulation Systems and Climates</u>, Oriental Publ. Co., Honolulu, Table 2, p. 105.)

(5) Refer to Figure 6.19 of your textbook. Fasten a sheet of tracing paper over the map and draw the rectangular map outline.

    (a) Insert the number of each region in its correct place.

    (b) Insert the months of maximum activity below or adjacent to the number of the region.

    (c) Label the following: North Pacific Ocean, South China Sea, North Atlantic Ocean, Bay of Bengal, South Pacific Ocean

(6) What three regions of tropical cyclone occurrence are not included in the table? Assign regional names and label them on the map.

(a) Arabian Sea. (b) South Indian Ocean (off Madagascar and southern Africa).

(c) South Indian Ocean, northwest of Australia.

(7) Judging from storm tracks shown on the North Atlantic map, and Figure 6.20, what is the lowest latitude at which tropical cyclones originate in the waters off North America?

    About 12°N.

(8) What major ocean has no tropical cyclones? Label the locality on the map. Can you think of one reason why they do not occur here?

South Atlantic Ocean. One posssible reason is that the sea-surface temperatures are lower here than in other oceans in the same latitude zone. Another possible reason, clearly shown in Figure 4.10, is that the ITC does not move south of the equator in January.

(Attach tracing sheet here.)

```
North                                North
                                     Atlantic
North                         2      Jul-Oct
Pacific        3                                               1   Jul-Oct
            Jul-Oct                       Arabian     So.
                                          Sea      4  China Sea
5                                  S. Atlantic    Bay of
          South Pacific            Ocean    So.   Bengal          5
          Dec-Mar                           Indian Aug-Nov
                                            Ocean
```

Name _____   Date _____

_____   _____

Exercise 7-A   Understanding Thermal Regimes
        [Text p. 144-45, Figure 7.2]*

In text Figure 7.2, the annual cycle of monthly mean temperature is displayed for eight stations ranging widely in latitude. The graphs are reproduced here as Figure A. The alphabetical list of stations is as follows, preceded by a code designation to be used in plotting the data and answering questions:

Code	Station Name	Mean Annual Temp. (°C)	Annual Range (C°)
DOU	Douala, Cameroon	25	2
EIS	Eismitte, Greenland	-30	32
FTV	Fort Vermilion, Alberta	-4	41
INS	In Salah, Algeria	26	22
MON	Monterey, California	13	7
OMA	Omaha, Nebraska	10	30
SIT	Sitka, Alaska	8	12
WAL	Walvis Bay, S.W. Africa	17	6

Using a clear plastic straightedge, laid horizontally across each graph, estimate the mean annual temperature for each station and enter in the space provided above. Estimate the annual temperature range for each station and enter in the space provided above. (Horizontal lines drawn through the highest and lowest months of each station will aid you in determining these values.)

Next, plot each of these stations on the blank graph provided, making a precise dot surrounded by a tiny circle. Enter the code symbol to the right of the dot.

(1) Disregarding for the moment the point for EIS, which is obviously in a position distant from the others, describe the arrangement of the remaining 7 points.

They appear to be roughly aligned in two groups. Straight lines drawn to fit

each group slant downward to the right. _____

_____

_____

_____

*Modern Physical Geography, 3rd Ed., p. 146-47, Figure 9.2, 9.3.

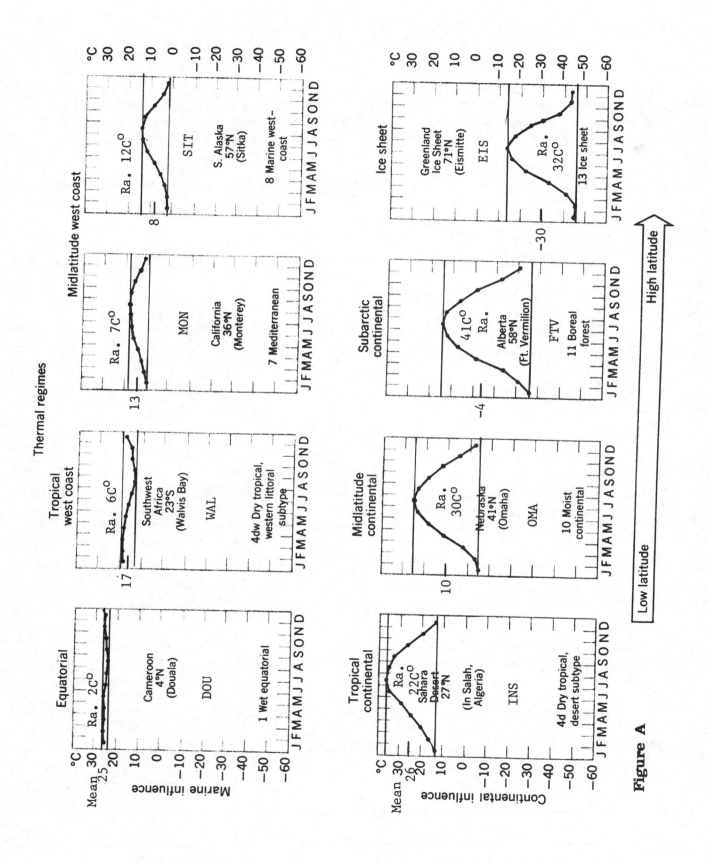

Figure A

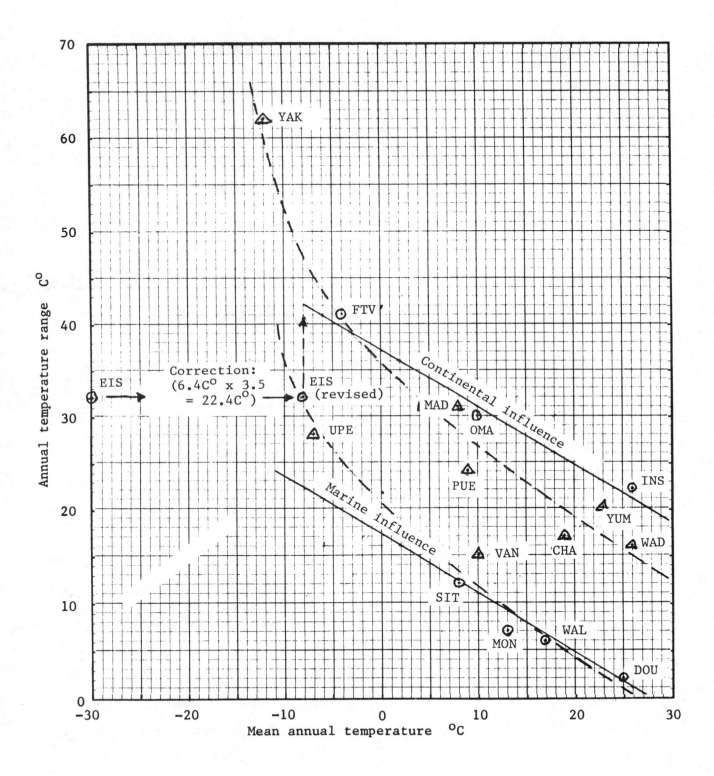

Using the straightedge, fit a slanting straight line to each of the two groups of points. Adjust each line so that the points lie as close as possible to it, evenly distributed on both sides of the line.

(2) Referring back to Figure 7.2, what is the obvious significance of the two lines of points on your graph?

The lower line consists of stations of marine influence; the upper line of those of

continental influence.

_____

Label the two sets of lines with the key words in your answer.

(3) Turning to the isolated point for EIS, suggest one or more reasons why it lies so far to the left (For inspiration, refer to Figure 11.18).

Eismitte station (literally "middle of the ice") has an elevation of about 3,500 m,

explaining why the mean annual temperature is so extremely cold.

_____

_____

(3) Using the environmental temperature lapse rate (Figure 3.6), revise the mean annual temperature for EIS to a sea-level value. Plot the new value on the graph. Enter the data below:

Number of degrees warmer at sea level: ___22.4___ C°

Revised sea-level value: ___-7.6___ °C

(4) The surface of the Greenland Ice Sheet consists entirely of snow throughout the year. Direct reflection of a large part of the incoming solar radiation prevents surface heating. Use this information to explain why the annual range at EIS would be expected to be considerably smaller than for a station at the same latitude in which vegetation and soil are snow-free during a brief summer (as on the arctic tundra).

Solar heating of a snow-free ground surface during the summer would greatly

boost the summer-month temperatures, increasing the annual range. This would

have the effect of moving the plotted point upward on the graph and bringing it

more nearly into line with the other three continental stations.

_____

Extra Credit Project

For good reason, scientists are skeptical of the significance of very small data samples. Eight additional stations, available in chapters 8 and 9 of your textbook, are listed below with references to text figures. Estimate the mean annual temperatures and temperature ranges. Plot the points on your graph, as before, using a different color for these dots.

Code	Station Name	Mean Annual Temp. (C)	Annual Range (C)
CHA	Charleston, S.Carolina (9.4)	19	17
MAD	Madison, Wisconsin (9.26)	8	31
PUE	Pueblo, Colorado (9.20)	9	24
UPE	Upernivik, Greenland (9.34)	-7	28
VAN	Vancouver, Br. Columbia (9.16)	10	15
WAD	Wadi Halfa, Sudan (8.16)	26	16
YAK	Yakutsk, USSR (9.29)	-12	62
YUM	Yuma, Arizona (8.17)	23	20

(5) What effect does the addition of eight more stations have on the fields of points?

The marine and continental groups remain separated, but scatter is substantially increased.

(6) With the doubled sample, does a slanting straight line still appear to be the best-fitting line? If not, sketch in a different curve for each group.

Curving lines, steepening toward the upper left, would appear to be better fitted to the data. These bring UPE and YAK on line.

For the Mathematicians

Research in physical geography makes use of a statistical procedure called _regression_ _analysis_, in which an equation is derived to show the relationship between the dependent variable, Y (which in this case is temperature range), and the independent variable, X (which is mean annual temperature). For a straight regression line (arithmetically linear), the regression equation takes the following form:

$$Y = a \pm bX, \text{ where a and b are constants.}$$

The regression coefficient _b_ describes the slope of the regression line. In this case, it has a negative sign, so our equation is $Y = a \pm bX$. You can evaluate _b_ by using a right triangle on the map and measuring its two legs, calculating the ratio of the vertical leg to the horizontal leg. To establish the coefficient _a_, you need to establish the y-intercept where X = 0.

(7) State your two regression equations:

For the continental stations: $\underline{Y = 37 - 0.625X}$ (where b = 0.625,

or tan 32°)

For the marine stations: $Y = 17 - 0.625X$

As an added project, determine the standard deviations for each equation. Then test each equation for significance.

Name _____ Date _____

_____   _____

Exercise 7-B   Variability of Rainfall from Year to Year
[Text p. 149-51, Figure 7.5.]*

From the news media we know that in a given year severe drought can strike a particular state or group of states, while at the same time other parts of the country are getting much more than their average amount of precipitation. In this exercise we study the concept of <u>variability</u> <u>of</u> <u>precipitation</u>. This is a concept not treated in your textbook, which presents the data of climate in terms of averages calculated over long periods of record. The concept of variability is thus a new climate dimension to investigate.

As a data base, we will use three rainfall graphs (Figure A, below). Each bar tells the actual amount of rainfall measured by rain gauge in a single month of a given year at a weather station. The period of consecutive years of record is stated on each graph. The data were compiled many years ago by H.H. Clayton, a scientist of the Smithsonian Institution. The study of global precipitation variability was investigated by geographers Erwin Biel and William Van Royen. The latter's world map of precipitation variability is reproduced as Figure A of Exercise 7-D.

On the graph, the average (mean) value for the station is given and is shown by a horizontal dashed line. The month selected for this study is that month which, on the average, is the rainiest month at the station.

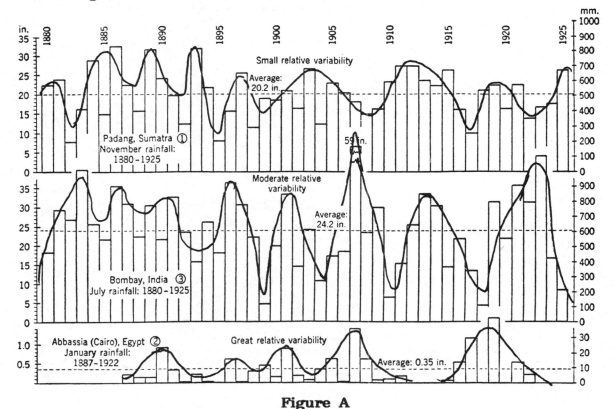

Figure A

---

*Modern Physical Geography, 3rd Ed., p. 149-50, Figures 9.5, 9.6.

(1) Identify the climate represented in each graph. To do this, find the stations on the world climate map, Figure 7.9 of your textbook.

		Climate name and number
Graph A	Padang, Sumatra	Wet equatorial climate (1)
Graph B	Bombay, India	Wet-dry tropical climate (3)
Graph C	Abbassia, Egypt	Dry subtropical climate (5d)

(2) From each graph, select the years of the three wettest months and the three driest months. Measure the <u>difference</u> between the monthly value and the average value (dashed line). Use inches as the scale. A pair of dividers will help you do this quickly. (Read to nearest whole inch for A and B; to nearest tenth-inch for C.) Record the differences in the following tables. Then carry out the indicated calculations to determine the <u>relative variability</u>.

Graph A Padang, Sumatra

	Difference, inches	Quantity squared
1	+13	169
2	+12	144
3	+12	144
4	-12	144
5	-12	144
6	-10	100
Sum of squares:		845

Sample variance (divide by 6):   140

Square root of variance: 11.8

Ratio of square root to
average (relative variability):   0.6 , or   60   percent

Graph B Bombay, India

	Difference, inches	Quantity squared
1	+35	1225
2	+19	361
3	+16	256
4	-19	361
5	-20	400
6	-17	289

Sum of squares: 2892

Sample variance (divide by 6): 482

Square root of variance: 22

Ratio of square root to average (relative variability): 0.92 , or 92 percent

Graph C Abbassia, Egypt (near Cairo)

	Difference, inches	Quantity squared
1	+1.40	1.96
2	+1.10	1.21
3	+0.80	0.64
4	-0.35	0.12
5	-0.35	0.12
6	-0.35	0.12

Sum of squares: 4.17

Sample variance (divide by 6): 0.70

Square root of variance: 0.83

Ratio of square root to average (relative variability): 2.4 , or 240 percent

(3) Which station has the smallest relative variability?   (A) Padang

Which the largest?   (C) Cairo

(4) What general principle or rule might be tentatively drawn from this relationship?

Increasing aridity is associated with increasing relative variability of

rainfall.

(5) Does relative variability consistently increase as monthly average rainfall decreases?

No. Station B, Bombay, which is intermediate in relative variance, has the

largest monthly rainfall.

(6) Close scrutiny of these graphs might lead you to suspect they show distinct rhythms of increase and decrease. Using a soft pencil (easy to erase and redraw repeatedly), attempt to draw a smooth curve over the tops of the bars, revealing a precipitation cycle.

(7) Describe in a very general way the cycles you read into the graph.

For both Padang and Bombay, cycles with peaks ranging from 5 to 9 years apart

are clearly present after about 1890. For Bombay, especially, the rhythm is well

developed.

(8) For crop farming, which of the three stations presents the greatest hazard? Explain.

For Abbassia, the climate is much too dry for any crop farming without

irrigation. Here the hazard lies in the reliability of the seasonal flood of the Nile.

For Padang, the risk is relatively small for rice cultivation; years of very low

rainfall are few and far between. For the region near Bombay, the hazard is

high--feast or famine--with very low rainfalls every few years (July is the peak

month of the monsoon).

Exercise 7-C   Global Precipitation Variability and Climate
[Text Figures 7.5, 7.7, 7.9.]*

From rainfall variability at a station for a given month over a period of years, we move to the scale of mean annual precipitation. Because snowfall is important in the colder regions, the term "precipitation" is appropriate here.

The accompanying world map, Figure A, shows the percentage departure of the total annual precipitation from the mean value. The lines on this map are a special kind of isopleth, scaled in percentage, according to the legend shown. You are to use this map in conjunction with the world climate map, Figure 7.9, in your textbook. Keep in mind that the variability map is highly generalized and largely ignores major mountains and plateaus.

(1) Locate on the variability map each of the 12 stations listed in Table A. Show the station by a dot and the station symbol. Enter the percentage range in the appropriate space in the table. Then use the world climate map to determine the corresponding climate and climate subtype.

(2) Do the data you have recorded show the same trends as those you obtained in Exercise 7-b?

Yes. Variability is low for the wet equatorial climate (Padang), intermediate for

the wet-dry tropical climate (Bombay, Kaduna), and greatest for the low-latitude

deserts (Abbassia, Cairo) _____

_____

(3) As shown on the variability map, which climates register areas of variability less than 10 percent? Where are these areas located?

Marine west-coast climate (8p, 8h) in western Europe, Tierra del Fuego, and

South Island of New Zealand. Wet equatorial climate (1) of Sumatra. Monsoon

climate (2) of Burma coast. Small area of moist continental climate (10h) in the

low countries of Europe. _____

_____

*Modern Physical Geography, 3rd Ed., Figures 9.5, 9.6, 9.8, Pl. C-2.

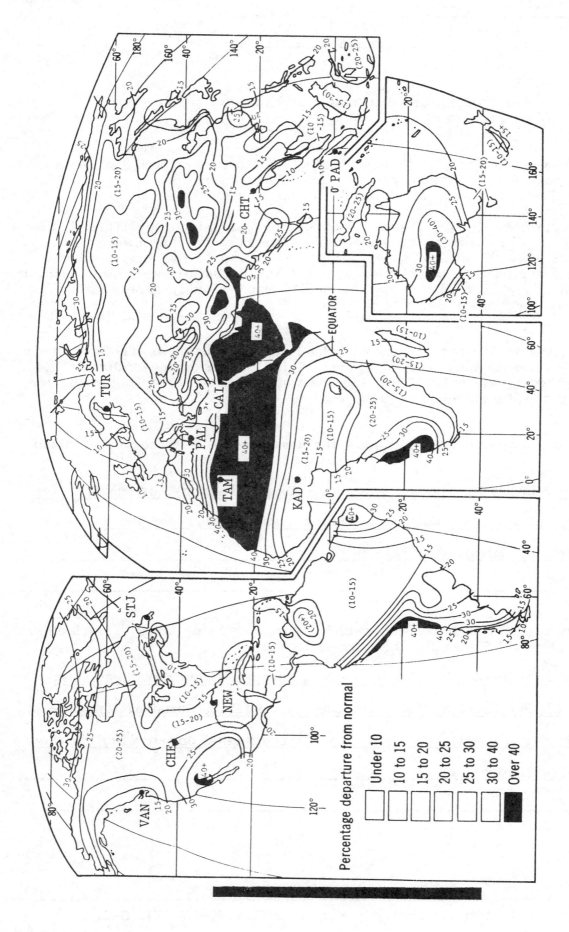

Figure A Precipitation variability map of the world. (After William Van Royen, Atlas of the World's Agricultural Resources, Prentice-Hall, Englewood Cliffs, N.J. Based on Goode Base Map. Copyright by the University of Chicago, Used by permission of the Geography Department.)

Percentage departure from normal

Under 10
10 to 15
15 to 20
20 to 25
25 to 30
30 to 40
Over 40

Table A

Station		Mean annual precip.(cm)	Variability (%)	Climate number and name	
STJ	St.Johns, Newfound-land 47°N, 53°W	137	<10	11p	Boreal forest
PAD	Padang, Sumatra 1°S, 100°E	241	10-15	1	Wet equatorial
CHT	Chittagong, Bangla-desh 22°N, 92°E	274	10-15	2	Monsoon
SHA	Shanghai, China 31°N, 121°E	114	10-15	10h	Moist continental
VAN	Vancouver, B.C. 46°N, 122°W	145	10-15	8h	Marine w.-coast
NEW	New Orleans, LA 30°N, 90°W	146	15-20	6h	Moist subtropical
TUR	Turku, Finland 60°N, 22°E	61	15-20	11h	Boreal forest
KAD	Kaduna, Nigeria 11°N, 7°E	127	15-20	3	Wet-dry tropical
PAL	Palermo, Sicily 38°N, 13°E	71	15-20	7	Mediterranean
CHE	Cheyenne, WY 41°N, 105°W	38	20-25	9s	Dry midlatitude
CAI	Cairo, Egypt 30°N, 31°E	3	30-40	5d	Dry subtropical
TAM	Tamanrasett, Algeria 23°N, 3°W	4	>40	4d	Dry tropical

(4) In terms of belts of prevailing winds, and air masses, what is the reason for the very small variability in western Europe?

Prevailing westerlies, bringing moist air masses from the North Atlantic throughout the year, provide ample precipitation throughout the year. The warm North Atlantic current intensifies this effect.

(5) How do you explain the low variability in Nova Scotia and New Brunswick?

Here, the marine influence of the North Atlantic is very strong. Cyclones over the coastal waters bring in moist marine arimasses in winter, thus keeping monthly precipitation high through the winter months. (Precipitation at Halifax is remarkably uniform throughout the year).

(6) Notice the small area of extremely high variability in northeastern Brazil, lat. 5 to 10°S. With what climate is this area associated? Where else can you find such extreme variability in the same latitude zone? Can you explain this area of aridity so close to the equator?

The climate here is the semiarid subtype of the dry tropical climate (4s). A similar area of extreme variability is found in Somalia on the "Horn" of Africa where the tropical desert climate (4d) occurs. (In Brazil, this semiarid area experiences a very dry season of nearly six months' duration and a short wet season. No easy explanation can be expected for this anomaly in terms of general patterns of pressure, winds, and airmasses.)

Exercise 7-D   Analyzing the Soil-Water Budget
          [Text p. 161-62, Figure 7.13.]*

The annual soil-water budget, making use of the data for each of the twelve months of the year, is much like a personal financial budget in which a checking account is combined with a loan account (i.e., a credit-card account). Your monthly statements tell what you put in, what you spent, and what you borrowed and repayed. Some months see a surplus, others an overdraft carried forward as a loan. Like ordinary people, the various climates differ greatly in the styles of their budgets. Some manage to keep on an even keel, others are always in debt, and a very few have more wealth than they know what to do with. One difference is that with the soil-water budget, excess income (as precipitation) can't be held in savings; it's just lost.

The terms of the soil-water budget, all in centimeters, are as as follows:

P   =   Precipitation

Ep =   Water need (potential evapotranspiration)

Ea =   Water use (actual evapotranspiration)

D   =   Soil-water shortage

-G =   Storage withdrawal

+G =   Storage recharge

R   =   Water surplus

Table A gives data for for a model soil-water budget in which all figures have been rounded off to the nearest 0.5-cm for easy addition and subtraction. Note that the annual total of -G must always equal the total of +G. Storage recharge, +G, will usually be completed midway through one of the months. In that case, assign to +G only enough of the monthly quantity needed to bring the annual total of +G to the annual total of -G. Assign the balance of that month to surplus, R. In the example below, recharge was completed during the month of November, so that 5.5 cm was assigned to +G and 3.5 cm to R.

(1) On the blank graph provided, plot monthly values of P, Ep, and Ea, following the example in Figure 7.13 of your textbook. Note that the points are plotted along the midline of the month. Use the following symbols:

P   Solid dot
Ep   Open dot (circle)
Ea   Triangle

---

*Modern Physical Geography, 3rd Ed., p. 171-72, Figure 10.11.

Table A

	J	F	M	A	M	J	J	A	S	O	N	D	Year
P	11.0	9.0	6.0	3.0	2.5	2.0	2.5	4.0	7.0	9.0	10.5	12.0	78.5
Ep	1.0	2.0	3.5	6.0	8.5	9.5	9.0	7.0	4.5	3.0	1.5	1.5	57.0
Ea	1.0	2.0	3.5	5.5	7.0	6.0	5.0	4.5	4.5	3.0	1.5	1.5	45.5
D	0.0	0.0	0.0	0.5	1.5	3.5	4.0	2.5	0.0	0.0	0.0	0.0	12.0
-G	---	---	---	2.5	4.5	4.0	2.5	0.5	---	---	---	---	14.0
+G	---	---	---	---	---	---	---	---	2.5	6.0	5.5	---	14.0
R	10.0	7.0	2.5	---	---	---	---	---	---	---	3.5	10.5	33.5

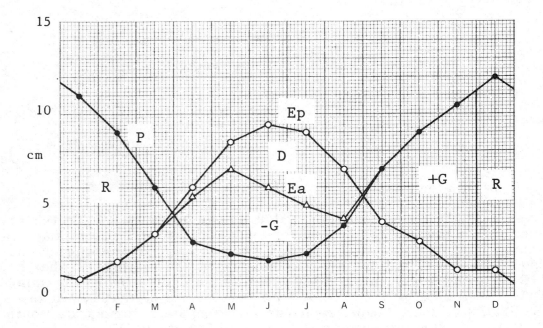

Connect the points with straight-line segments (not curved lines, as in text Figure 7.13). Use a vertical line to separate that month in which recharge is completed from the following month in which all the excess precipitation is a surplus quantity. Label the three curves and the areas between the curves, as on the example, Figure 7.13.

To enhance the impact of this graph, you may color the areas between plotted lines, using the following code: D = red, R = blue, G = yellow, +G = diagonal blue lines.

(2) What precipitation type is depicted in this soil-water budget? Refer to text Figure 7.5 and select a type.

Mediterranean type: west-coast with dry summer and winter maximum.

(3) Why does actual evapotranspiration, Ea, which we call "water use," quickly reach a peak and then decline steadily?

When water need, Ep, exceeds precipitation, P, plants begin to compensate by transpiring less water. As the soil becomes more dry, evapotranspiration goes on more slowly.

(4) What is the soil-water condition during the period from September through November?

Because precipitation, P, exceeds water need, Ep, the soil stays fully at its water-holding capacity (known as "field capacity").

(5) How does the total runoff, R, compare with the total precipitation, P?

Total runoff is somewhat less that half the total precipitation (actually about 43%).

(6) If you were operating a vegetable garden or keeping up the greens of a golf course in this area, would irrigation be needed in an average year? If so, during what months? How much water would need to be applied during this season?

Irrigation would be needed, starting about in May (in an average year) and continuing through August. About 12 cm of water would be needed for full irrigation during the entire deficit season.

(7) Hazard a guess as to the actual location of this station. Name more than one possible location.

Northern California coastal zone; Spain, Italy, or Israel in the Mediterranean region.

Name _____ Date _____

_____   _____

Exercise 7-E Identifying Köppen Climates
[Text p. 164-69, Figures 7.16-7.19.]*

Table A gives climate data for ten stations, designated by letters (a) through (j). Monthly mean temperatures (°F) and mean annual temperature are given in the upper line of figures. The lower line gives mean monthly precipitation in inches, and mean annual total.

Table A

		J	F	M	A	M	J	J	A	S	O	N	D	Year
(a)	°F	13.3	13.6	27.0	43.9	53.8	62.2	69.1	65.8	56.1	43.9	30.9	20.1	41.5
	In.	0.7	0.6	0.5	1.0	2.0	2.9	1.8	1.2	1.3	0.7	0.6	0.6	13.9
(b)	°F	−14.4	7.4	13.3	33.8	46.8	58.5	62.6	58.5	46.0	30.0	6.8	−9.2	27.1
	In.	0.1	0.1	0.1	0.4	0.9	1.7	2.4	3.0	1.3	0.3	0.2	0.2	10.3
(c)	°F	71.1	73.6	83.1	91.6	94.5	93.7	89.2	86.5	89.2	88.9	80.8	71.1	84.4
	In.	0	0	0.1	0	0.3	0.9	3.5	2.8	1.1	0.4	0	0	9.0
(d)	°F	33.4	35.2	41.9	53.2	64.0	72.1	76.6	75.0	68.0	57.7	46.2	36.9	55.0
	In.	3.4	3.7	3.6	3.4	3.5	3.9	4.6	4.3	3.4	2.8	2.6	3.3	42.5
(e)	°F	47.8	49.5	52.5	58.3	66.4	74.1	79.7	79.5	73.2	66.2	57.0	57.8	63.0
	In.	2.2	1.8	1.3	0.9	0.8	0.6	0.3	0.6	0.7	1.4	2.9	2.5	16.0
(f)	°F	77.9	78.4	79.3	79.9	80.6	79.9	80.2	79.7	79.5	79.7	79.0	78.3	79.3
	In.	9.7	7.1	7.3	7.8	6.5	7.0	6.7	7.8	6.9	7.9	10.1	10.5	95.1
(g)	°F	80.8	80.6	80.8	80.4	78.4	75.0	74.8	78.1	81.3	82.6	81.7	81.0	79.7
	In.	9.6	8.9	8.1	4.1	2.0	0.2	0.1	1.1	2.0	4.4	6.0	7.9	54.5
(h)	°F	−14.3	−12.8	−2.0	12.0	28.6	38.7	45.7	44.2	36.3	26.1	12.4	−3.8	17.6
	In.	1.1	1.3	0.8	1.5	1.5	1.1	2.6	2.0	1.2	1.5	2.2	1.5	18.3
(i)	°F	63.3	63.0	62.2	58.3	54.1	51.6	50.4	53.4	60.1	63.5	64.6	63.9	59.0
	In.	10.6	8.4	8.3	2.2	0.9	0.6	0.8	0.3	0.3	1.3	3.5	8.5	45.7
(j)	°F	22.6	23.9	31.6	42.4	52.5	61.9	67.5	65.7	59.2	49.6	37.9	27.7	45.1
	In.	3.9	4.3	3.8	3.4	3.3	3.3	3.2	3.1	3.1	3.1	3.5	3.9	41.9

Determine the Köppen climate to which each station belongs, referring to the climate list and symbols on text p. 165, 168-69. Enter the code symbols in the spaces provided. If possible, give the third letter of the code as well. For each, state the criteria used in assigning the station to that climate.

---

*Modern Physical Geography, 3rd Ed., p. 162-65, Figures 9.33, 9.34, Pl. C.4.

(a) Code ___BSk___ Criteria ___Falls in the BS area in middle graph of Figure 7.17. Mean annual temperature under 64.4°F. (Havre, Montana, 49°N.)___

(b) Code ___Dwc___ Criteria ___Warmest month mean over 50°F. Coldest month mean under 26.6°F. Fewer than four months with mean over 50°F. (Tunka, Siberia, 54°N.)___

(c) Code ___BWh___ Criteria ___Falls in BW area in middle graph of Figure 7.17. Mean annual temperature well over 64.4°F. (Tombouctou, or Timbuktu, Mali, 17°N.___

(d) Code ___Cfa___ Criteria ___Mean temperature of coldest month lower than 64.4°F, but above 26.6°F. Ample precipitation in all months. Warmest month mean over 71.6°F. (Baltimore, Maryland, 39°N.)___

(e) Code ___Csa___ Criteria ___Mean temperature of coldest month lower than 64.4°F, but above 26.6°F. Dry season in summer months. Warmest month mean over 71.6°F. (Athens, Greece, 38°N.)___

(f) Code ___Af___ Criteria <u>All monthly mean temperatures over 64.4°F. Ample</u> <u>precipitation in all months. Least is 6.5 inches in May. (Singapore,</u> <u>1°N.)</u>

_____

(g) Code ___Aw___ Criteria <u>All monthly mean temperatures over 64.4°F. Dry</u> <u>season in low-sun period, June-July. (Cuiaba, Brazil, 16°S.)</u>

_____

_____

(h) Code ___ET___ Criteria <u>Warmest month mean under 50°F. (Lake Harbor,</u> <u>Baffin Island, 63°N.)</u>

_____

_____

(i) Code ___Cwb___ Criteria <u>Mean temperature of coldest month below 64.4°F</u> <u>but above 26.6°F. Dry season in low-sun period. Warmest month mean under</u> <u>71.6°F. More than four months have means over 50°F. (Juliasdale, Zimbabwe,</u> <u>18°S., el. 6070 ft.)</u>

(j) Code ___Dfb___ Criteria <u>Mean temperature of coldest month below 64°F.</u> <u>Ample precipitation in all months. Warmest month mean under 71.7°F. More than</u> <u>four months have means over 50°F. (Portland, Maine, 44°N.)</u>

_____

Name _____   Date _____

_____   _____

Exercise 8-A Rainfall Regimes of the Low-Latitude Climates

Part 1 Seasonal Trends in Monthly Rainfall of Stations in the Wet Equatorial Climate
[Text, p. 171-72, Figures 8.1, 8.2.]*

Graph A shows mean monthly rainfall for five equatorial-zone stations having substantial rainfall in all months and no soil-water shortage in any month. Although all are classified as belonging to the wet equatorial climate, the rainfall regimes of these stations differ quite a lot among themselves both as to the total annual rainfall and in what seem to be unique seasonal patterns of variation in monthly rainfalls.

Station data are as follows:

Code	Station name	Location lat., long.	Mean annual total rainfall	Length of record, yrs.
BAS	Basoko, Zaire	1°N 23½°E	71 in. (180 cm)	42
BEL	Belém, Brazil	1½°S 48½°W	96 in. (244 cm)	20
GEO	Georgetown, Guyana	7°N 58°W	89 in. (226 cm)	35
JAL	Jaluit, Marshall Is.	6°N 170°E	159 in. (404 cm)	17
SIN	Singapore, Republic of Singapore	1°N 104°E	95 in. (241 cm)	64

Locate each of these stations on (a) the world climate map, text Figure 7.9, and (b) the world map of rainforest, text Figure 21.1. Make a dot at the station location and write in the station code.

Basoko, an important trading post in the former Belgian Congo, lies on the Congo River. Belém, founded by the Portuguese in 1616 at the mouth of the mighty Amazon River, was the gateway to the vast lowland rainforest of Amazonia with its rich yield of natural rubber and other forest products. Georgetown, port city of what was formerly British Guiana, gained notoriety in the American press as the site of evangelist Jim Jones's colony, wiped out by mass suicide in 1978. Jaluit, an atoll island in the central Pacific was occupied by the Japanese in World War II, then captured by U.S. forces in 1944. Singapore, a highly industrialized island city, is smaller in area than New York City but has a population of over 2 million.

Data source: Throughout the exercises for Chapters 8 and 9, station data are from Tables of Temperature, Relative Humidity and Precipitation for the World, Meteorological Office of Great Britain, Her Majesty's Stationery Office, London.

---

*Modern Physical Geography, 3rd Ed., p. 153-156, 177, Figures 9.9, 10.16.

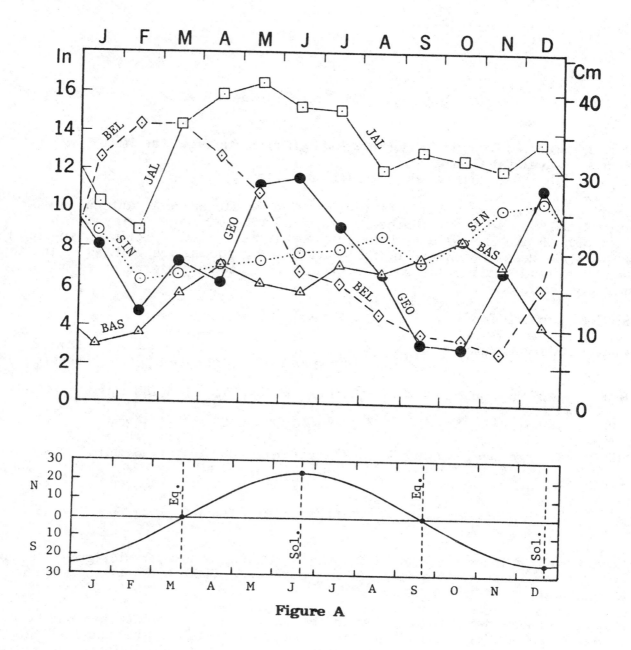

Figure A

(1) Enter the station code for each of the following: Largest total annual rainfall:
___JAL___ . Least annual rainfall: ___BAS___ . Greatest range in monthly rainfall:
___BEL___ (__12__ in.). Smallest range in monthly rainfall: __SIN__ (__4__ in.).

(2) Look for strong annual cycles in these rainfall records. Identify a station
with a particularly stong cycle consisting of one "crest"
and one "trough." ___BEL___ . Identify a station with two "crests"
and two "troughs." ___GEO___ .

We have included in Graph A the annual curve of the sun's declination. You can think of this graph as telling the parallel of latitude at which the sun at noon is directly overhead in the sky (i.e., in the zenith). Keep in mind also that the declination curve tells in a general way how the belts of pressures and winds shift alternately north and south. In this case, we are concerned with the shifting position of the intertropical convergence zone (ITC), and the trade-wind belts that flank it.

(3) Comparing the rainfall graph of each station with the declination curve, can you identify significant correlations between seasons of high (or low) rainfall with position of the ITC or the trade winds? Keep in mind that these rainfall seasons may lag two months or so behind the sun's declination. (Hint: Try JAL first, referring to the world maps of pressure and winds, text Figure 4.10.)

One possibility is that the strong rainfall minimum for JAL (lat. 6°N) in January and February is an effect of the southward migration of the northeast trades into this area and the distancing of the ITC toward the south. (This is a difficult question to answer and all kinds of statements can be expected. What counts is the intensity of inquiry, based on fundamentals of general global climatology.)

A Word About English and Metric Units of Measurement

In Chapters 8 and 9, we use English units--Fahrenheit degrees and inches--for the measurement of air temperature and precipitation, respectively. This practice is unacceptable today in college courses in the physical and biological sciences; instead, Celsius degrees and centimeters are required.

Not only are metric units the world standard for all science, but in the physical sciences Le Système Internationale d'Unités ("International System of Units," or "SI") is demanded of all students and research scientists. Under the SI, the correct units are meter, kilogram, and second (MKS) for mechanics, and the Kelvin degree (K) for temperature. Thus, one inch (2.54 cm) of rainfall would be expressed as 0.0254 m; a temperature of 50°F (10°C) as 313 K.

Why, then do we continue to use the English units in these exercises? There are two reasons. First and foremost, as Americans we associate Fahrenheit degrees with levels of bodily comfort or discomfort and with and many practical activities of daily life. We measure small lengths in inches and feet. Using these familiar units allows us to comprehend the statistics of climate in distant lands in a practical and down-to-earth way. Second, the National Weather Service continues to use the English units in passing along useful information on weather and climate to the general public. Because most college students of physical geography are not science majors, we continue to present the physical environment of humans in a way that is most easily understood and appreciated in the context of a general education.

(Note, however, that for exercises on the soil-water budget, we move one step closer to the SI by using centimeters exclusively.)

Part 2 Seasonal Rainfall Contrasts in the Wet-Dry Tropical Climate
[Text, p. 178-79, Figures 8.11, 8.12]*

The bar graph below shows seasonal rainfall contrasts for four stations in the wet-dry tropical climate (3). The rainfall of the wettest month is compared with the combined rainfall of the three driest months.

Station data are as follows:

Code	Station name	Location lat., long.	Mean annual total rainfall	Length of record, yrs.
BOM	Bombay, India	19°N 73°E	71.2 in. (181 cm)	60
DAR	Darwin, N.T., Australia	12½°S 132°E	58.7 in. (149 cm)	70
LUB	Lubumbashi, Zaire	12°S 27½°E	48.7 in. (124 cm)	17
PAR	Paraná, Brazil	12½°S 48°W	62.3 in. (158 cm)	19

Locate each of these stations on (a) the world climate map, text Figure 9.7, and the world map of tropical raingreen vegetation, text Figure 21.3. Make a dot at the station location and write in the station code.

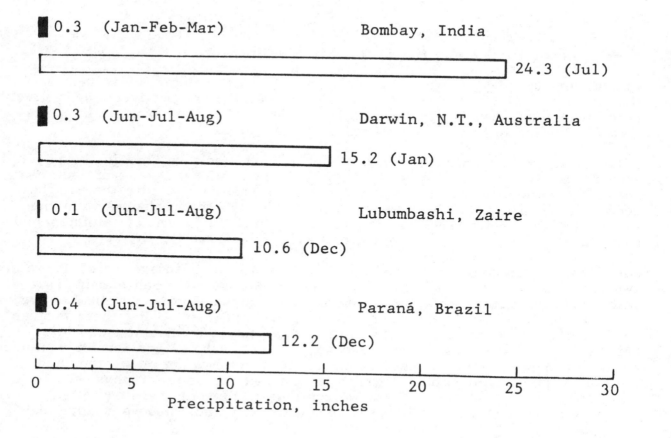

Figure B

The port city of Bombay, long known to colonial Britons bound for military or civil service in India as the "Gateway to India," lies next to the Arabian Sea. Darwin, on the northern coast of Australia, was originally called Palmerston but was renamed in the honor of the great evolutionary biologist whose ship, <u>H.M.S Beagle</u>, visited the spot in 1839. Lubumbashi was known during the colonial era as Elisabethville. A leading city in the Belgian Congo, it prospered as a smelting center for a booming copper-mining industry. Paraná lies only about 200 mi (300 km) north of Brazil's spanking new capital city, Brasilia; they share much the same kind of climate, although Brasilia is on higher ground.

(4) Clearly, Bombay has the largest rainfall in the wettest month. Offer an explanation.

<u>The Asiatic monsoon effect is the obvious cause of Bombay's greater rainfall, with</u>

<u>its favorable coastal location open to moist equatorial airmasses moving in from</u>

<u>the Arabian Sea.</u>

_____

(5) Darwin's wettest month is January. Explain. Why is the June-August period so dry? (Hint: study the world pressure/winds maps of text Figure 4.10.)

<u>In January, Darwin lies along the ITC in a large low-pressure region dominated</u>

<u>by the, maritime equatorial airmass, whereas in July it is in the belt of southeast</u>

<u>trades and is dominated by the continental tropical airmass from interior</u>

<u>Australia.</u>

_____

(6) The maximum-month rainfalls of both Lubumbashi and Parana are low, compared with the other two. Give an explanation in terms of air masses.

<u>A possible explanation is that these stations are in the deep continental interiors.</u>

<u>Moist air masses must travel long distances over land to reach these locations and</u>

<u>will have already undergone a substantial loss of moisture content.</u>

_____

Name _____   Date _____

_____   _____

Exercise 8-B   Air Temperatures in the Tropical Desert
        [Text, p. 143-44, Figure 7.2; p. 182-83, Figures 8.16, 8.17.]*

In the low-latitude deserts, where a rainstorm is a rare event, air temperature dominates the climate as a factor in the human environment. We think first of the dreadful heat of deserts such as the Sahara and the Kalahari of Africa, giving little thought to near-freezing chill of the predawn hours during the season of low sun, especially where the surface elevation is high.

For this exercise you will use the data of text Figure 8.17, reproduced below as Graph A. It shows the spectrum of air temperatures observed at Bou-Bernous, Algeria, over a period of 15 years, during the time when Algeria was French territory. Remote fortified military outposts, such as Bou-Bernous, were then manned by the legendary French Foreign Legion. Bou-Bernous may well be among the leading contenders for the hottest climate on earth, although Azizia, Libya, still claims the official world's record for highest observed air temperature--136°F (58°C) in 1922. Death Valley, California, ranks not far behind with 134°F (57°C).

(1) Examine carefully the five sets of data for which points on the graph are connected by straight-line segments. What thermal regime is represented by this record? (See text Figure 7.2.)

    tropical continental

(2) Which month has the highest value in all five data sets?   July

(3) Is January the month of lowest values in all five data sets? Give details.

Yes, for mean of daily means. Not for mean of daily maxima and mean of highest in month, for which December is the lowest.

(4) The mean annual temperature, calculated from data of this period of record, is 79°F. Draw a horizontal line across the graph for this value and label in F and C degrees.

---

*Modern Physical Geography, 3rd Ed., p. 146-47, Figures 9.2, 9.3; p. 155-56, Figures 5.11, 9.14.

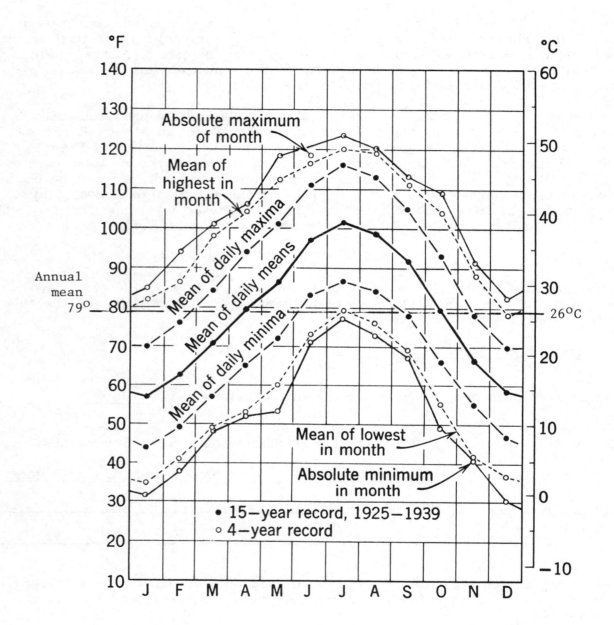

°F

°C

Annual
mean
79°

Absolute maximum
of month

Mean of
highest in
month

Mean of daily maxima

Mean of daily means

Mean of daily minima

Mean of lowest
in month

Absolute minimum
in month

• 15—year record, 1925—1939
○ 4—year record

26°C

Graph A

(5) Connect the points showing absolute maximum of month, and for the absolute minimum. Compare the regularity (smoothness) of curve of the annual cycle for the absolute maximum (minimum) with that of mean of highest (lowest) in month and mean of daily maxima (minima). Give an explanation for your conclusion. (Note: These data are for only 4 years of record.)

Points for the absolute maximum and minimum show a high degree of irregularity as compared to a smooth annual curve. The curves of mean of highest and lowest in month are more regular; those for the mean of daily maxima and minima give the smoothest curve. Averaging has a smoothing effect. Moreover, each graph point for the absolute maximum (minimum) represents only one thermometer reading in the 4-year period of record. Points for the mean of highest (lowest) in month are computed from only 4 readings during the period of record.

(6) Approximately how many individual thermometer readings went into the calculation of the mean daily maximum temperature?

Using 30 as the number of days per month and multiplying times 15 years, we arrive at 450 readings for each point.

(7) Calculate the annual range of temperature at Bou Bernous, (Use mean of daily means.) Compare this value with the range for In Salah, Algeria, text Figure 7.2, and for Wadi Halfa, Sudan, Figure 8.16.

Bou-Bernous:	44	F°,	79	C°
In Salah:	45	F°,	25	C°
Wadi Halfa:	29	F,°	16	C°

(8) Speculate on the differences in these ranges. Latitudes are as follows: Bou-Bernous and In Salah, 27°N; Wadi Halfa, 22°N. (Hint: Look up in an atlas the location of Wadi Halfa.)

Wadi Halfa's lower range may be related to its lower latitude, placing it about 350 miles farther south and therefore closer to the wet-dry tropical climate zone, with its smaller temperature range. Because Wadi Halfa is a city on the Nile River, perhaps the thermometer shelter was situated in or near irrigated fields where the soil was moist, allowing evaporation to lower the air temperatures in the high-sun months.

Exercise 8-C   Climographs of Low-Latitude Climates
        [Text p. 171-84, Figures 8.1, 8.4, 8.7,
        8.11, 8.16, 8.19, 8.20.]*

Plotting climographs for stations scattered around the world and analyzing their contents is a good way to strengthen your understanding of climate types. The station data given on following pages are arranged in the same order as in the textbook. To identify the climate and subtype, you would need only to find the station on the world climate map, text Figure 7.9, and read from the map legend. However, you will be asked to compare the station with the example shown in the textbook and to explain the differences between these and the other plotted climographs.

Using the latitude and longitude given, find the location of each station on the world climate map, Figure 7.9, and show it with a dot. Write in the station name next to the dot.

Plotting the data: As in your textbook climographs, monthly temperatures (°F) are plotted as points centered in each monthly column and connected by straight-line segments. Mean monthly precipitation amounts (inches) are plotted as bars, forming a step-graph.

Beside each climograph is a list of quantities to be entered in the blanks provided. The <u>mean</u> <u>annual</u> <u>temperature</u> (sum of monthly means divided by 12) is given with the station data. You are asked to calculate and enter the following statistics: <u>annual</u> <u>temperature</u> <u>range</u> (difference between highest and lowest monthly means); <u>annual</u> <u>total</u> <u>precipitation</u> (sum of the monthly values).

Enter the number and name for the climate type and subtype in the blank spaces provided. These should follow the system given in the legend of the world climate map, text Figure 7.9.

If your instructor requires the Köppen climate system, identify the climate by use of the Köppen-Geiger world map, text Figure 7.17, and text definitions and boundary graphs, p. 164-69 and Figures 7.18 and 7.19. Enter the Köppen code symbol and Köppen climate name in the blank spaces provided beside each climograph.

---

*Modern Physical Geography, 3rd Ed., p.153-56, Figures 9.9-9.15.

Station Data

(a) Cristobal, Panama (Canal Zone) 9½°N 80°W

	J	F	M	A	M	J	J	A	S	O	N	D	Mean
T (°F)	80	80.5	81	81.5	81	81	80.5	80.5	80.5	80.5	79.5	80.5	80.5
P (in.)	3.4	1.5	1.5	4.1	12.5	13.9	15.6	15.3	12.7	15.8	22.3	11.7	

(b) Saigon, Vietnam 11°N 107°E

	J	F	M	A	M	J	J	A	S	O	N	D	Mean
T (°F)	79.5	81	83.5	85.5	84	82	81.5	81.5	81	81	80	79	82
P (in.)	0.6	0.1	0.5	1.7	8.7	13.0	12.4	10.6	13.2	10.6	4.5	2.2	

(c) Cairns, Queensland, Australia 17°S 146°E

	J	F	M	A	M	J	J	A	S	O	N	D	Mean
T (°F)	82	81.5	80	77.5	73.5	71.5	69.5	71	73.5	77	79	81.5	76.5
P (in.)	16.6	15.7	18.1	11.3	4.4	2.9	1.6	1.7	1.7	2.1	3.9	8.7	

(d) Nagpur, India 21°N 79°E Elevation 1025 ft (410 m).

	J	F	M	A	M	J	J	A	S	O	N	D	Mean
T (°F)	69.5	74.5	83	90.5	95.5	88.5	81.5	81	81.5	79	72.5	67.5	80.5
P (in.)	0.4	0.7	0.6	0.6	0.8	8.8	14.6	11.4	8.0	2.2	0.8	0.5	

(e) Alice Springs, N.T., Australia 23½°S 133½°E Elevation 1900 ft (580 m).

	J	F	M	A	M	J	J	A	S	O	N	D	Mean
T (°F)	83.5	82	76.5	67.5	59.5	54	53.5	59.5	65	73	78.5	82	68
P (in.)	1.7	1.3	1.1	0.4	0.6	0.5	0.3	0.3	0.3	0.7	1.2	1.5	

(f) Lima, Peru 12°S 77°W

	J	F	M	A	M	J	J	A	S	O	N	D	Mean
T (°F)	74	75	74.5	71.5	67	63	61	62	62.5	64.5	67	70	68
P (in.)	0.1	<0.1	<0.1	<0.1	0.2	0.2	0.3	0.3	0.3	0.1	0.1	<0.1	

(g) Quito, Ecuador ¼°S 78½°W Elevation 9,500 ft (2,900 m)

	J	F	M	A	M	J	J	A	S	O	N	D	Mean
T (°F)	59	59	59	58.5	58.5	58	58	59	59	59	58.5	59	59
P (in.)	3.9	4.4	5.6	6.9	5.4	1.7	0.8	1.2	2.7	4.4	3.8	3.1	

CLIMOGRAPHS

(a) Cristobal, Panama

Mean Ann. Temp.   80.5

Ann. Temp. Ra.   2

Ann. Precip.   130.3

Climate number   1

Climate name   Wet equatorial

Köppen code   Am

Köppen name   Monsoon var. of
                    Af

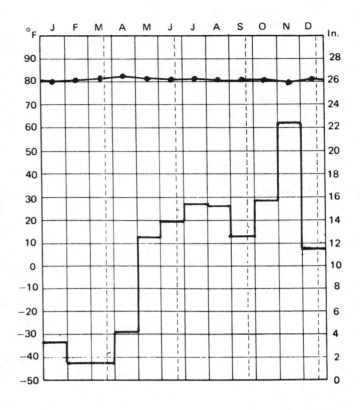

(b) Saigon, Vietnam

Mean Ann. Temp.   82

Ann. Temp. Ra.   6.5

Ann. Precip.   78.1

Climate number   2

Climate name   Monsoon

Köppen code   Aw

Köppen name   Tropical savanna

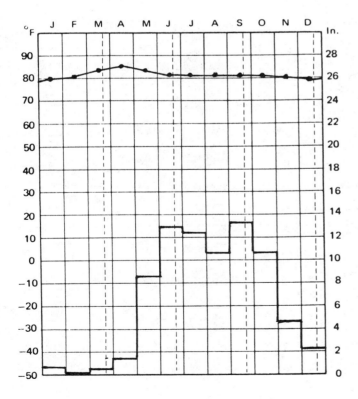

(c) Cairns, Australia

Mean Ann. Temp.	80.5
Ann. Temp. Ra.	12.5
Ann. Precip.	88.7
Climate number	2
Climate name	Trade-wind littoral
Köppen code	Af
Köppen name	Tropical rainforest climate

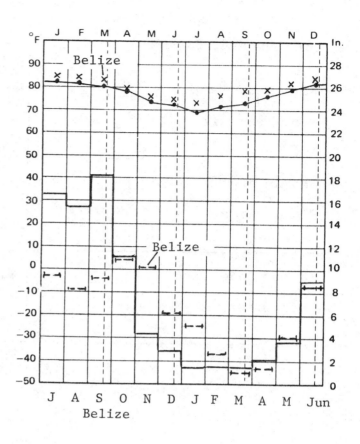

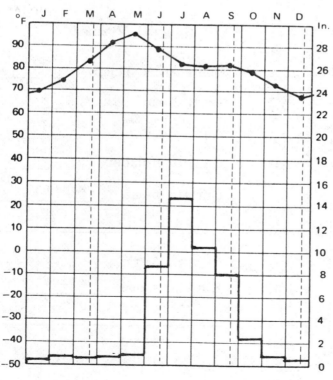

(d) Nagpur, India

Mean Ann. Temp.	80.5
Ann. Temp. Ra.	28
Ann. Precip.	49.4
Climate number	3
Climate name	Wet-dry tropical
Köppen code	Aw
Köppen name	tropical savanna

(e) Alice Springs, Australia

Mean Ann. Temp.	68
Ann. Temp. Ra.	30
Ann. Precip.	9.9
Climate number	4 d
Climate name	Dry tropical desert
Köppen code	BWh
Köppen name	Desert climate; dry-hot

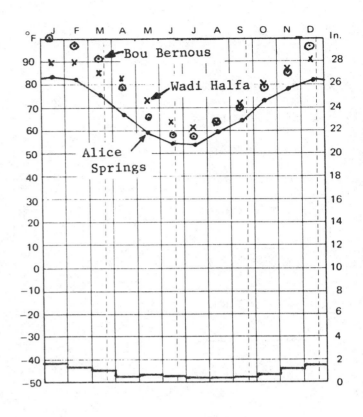

(f) Lima, Peru

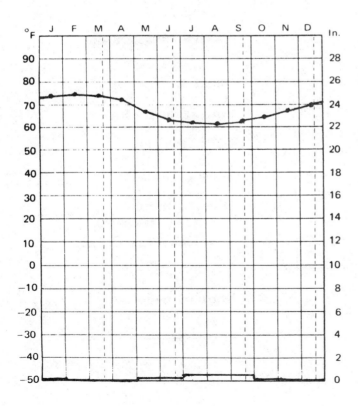

Mean Ann. Temp.	68
Ann. Temp. Ra.	14
Ann. Precip.	1.6
Climate number	4dw
Climate name	Dry tropical, desert, western littoral
Köppen code	BWh
Köppen name	Desert climate, dry-hot

(g) Quito, Ecuador

Mean Ann. Temp.   59

Ann. Temp. Ra.   1

Ann. Precip.   43.9

Climate number   _____

Climate name   Highland

_____

Köppen code   _____

Köppen name   _____

_____

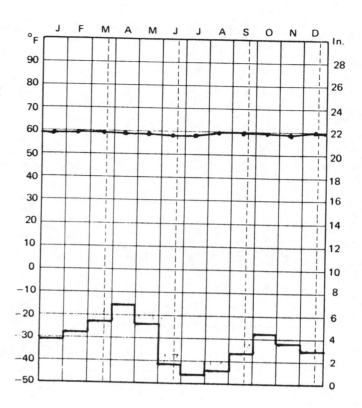

(Extra climograph)

Mean Ann. Temp.   _____

Ann. Temp. Ra.   _____

Ann. Precip.   _____

Climate number   _____

Climate name   _____

_____

Köppen code   _____

Köppen name   _____

_____

QUESTIONS

(a) Cristobal, Panama

Cristobal is located near the northern (Atlantic) end of the Panama Canal, where the Gatun Locks allow ships to ascend from sea level to the 85-foot (28-m) level of Lake Gatun. From these data you can begin to appreciate the horrendous climatic difficulties that beset the builders of the canal. The two-month dry period is a mere interruption of a succession of wet months in which rain falls in about two-thirds of the days of every month, amassing a seasonal total of about 120 inches. Small wonder that water-saturated soil and rock gave way in great mudslides.

(1) Compare the climograph of Cristobal with those of Iquitos, text Figure 8.1, and Belize, Figure 8.4, particularly as to the rainfall regime.

In terms of total annual rainfall, Cristobal (130 in.) compares more closely with Iquitos (103 in.) than with Belize (79 in.). In terms of an annual rainfall cycle, Cristobal resembles Belize in having a sharply defined but brief period of low rainfall in the early months of the year (Feb-Mar-Apr).

(2) What is the cause of the long wet season at Cristobal?

During and following the June solstice period, the ITC lies over or close to this station, bringing copious convectional rainfall for eight consecutive months.

(3) Explain the low-rainfall period in the rainfall cycle of Cristobal.

Here at latitude 9° N, the winter solstice causes a southerly shift in the trade-wind belt, bringing with it the aridity of the subtropical high that lies farther to the north. (See this same effect more strongly in the climograph for Belize.)

(b) Saigon, Vietnam

Familiar to most Americans through the news media during the Vietnam War, the city of Saigon (renamed Ho Chi Minh City), its great military airport, and its environs are well established in a tropical setting of alternating wet and dry seasons. Many will recall the devastating Tet offensive mounted by the North Vietnamese army and the Viet Cong in February, 1968, which appears in our data table as the month of least rainfall--only 0.1 inch, on the average.

(1) Compare the precipitation cycle of Saigon with that of Cochin, text Figure 8.7.

Saigon's cycle matches closely that of Cochin, with similar timing of dry and wet seasons, but Saigon has lower monthly and annual totals in the wet monsoon period. Both are typical monsoon climates of southeast Asia.

(2) Compare the precipitation cycle of Saigon with that of Aparri, text Figure 8.5.

Aparri's dry period comes two months later, as does its sharply-peaked rainfall maximum. This is the trade-wind littoral pattern, similar to that for Belize (Figure 8.4). The explanation lies in the more northerly latitude of Aparri and its greater distance from the Asiatic mainland.

(c) Cairns, Australia

Cairns is situated on Trinity Bay in Queensland. Along this tropical east coast of Australia the Great Barrier Reef lies offshore. Cairns now serves as a center for tourists visiting the Barrier Reef. With a hinterland belt of tropical wet-dry (savanna) climate, favorable to cultivation of sugar cane, Cairns is the principal sugar port of Australia. This climograph will give you a chance to become accustomed to the seemingly "inverted" cycle of climate seasons in the southern hemisphere.

(1) Plot the rainfall data of Belize, text Figure 8.4, on the same graph as for Cairns, but "synchronize" the seasons by advancing the Belize calendar by 6 months. This means that January in Belize coincides with July in Cairns. (To keep things neat and legible, enter the Belize data on a small rectangle of tracing paper over the Cairns graph.) Compare the two sets of data in terms of temperature and rainfall.

The temperature cycles track quite well, but with Belize having readings 2 to 4 degrees warmer than Cairns in all months. The rainfall cycles are generally in phase, but Cairns has substantially greater monthly values in the wet season and these drop more quickly to low values in the dry season.

(2) Find Cairns and Belize on the world maps of pressure and winds, Figure 4.10. Mark the locations of each on both January and July maps. Can you find a cause of the greater monthly rainfall values at Cairns in the high-sun months?

The ITC comes close to Cairns in January and a large sub-equatorial low-pressure cell is developed in this region. In contrast, Belize in July is farther from the ITC, which is here associated with only a weak equatorial pressure trough. Convectional activity would thus be much more intense in the region of Cairns.

(3) Refer to the special section on El Niño, text Chapter 6, p. 139-40. During the El Niño event of 1982-1983, how would the climograph of Cairns have been different than for the long-term record? Speculate on how the change in track of the subtropical jet stream might have affected the climograph of Belize during the mature phase of El Niño.

The equatorial low was replaced by persistent high pressure and the trade winds were inactive. Severe drought resulted over northern Australia. The climograph of Cairns would have shown low rainfall in all months. For Belize, the southward shift of the jet stream, shown in Figure 6.23, might have caused a greatly increased rainfall in the low-sun season (normally dry) because of unusual increase in the number and intensity of easterly waves.

(d) Nagpur, India

Nagpur, a city of about one million inhabitants in the Indian state of Maharashtra, lies in the heart of the Indian peninsula, an ancient landmass that was once part of the supercontinent of Gondwana. The city name, freely translated, means "cobra-town." Nagpur lies strategically on the main west-east rail line from Bombay to Calcutta, and the main line from south India to the northern cities of Agra and Delhi.

Note: The temperature and rainfall data for Raipur, located only a short distance east of Nagpur, are given in Figure 8.12. The two data sets are almost identical.

(1) How does the rainfall regime of Nagpur compare with that of Timbo, Figure 8.11? Cite specific differences in the two climographs.

Nagpur's dry season is somewhat longer (7 months) than that of Timbo (5 months), but Timbo has three months of virtually no rainfall, whereas Nagpur shows roughly one-half inch of rain in all the dry-season months. Rainfalls of the three wettest months are almost identical for both stations, but the total rainfall for Timbo is greater.

(2) Compare the temperature cycle of Nagpur with that of Timbo. Which place has a hotter hot-season? How does the effect of the rains show in the temperature graph?

Nagpur's annual temperature range is substantially greater, because of its more northerly latitude and the cooling effect of the dry low-sun monsoon winds arising in the cold Asiatic interior. Nagpur has a substantially higher peak temperature (95°F) in May. Once the rains begin (June) temperature fall sharply at Nagpur; this effect is not so marked at Timbo.

(3) Nagpur is about at the same latitude as Bombay, for which the seasonal rainfall contrast is presented in Exercise 8-A, Part 2. Compare the Bombay data with those of Nagpur. Explain the differences.

The month of greatest rainfall is July for both stations, but Bombay's July rainfall is almost double that at Nagpur. The inland location of Nagpur explains this difference. Nagpur has more rain in its three driest months, possibly because some moist air produced over the Bay of Bengal penetrates the interior during the otherwise dry monsoon season.

(e) Alice Springs, Australia

Alice Springs lies squarely in the center of Australia's great interior desert and nearly on the Tropic of Capricorn. It is the end of the line for a railroad from the south coast and lies on a north-south transcontinental highway connecting Darwin on the north coast with Adelaide and other cities of the south coast. Some of you will recall scenes from the epic Australian film "A Town Named Alice," in its World War-II setting, depicting life in the "outback" desert environment.

(1) On the climograph for Alice Springs, plot monthly mean temperatures for Wadi Halfa (Figure 8.17) and Bou-Bernous (Figure 8.17) in the same phase as Alice Springs (i.e., offset by 6 months in the calendar). Compare the temperature cycle of Alice Springs with those of the two North African stations. Offer explanations for the differences. (See Exercise 8-B, questions 6 and 7.)

Although their cycles are similar, with annual ranges almost equal, Alice Springs runs about 10 degrees cooler than Wadi Halfa throughout the year and the annual means differ by 10F°. Bou-Bernous is much hotter than either in the high-sun months. Reasons for the cooler temperatures of Alice Springs are not obvious. Its greater elevation is most likely a factor. That Alice Springs has some rainfall in all months (but Wadi Halfa does not) suggests that cloud cover is a cooling factor and that the evaporation of soil moisture may introduce a significant cooling factor.

(2) Describe the rainfall cycle at Alice Springs. Why is there more rain from October through March?

The October-March period is in the summer season of this hemisphere and suggests that invasions of the maritime tropical airmass may occur from the north or northeast. The upland topography may be effective in triggering convectional precipitation.

(f) Lima, Peru

Lima, capital city of Peru, lies on the Pacific shore on a narrow coastal plain footing the soaring Cordillera Occidental, a range of the Andean mountain chain. Already a city of over 2.5 million inhabitants by 1970, it has witnessed the growth of great peripheral shanty slums as country folk come to seek employment and relief from intense rural poverty--all of this in a nearly rainless, cool (for this latitude), and foggy climate. But fresh water is not lacking, for the Rimac River flows close by, draining a vast area of high, snowcapped Andean peaks.

(1) Compare the climograph of Lima with that of Walvis Bay, Figure 8.20. Both are in the southern hemisphere, but some confusion can arise because in Figure 8.20, the calendar has been advanced by six months. (Note: The legend of Figure 8.20 incorrectly assigns Walvis Bay to the tropical desert climate, 4wd, whereas it should be assigned to the subtropical climate 5wd. The difference between the two is quite trivial, close to their common boundary.)

The temperature cycles are quite similar, but the annual range of Lima is much larger--double that of Walvis Bay. Quite possibly this difference is due to the positioning of the observing station, since the littoral effect weakens landward in a very short distance. The periods of appreciable (measurable) rainfall occur at different times--much earlier in the year at Walvis Bay than at Lima.

(2) What significant change in the precipitation cycle at Lima might be expected during a strongly developed El Niño event? Explain.

During the El Niño event of 1982-1983, torrential rainfalls occurred in the Andes of Peru, Bolivia, and Colombia. If this effect extended as far south as Lima, significant increases in monthly rainfall would have been recorded. Cessation of the upwelling of cold water off the coast could have temporarily destroyed the stable temperature inversion repsonsible for the coastal dryness.

(g) Quito, Ecuador

Quito is our example of a highland station of the low-latitude climates. Quito, with over one million inhabitants, is the capital city of Ecuador. It lies almost on the equator. In this part of Ecuador the Andean range is narrow, but with summits reaching over 18,000 ft (5,500 m). Quito lies in a narrow but fertile valley at the foot of a lofty active volcano named Pichincha. To the east, within a distance of only about 100 mi (160 km), the land surface descends steeply to the rainforest lowland of Amazonia.

(1) Compare the climograph of Quito with that of Iquitos, Peru, text Figure 8.1. Iquitos is located at lat. 3½°S, long. 73°W. (Locate and label Iquitos on the world climate map, Figure 7.9.)

Temperatures of both stations follow closely the wet equatorial regime, with very small annual range, but the annual mean temperature of Quitos is about 20F° lower than for Iquitos. The rainfall regimes are similar in pattern, with a marked dip in July and August, but the annual total for Quito is less than half that of Iquitos.

(2) Can the difference in mean annual temperatures of Iquitos and Quito be fully accounted for by the altitude effect alone, according to the environmental lapse rate? (See text p. 66 and Figure 3.6.) Calculate the difference by lapse-rate and explain your results.

Using an altitude difference of 9,000 ft and a lapse rate of $3\frac{1}{2}$F° per 1000 ft, the air should be about $31\frac{1}{2}$° cooler at Quitos, whereas the actual difference in mean annual temperatures is only 18° (77 - 59 = 18). Perhaps this discrepancy is accounted for by the much higher intensity of incoming solar radiation received at the ground surface at Quito than at Quitos. At high altitude the overlying air is less dense and has a lower water-vapor content. (Would not the outgoing radiation also be greater at Quito, causing more intense nocturnal cooling?

(3) Repeat this calculation, using the data of Figure 3.7 to compare the July mean temperatures of Mollendo and Vincocoya. Compare your results with those from Question 2.

Subtracting 31°F for Vincocoya from 61° for Mollendo, the difference is 30°. The altitude difference, rounded to 14,000 ft, multiplied by 3.5° per 1000 ft, gives a temperature difference of 49°. The ratios are very nearly the same as for Question 3, i.e., about 60%.

(4) Does Quito show a clearly-defined seasonal cycle of rainfall? Describe and explain what you observe.

Two maxima occur in the month following equinox; two minima occur in the month following solstice. This suggests that the ITC may pass over this equatorial position twice each year, bringing heavier rainfall just after each equinox. At solstices, the ITC is drawn away from the equator, and rainfall is reduced. (This suggestion is not in agreement with the world maps of pressures and winds, Figure 4.10.)

(5) Compare the climograph of Quito with that of Simla, India, text Figure 8.24. What are the major differences? Explain those differences.

Two major differences are present. The annual temperature cycle at Simla is strongly developed, with an annual range of about 25F°; that of Quito is only about 1°. The greater range is keyed to Simla's greater latitude (31°N) and a relatively large seasonal swing in the cycle of solar radiation. The strong Asiatic monsoon rainfall cycle, peaking in July and August at Simla, is not found in the equatorial zone of South America.

Exercise 8-D   Soil-Water Budgets of Low-Latitude Climates
[Text p. 171-73, 179, 163; Figures 8.1, 8.5, 8.18.]*

In this exercise, you build on Exercise 7-D, in which you learned to plot the annual soil-water budget. Refer back to that exercise for instructions as to how the data are to be entered on the blank graphs.

In this exercise, the unit used is the centimeter, rather than the inch. Scientists of the Laboratory of Climatology at Centerton, New Jersey, who amassed a great data bank of world soil-water budgets, used only the metric units. Temperature is not needed in the soil-water budget, because that essential factor has already been included in the calculation of evapotranspiration.

In the data tables given you, four lines are left blank. You will need to calculate the missing figures and enter them into the table. Instructions for carrying out these calculations are given in detail below.

The data given consist of monthly values of precipitation (P), water need (Ep), and water use (Ea). From these data, calculate monthly values of the following terms and enter the figures in the blank spaces provided in the tables.

D   Soil-water shortage (Subtract Ea from Ep.)

-G   Storage withdrawal (Subtract P from Ea, where Ea is the larger quantity.)

+G   Storage recharge (Subtract Ep from P, where P is the larger quantity.)

R   Water surplus (Subtract Ep from P, but only for those months during and following completion of the soil-water recharge.)

Credit: Data used in this exercise are from C.W. Thornthwaite Associates, <u>Average Climatic Water Balance Data of the Continents</u>, Laboratory of Climatology, Publications in Climatology, Centerton, N.J., Vols. 15-18, 1962-1965.

---

*Modern Physical Geography, 3rd Ed., p. 177-80, Figures 10.16-10.20.

Station Data

(a) Iauarete, Brazil   ½°N   69°W

	J	F	M	A	M	J	J	A	S	O	N	D	Year
P	24.4	21.7	25.9	36.2	35.8	35.1	29.9	27.2	26.4	23.2	21.3	20.4	327.5
Ep	11.2	11.0	11.9	11.5	11.2	10.3	10.3	10.6	10.9	12.2	11.8	11.9	134.8
Ea	11.2	11.0	11.9	11.5	11.2	10.3	10.3	10.6	10.9	12.2	11.8	11.9	134.8
D	0.0	0.0	0.0	0.0	0.0	0.0	0.0	0.0	0.0	0.0	0.0	0.0	0.0
-G													
+G													
R	13.2	10.7	14.0	24.7	24.6	24.8	19.6	16.6	15.5	11.0	9.5	8.5	192.7

(b) Kaduna, Nigeria   10½°   7½°E

	J	F	M	A	M	J	J	A	S	O	N	D	Year
P	0.0	0.2	1.3	6.9	14.7	18.0	21.8	31.2	27.9	7.6	0.5	0.0	130.1
Ep	8.4	10.0	14.3	15.7	15.6	11.9	10.0	9.3	10.0	11.4	9.4	8.2	134.2
Ea	3.6	3.4	4.2	8.2	14.8	11.9	10.0	9.3	10.0	11.2	7.3	4.7	98.6
D	4.8	6.6	10.1	7.5	0.8	0.0	0.0	0.0	0.0	0.2	2.1	3.5	35.6
-G	3.6	3.2	2.9	1.3	0.1					3.6	6.8	4.7	26.2
+G						6.1	11.8	8.3					26.2
R								13.6	17.9				31.5

(c) Hyderabad, Pakistan   25½°N   68½°E

	J	F	M	A	M	J	J	A	S	O	N	D	Year
P	0.5	0.8	0.5	0.2	0.5	1.0	7.9	5.3	1.5	0.0	0.2	0.2	18.6
Ep	2.5	3.3	11.8	17.5	20.4	20.5	20.3	18.4	16.5	15.3	7.2	2.6	156.3
Ea	0.5	0.8	0.5	0.2	0.5	1.0	7.9	5.3	1.5	0.0	0.2	0.2	18.6
D	2.0	2.5	11.3	1.73	19.9	19.5	12.4	13.1	15.0	15.3	7.0	2.4	137.7
-G													
+G													
R													

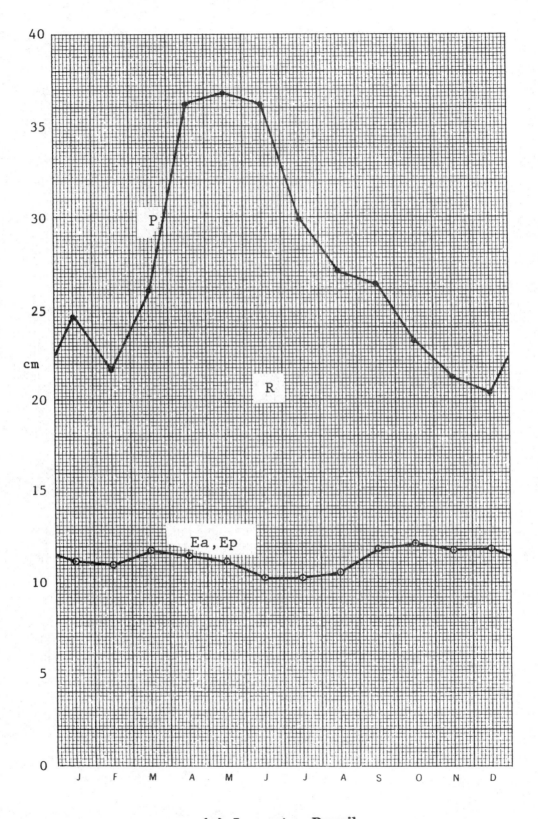

(a) Iauarete, Brazil

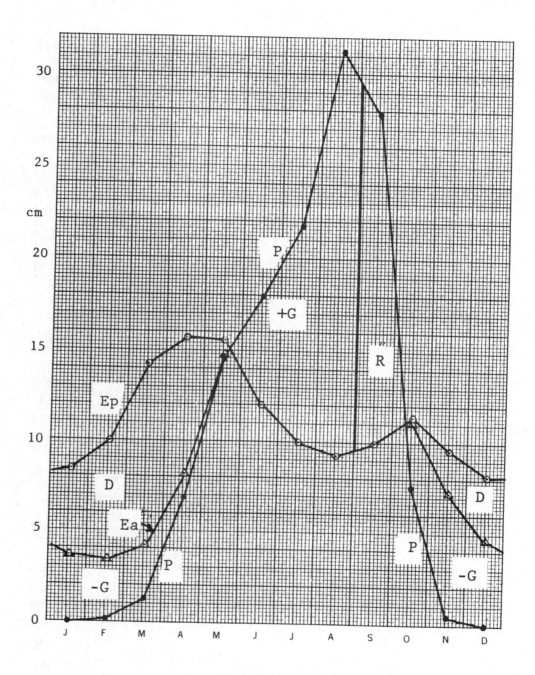

(b) Kaduna, Nigeria

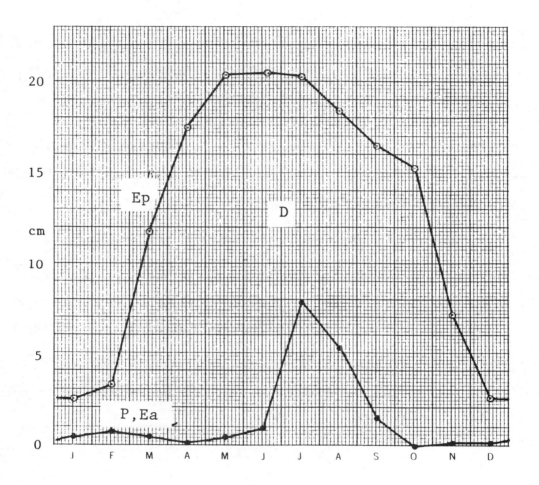

(c) Hyderabad, Pakistan

Questions

(a) Iauarete, Brazil

Iauarete is a small river settlement in the heart of Amazonia. It lies on the Uaupes River, an obscure tributary to the great Negro, which in turns feeds a large discharge into the Amazon River at Manaus. The elevation of Iauarete is only 400 ft (122 m) but it lies (1400 mi) 2300 km airline distance from the Atlantic Ocean. This gives you some idea of the flatness and vastness of the Amazon Basin.

(1) What climate does this soil-water budget illustrate?

Wet-equatorial climate (1)

(2) Describe the most striking features of this soil-water budget.

Precipitation (P) is very heavy in all months and is double the water need (Ep), or more, in very month. Water use (Ea) equals water need (Ep) in every month. There is a large water surplus (R) in all months.

(3) Compare the soil-water budget of Iauarete with that of Singapore, text (Figure 8.2).

The budgets are generally similar, but Iauarete has about one-third greater annual precipitation. The annual precipitation cycles are in almost opposite phase. Singapore shows one month of precipitation equal to water need, whereas Iauarete has a large surplus in all months.

(4) What form of vegetation would you expect to find in this region? How does the soil show a response to the great water surplus?

Broadleaf evergreen rainforest is typical of this water budget and the warm uniform temperature regime that goes with it. Soils are leached of nutrients, and these must be recycled directly from plant remains back into the trees.

(5) What conditions of river flow would you predict for this region?

Rivers flow copiously and fairly uniformly throughout the year, but with flooding to be expected in the three rainiest months (April, May, and June.

_____

(b) Kaduna, Nigeria

Kaduna is an important commercial and industrial center of northern Nigeria, from which cotton, peanuts, and sorghum are shipped south by rail to port cities on the Gulf of Guinea. The altitude here is around 2000 ft (600 m), which may help account for the generous amount of rainfall in the wet season.

(1) To what climate type should this station be assigned? Give name and number.

Tropical wet-dry climate (3)

(2) Describe the outstanding features of this soil-water budget.

Kaduna has a large soil-water shortage (D) in one season and a large water surplus (R) in the other. Precipitation (P) shows a very strong annual cycle with total drought at the time of low sun and a very wet season at time of high sun. Water need (Ep) drops off in June, when rain and cloud cover cause a drop in air temperature.

_____

(3) Compare the budget of Kaduna with that of Raipur, India, text Figure 8.12. The two budgets are strikingly similar in many ways, but there are minor differences. Water need (Ep) peaks much more strongly at Raipur just before the rains. The strong saddle in water need (Ep) during the height of the rains seen in the Kaduna graph is only weakly shown for Raipur.

_____

(4) What forms of native vegetation (cultivated crops excluded) would you expect to find in the area surrounding Kaduna?

Plants in this climate must be capable of withstanding a long drought. Coarse grasses would survive the drought and thrive during the rains. Trees would be specialized, for example, bearing thick or thorny leaves, or shedding leaves in the dry season.

(5) What flow conditions can you predict for rivers in this region?

Rivers will rise to flood stage in August and September, but will fall to low stage and all but the largest may dry up completely from January through May, which is the dry season.

(c) Hyderabad, Pakistan

The city of Hyderabad, third largest in Pakistan, lies on the great Indus River not far from its mouth on the Arabian Sea. Like the Nile of Egypt, the Indus is an exotic river; it is fed by melting snows of the mighty Himalayas and Hindu Kush and flows more than 1000 km (600 mi) across the Thar Desert.

(1) What climate type is represented by Hyderabad? Name and number.

Dry tropical climate, desert subtype (4d).

(2) Describe the outstanding features of this soil-water budget.

Water need (Ep) greatly exceeds precipitation (P) in all months. All rain that falls is returned to the atmosphere through evapotranspiration. The total annual soil-water shortage (D) is very large.

(3) Compare the soil-water budget of Hyderabad with that of Khartoum, Sudan, text Figure 8.18.

The two budgets area almost identical in form. One difference is that water need (Ep) at Khartoum remains higher (7 cm) in the low-sun months than at Hyderabad. This is explained by the lower latitude of Khartoum.

(4) Explain the strong annual cycle of water need (Ep).

During the high-sun season, air temperatures rise to high values. During the low-sun season, air temperatures drop to moderate levels. The cycle of water need follows this temperature cycle.

(5) What forms of natural vegetation would you expect to find in this region?

Only drought-resisant plants could survive in this climate. Desert grasses and shrubs would be typical. Plant growth would be limited to the short rainy season in July and August.

(6) Under what conditions might field crops, such as grains, be grown in this climate?

Irrigation would be essential for crop production. The Indus Valley irrigation system has been developed for this purpose, but suffers from salinization and waterlogging.

Name _____   Date _____

_____   _____

Exercise 9-A The Subtropical Desert Temperature Regime
[Text p. 191-92, Figures 9.1, 9.2.]*

In this exercise we follow up on the theme of Exercise 8-B, investigating the desert temperature regime. The vast subtropical desert of the southwestern United States attracts not only large numbers of recreation seekers, but also those retired persons looking for permanent residence in a land without a cold winter. For the latter group, the extremes of temperature are the most important climate factor. Air conditioning and space heating are an expensive necessity for much of the year. Water is a scarce commodity and must be imported from distant sources. Evaporation of soil moisture is intense in the hot months, disposing of much of that precious water with little benefit received.

Part 1 July and January in Phoenix, Arizona

Phoenix, the burgeoning capital city of Arizona, lies in the Sonoran Desert (33½°N, 112°W), close to the Gila River. Only a few miles away is Sun City, one of the largest and most popular retirement communities in the United States, it population approached the 50,000 mark in 1988. Would your parents like to spend their "golden years" in this environment? If they have such ideas, the temperature record for a typical July and January in Phoenix is well worth studying closely. Perhaps, instead, Sun City would be just "a nice place to visit."

Over a 52-year period of record, Phoenix had the following temperature statistics: Average daily maximum in July: 104°F. Average daily minimum in January: 39°. Highest recorded temperature: 118°. Lowest: 16°.

Graph A shows the daily high and low temperature for each month. These would be readings of the maximum-minimum thermometer in the standard shelter. The mean of the month and the average daily range are also shown.

(1) Fill in the following data for July: (Use °F.)

    Most consecutive days above 100°:   20

    Highest temperature recorded: 112°

    Lowest temperature recorded: 68°

    Greatest range in one day: 30°

(2) Fill in the following data for January:

    Number of days with maximum 70° or over: 5

    Number of nights below 35°:   20

---

*Modern Physical Geography, 3rd Ed., p. 156-57, 180, Figures 9.17, 10.21.

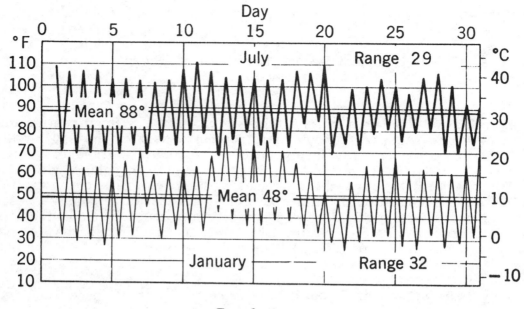

Graph A

Number of nights below 30°: ___10___

Greatest range in one day: ___39°___

(3) Using the figures on the graph, about how great is the annual temperature range at Phoenix? How does this compare with the annual range at Yuma (text Figure 9.1) and at Wadi Halfa (Figure 8.16). Explain.

The range in Phoenix is about 40F° (acutally, 38° from long-term record). That in Yuma is somethat less (36°), but for Wadi Halfa only 29°. Decreasing latitude is largely responsible for this decrease in range, as the amplitude of the insolation cycle diminishes.

(4) For retirees owning a house in Sun City, might energy costs be a financial burden? Explain.

Air conditioning would be a must--probably a central system combined with central space heating. By far the greater share would go space cooling. Depending on electricity costs, this item might prove a major part of the monthly budget.

(5) In the environment of Phoenix, how would temperature restrict summer outdoor physical activity, such as golf, tennis, or running?

Early morning hours are best, perhaps even the only wise choice for vigorous aerobic workouts.

(6) When air temperatures are over 100°, what level of relative humidity might be expected?

Relative humidity readings of from 12% to 20% would be typical in midafternoon in summer, accompanying the highest air temperatures.

Part 2 Maximum Temperatures in the Western United States

Figure A is a map of the western United States showing the highest air temperature observed by the U.S. Weather Service during a 40-year period. Isotherms (°F) are used to display the information, but they are highly generalized because observing stations are few and far between over large parts of the West. The problem here is to make inferences as to the tie-in between air temperatures and major features of the terrain.

(1) Begin by identifying each state on the map and entering its code letters at a convenient location near its upper-right corner. Using a red crayon, color all areas enclosed by the 125-degree isotherm. Use orange for areas between 120 and 125; yellow for areas between 115 and 120.

(2) How many areas can you find enclosed by the 125-degree isotherm? Using an atlas or other detailed map showing both cultural and terrain features, name each of these "hot-spots." Label them on the map. (Labels can be written in the page margin, with a straight line drawn to the center of the feature.) How do these spots relate to terrain?

Four "hot-spots" are shown: Death Valley, Lower Colorado River Valley, Imperial Valley, Gila Valley. All are valley bottoms of low elevation.

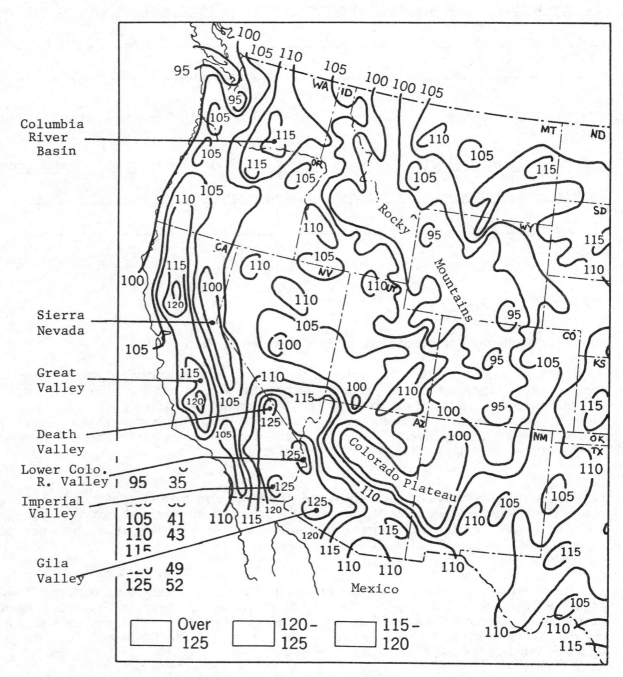

Figure A

Columbia River Basin	
Sierra Nevada	
Great Valley	
Death Valley	
Lower Colo. R. Valley	95 35
Imperial Valley	105 41
	110 43
	115
Gila Valley	49
	125 52

Over 125 | 120–125 | 115–120

(3) Locate two small areas enclosed by the 120-degree isotherm. (Disregard areas surrounding the hot-spots located in Question 2.) Name and label the terrain feature defined by these spots and the 115-degree isotherm that surrounds them. What terrain feature lies adjacent to this zone on the east? Name and label it.

These centers of 120+ occupy the Great Valley of California, a continuous, broad

lowland lying west of the Sierra Nevada, a lofty mountain range roughly

delineated by a 100-degree isotherm.

(4) Identify the two centers of high temperature in Washington and Oregon. Label the terrain feature associated with them.

Can be labeled "Columbia River basin." This is a topographic basin or lowland

formed by the valleys of the Columbia and Snake rivers east of the massive

mountain barrier of the Cascades. It is a rainshadow zone.

(5) Extending across north-central Arizona and western New Mexico is an area of temperatures below 100°. Identify this terrain unit and label it directly on the map. Explain the lower temperatures.

This is the Colorado Plateau, a large upland area that contains the Grand Canyon

and other canyonland features. Surface elevations are generally high, accounting

for the lower temperatures.

Name _____ Date _____

_____   _____

Exercise 9-B   The Boundary Between Subhumid and Semiarid Climate Subtypes
[Text p. 202-3, Figures 9.20, 9.21.]*

American pioneers on the long trek across the great interior plains could not help but notice that week after week the native vegetation, soils, and animal life showed a distinct change. Tall-grass prairie with its flowering forbs--such as the black-eyed susan--gave way to short-grass prairie with bare soil showing between the clumps of buffalo grass. At the same time, the rich almost-black prairie soil gave way to brown soil. Here, vast herds of bison grazed the grassland, and prairie-dog colonies were a common sight. Potable surface water became more difficult to find. Surely a climate boundary had been crossed, but where do you draw that line? Drawing the line between the subhumid climate on the east and the semiarid climate on the west has always been an important problem for geographers and agronomists. Here, we tackle that problem, using climate data.

Figure A is a map showing by isotherms the mean annual temperature (°F) and by isohyets the mean annual precipitation (inches) for the Great Plains region of the United States. A bold line running roughly north-to-south is the climate boundary between the semiarid subtype of climates 9 and 5 (to the west) and the subhumid subtype of climates 10 and 6 (to the east). Find this climate boundary on the world map of climates, text Figure 7.9).

Interpretation of the map data is focused on the changing values of temperature and precipitation in north-south and east-west directions. We need to establish the regional pattern present and then to explain it in terms of factors that affect temperature and precipitation.

(1) What is the general trend of the isotherms in the region of the climate boundary? Explain.

Isotherms trend generally east-west, if one looks along the zone to the east of

the climate boundary. This is the normal, or typical trend resulting from the

diminishing of mean annual insolation with increasing latitude.

_____

*Modern Physical Geography, 3rd Ed., p. 159-60, 183, Figures 9.23, 10.26.

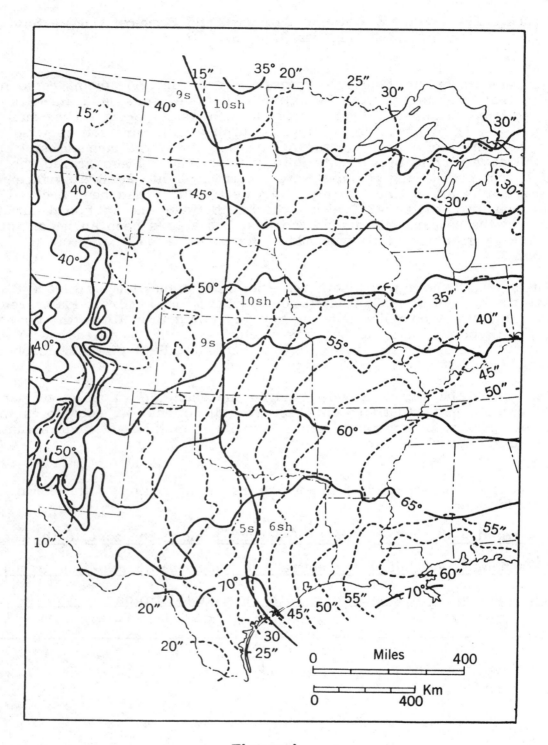

Figure A

(2) What is the general trend of the isohyets in the vicinity of the climate boundary?   Explain.

Trend of the isohyets is northeast-to-southwest in the northern part, and more nearly north-south in the southern part. Eastward increase in precipitation is to be expected as the effect of the rainshadow created by the Rocky Mountains dies out with distance. The diagonal trend may be the result of diminishing distance of inland penetration of moist tropical air masses into the continent from the Gulf.

(3) In the space below, make a schematic map-diagram of the trends of isotherms and isohyets. Use straight lines in parallel sets. Show the climate boundary as a straight north-south line.

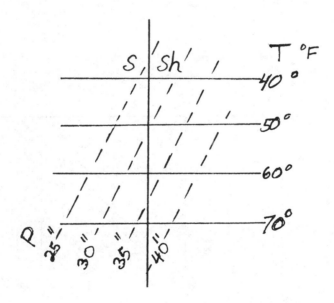

(4) What is the general relationship between temperature and precipitation as the climate boundary is followed from north to south?

Along the boundary, both temperature and precipitation increase from north to south.

(5) Using the blank graph, Figure B, plot temperature (vertical axis) against precipitation (horizontal axis). First, locate the point at which a given isotherm crosses the climate boundary. Second, interpolate (estimate) the precipitation at the same point, based on proportional distance between two successive isohyets.

(6) Describe the trend of the plotted points. Using a straightedge, fit a line that seems by visual inspection best to illustrate the relationship.

A straight-line trend seems to fit the data, despite the scatter of the points from

such a line. The line slopes upward from left to right at about a 45-

degree angle.

(7) Refer to the world map of vegetation, text Figure 20.12. Find the location of the s/sh climate boundary on this map and determine whether it is associated with an important vegetation boundary. What do you find?

The s/sh climate boundary is remarkably coincident with the boundary between

short-grass prairie (to the west) and tall-grass prairie (to the east).

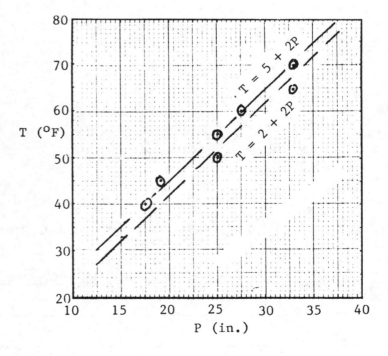

Figure B

Mathematical problem for special credit:

(8). Fit a simple arithmetic linear regression equation to the data. As a first approximation, assume the regression equation to have a 45-degree slope on the graph paper. Show your results below.

Equation of form $Y = a + bX$: $T = a + bP$. Proportionality coefficient, "b",

for 45° is 2 and $T = a + 2T$.   The constant "a" will range from 2 to 3, depending

on where the line is drawn.

Name _____   Date _____

_____   _____

Exercise 9-C   Climographs of Midlatitude and High-Latitude Climates
[Text p. 189-217, Figures 9.4, 9.10, 9.16, 9.20, 9.26
9.29, 9.34, 9.35.]*

This exercise is a continuation of Exercise 8-C and the same instructions apply. Using the latitude and longitude given, find the location of each station on the world climate map, Figure 7.9, and show it with a dot. Write in the station name next to the dot.

Plotting the data: As in your textbook climographs, monthly temperatures (°F) are plotted as points centered in each monthly column and connected by straight-line segments. Mean monthly precipitation amounts (inches) are plotted as bars, forming a step-graph.

Beside each climograph is a list of quantities to be entered in the blanks provided. The mean annual temperature (sum of monthly means divided by 12) is given with the station data. Calculate and enter the following statistics: annual temperature range (difference between highest and lowest monthly means); annual total precipitation (sum of the monthly means).

Enter the number and name for the climate type and subtype in the blank spaces provided. These should follow the system given in the legend of the world climate map, text Figure 7.9.

If your instructor requires the Köppen climate system, identify the climate by use of the Köppen-Geiger world map, text Figure 7.17, and text definitions and boundary graphs, p. 164-69 and Figures 7.18 and 7.19. Enter the Köppen code symbol and Köppen name in the blank spaces provided beside each climograph.

---

*Modern Physical Geography, 3rd Ed., p. 156-62, Figures 9.17-9.31.

Station Data

(a) Brisbane, Queensland, Australia   27½°S 153°E

	J	F	M	A	M	J	J	A	S	O	N	D	Mean
T (°F)	77	76.5	74	70	65	60	58.5	60.5	65.5	70	73	76	69
P (in.)	6.4	6.3	5.7	3.7	2.8	2.6	2.2	1.9	1.9	2.5	3.7	5.0	

(b) Zakinthos, Greece   38°N 21°E

	J	F	M	A	M	J	J	A	S	O	N	D	Mean
T (°F)	52	57.5	55.5	60.5	67.5	74.5	79.5	79.5	75.5	68.5	61	55.5	65
P (in.)	7.2	5.3	3.4	2.2	1.2	0.3	0.1	0.4	1.4	5.1	8.1	9.2	

(c) Oporto, Portugal   41°N 8½°W

	J	F	M	A	M	J	J	A	S	O	N	D	Mean
T (°F)	48	49.5	53	56	58.5	64.5	67	67.5	65.5	60	53.5	48.5	57.5
P (in.)	6.0	4.6	5.4	4.1	3.3	1.7	0.9	0.7	2.1	4.3	5.9	6.5	

(d) Volgograd (Stalingrad), U.S.S.R.   49°N 44½°E

	J	F	M	A	M	J	J	A	S	O	N	D	Mean
T (°F)	9.5	15.5	25.5	44	60.5	69.5	74.5	72	59.5	45	30.5	20	43.5
P (in.)	0.9	1.0	0.6	0.6	1.0	1.9	0.9	0.8	0.7	1.0	1.5	1.3	

(e) Inch'on, Korea   27½°N 127°E

	J	F	M	A	M	J	J	A	S	O	N	D	Mean
T (°F)	26	29	38	50	59.5	68.6	75	77.5	68.5	58	43.5	30.5	52
P (in.)	0.8	0.7	1.2	2.6	3.3	3.9	10.9	8.8	4.3	1.6	1.6	1.1	

(f) Vardo, Norway,   70½°N 31°E

	J	F	M	A	M	J	J	A	S	O	N	D	Mean
T (°F)	23	22	24.5	30	36	42.5	48.5	48.5	43.5	35	29.5	26	34
P (in.)	2.5	2.5	2.3	1.5	1.3	1.3	1.5	1.7	1.9	2.5	2.1	2.4	

(g) Point Barrow, Alaska   71°N 157°W

	J	F	M	A	M	J	J	A	S	O	N	D	Mean
T (°F)	-15.5	-18.5	-15	-0.5	18.5	34	39.5	38.5	30.5	17	1	-10.5	10
P (in.)	0.2	0.1	0.1	0.1	0.1	0.3	0.9	0.8	0.5	0.5	0.3	0.2	

CLIMOGRAPHS

(a) Brisbane, Australia

Mean Ann. Temp.	69
Ann. Temp. Ra.	18.5
Ann. Precip.	44.7
Climate number	6h
Climate name	Moist subtropical, humid subtype
Köppen code	Cfa
Köppen name	Mild, humid climate, no dry season, hot summer

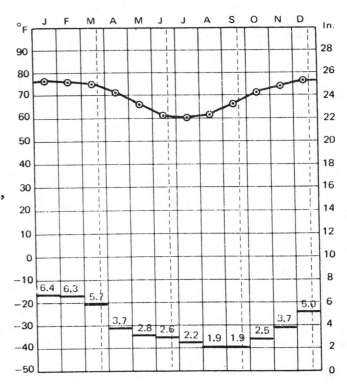

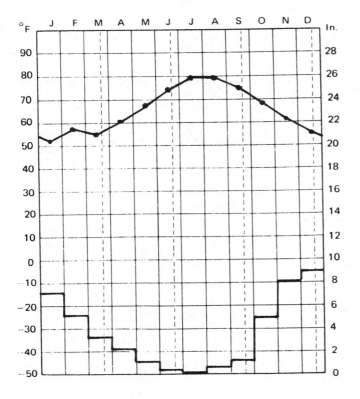

(b) Zakinthos, Greece

Mean Ann. Temp.	65
Ann. Temp. Ra.	27.5
Ann. Precip.	43.8
Climate number	7h
Climate name	Mediterranean, humid subtype
Köppen code	Csa
Köppen name	Mild humid climate with dry summer, with hot summer

(c) Oporto, Portugal

Mean Ann. Temp.   57.5

Ann. Temp. Ra.   19.5

Ann. Precip.   45.5

Climate number   8sh

Climate name   Marine west-coast,

    subhumid subtype

Köppen code   Csb

Köppen name   Mild humid

climate, dry summer, with warm

    summer

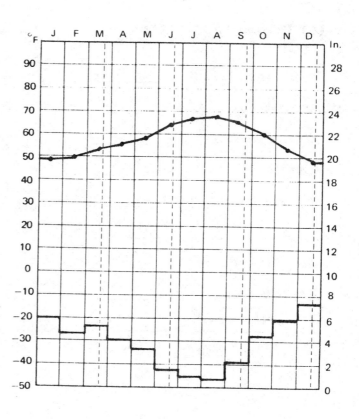

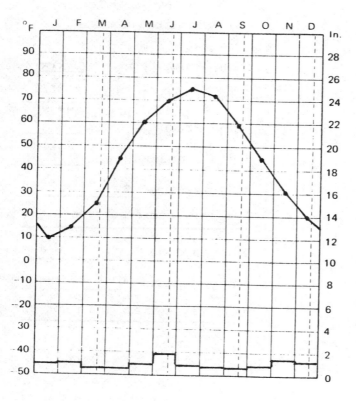

(d) Volvograd, U.S.S.R.

Mean Ann. Temp.   43.5

Ann. Temp. Ra.   65

Ann. Precip.   12.2

Climate number   9s

Climate name   Dry midlatitude,

    semiarid subtype

Köppen code   BSk

Köppen name   Steppe climate,

    dry-cold

(e) Inch'on, Korea

Mean Ann. Temp.	52
Ann. Temp. Ra.	51.5
Ann. Precip.	40.8
Climate number	10h
Climate name	Moist continental, humid subtype
Köppen code	Dwa
Köppen name	Snowy-forest climate, dry winter

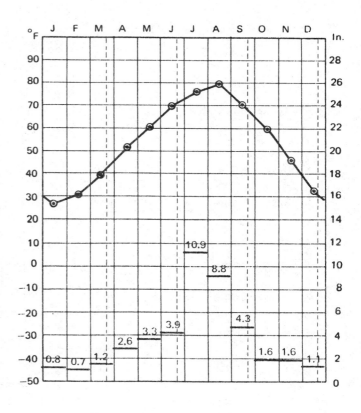

(f) Vardo, Norway

Mean Ann. Temp.	34
Ann. Temp. Ra.	$26\frac{1}{2}$
Ann. Precip.	23.5
Climate number	11h
Climate name	Boreal forest, humid subtype
Köppen code	Cfc
Köppen name	Mild humid climate, no dry season, cool short summer

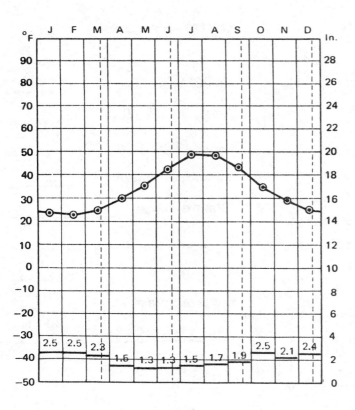

(g) Point Barrow, Alaska

Mean Ann. Temp.   10

Ann. Temp. Ra.   58

Ann. Precip.   4.1

Climate number   12sh

Climate name   Tundra

      subhumid subtype

Köppen code   ET

Köppen name   Tundra climate

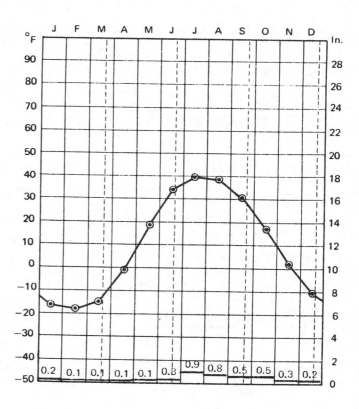

(Extra climograph)

Mean Ann. Temp.   _____

Ann. Temp. Ra.   _____

Ann. Precip.   _____

Climate number   _____

Climate name   _____

Köppen code   _____

Köppen name   _____

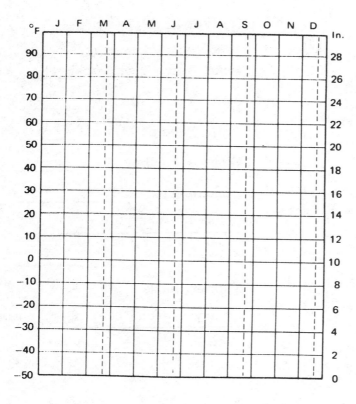

QUESTIONS

(a) Brisbane, Australia

Brisbane, capital city of Queensland, is located on Moreton Bay on the east coast of Australia. Founded in 1828 as a penal colony, it is now an industrial city that exports wool, meat, fruit, and sugar, along with coal and other minerals brought from the arid hinterland lying west of the Great Dividing Range and the Queensland coast that stretches far to the north into the tropical zone.

(1) Compare the climograph of Brisbane with that of Charleston, South Carolina, text Figure 9.4, as to the thermal regimes shown.

Charleston has a markedly greater temperature range (31.4F° vs. 18.5), with both a warmer summer and a cooler "winter." One reason is that Charleston lies at a higher latitude, equivalent to about 380 miles. Perhaps more important is that Charleston experiences the influx of polar continental air masses in winter, whereas Australia lacks a continental polar air mass source region.

(2) Compare the precipitation cycles of Brisbane and Charleston.

Brisbane's precipitation cycle is a simple wave-curve with well developed trough at low-sun and crest at high-sun. Charleston shows a secondary maximum in the winter months. (Explanation for this double maximum is complex.)

(b) Zakinthos, Greece

Zakinthos is the southernmost of the Ionian Islands, lying off the west coast of Greece. Zakinthos is also the name of the island's chief town and port. According to legendary history of ancient Greece, Zacynthus, son of the Arcadian chief Dardanus, led the first settlers to this island from the heart of the Peloponnesus. Intensively cultivated, the island produces wine, currants, citrus fruits and olive oil--a typically Mediterranean fare.

(1) Compare the climograph of Zakinthos with that of Monterey, California, text Figure 9.10, as to the temperature cycle. Explain the differences.

Zakinthos has a much greater annual temperature range (27.5 F°) than Monterey (12 F°). Zakinthos also has a much warmer summer. The explanation lies in the presence of the cold, upwelling California current lying offshore of Monterey. No such cold current flows in the Mediterranean Sea, which builds up warm surface water temperature in summer.

(2) Compare the precipitation regime of Zakinthos with that of Monterey. Explain the differences.

Zakinthos has a much greater annual total precipitation (43.8 in.) than Monterey

(17 in.). This very heavy winter precipitation suggests a possible orographic

effect acting at Zakinthos, together with an absence of the low-level coastal

inversion found at Monterey. (Explanation is complex.)

(c) Oporto, Portugal

Oporto (Pôrto), Portugal's second largest city, lies near the Atlantic at the mouth of the Douro River. The city is internationally known for its association with port wine--its most popular export. Oporto's long and checkered history includes occupation by the Moors between 716 and 1092. Climatologically speaking, this station is interesting as being transitional between the marine west-coast climate and the Mediterranean climate.

(1) Compare the climograph of Oporto with that of Vancouver, British Columbia, text Figure 9.16, as to the annual temperature cycle. Explain the differences.

Vancouver's winters are considerably colder, as would be expected with its much

higher latitude. The latitude difference is about 8 degrees, equal to about 550

miles. Summer temperatures are quite similar, however.

---

(2) Compare the precipitation regimes of Oporto and Vancouver. Explain any differences.

The precipitation cycles are quite similar, but with Vancouver having greater

amounts in the winter, and thus a greater annual total. Being farther north,

Vancouver intercepts more frequent and more intense winter cyclones. The

orographic effect of high mountains on both sides of Vancouver may also be a

major factor.

---

(3) Compare the precipitation cycle of Oporto with that of a Mediterranean climate station, such as Zakinthos or Monterey. What is the essential difference?

Oporto has precipitation amounts of 0.7 in. or more throughout the driest months

of summer, whereas the Mediterranean stations have at least one summer month of

less than 0.1 in. Extreme summer drought is the criterion here for the

Mediterranean climate.

(d) Volgograd, U.S.S.R.

Volgograd lies on the west bank of the mighty Volga River in the midst of a vast plain that extends far eastward to the Kirgiz Steppe. Named Stalingrad in 1925, the city was nearly destroyed in World War II when under seige by Hitler's armies. The city's name was changed to Volgograd in 1961, under the regime of Nikita Khruschev, who had denounced Stalin and his dictatorship. Today a city of heavy industry, Volgograd derives hydrolectric power from an enormous dam that spans the Volga just above the city.

(1) Identify the climate type and subtype for Volgograd. Compare the climograph with that of Pueblo, Colorado, text Figure 9.20, with respect to the temperature cycle. Explain the important differences.

Volgograd falls within the semiarid subtype of the dry midlatitude climate (9s). It

has both a greater annual range (65F° vs. 43) and a colder winter temperatures

(10°F in January) than does Pueblo (30)°. These differences are easily explained

by the higher latitude of Volvograd and the greater size of the Eurasian

landmass. Summer temperatures are almost identical.

(2) Compare the temperature cycle of Volgograd with that of Medicine Hat, Alberta, text Figure 9.21.

The two cycles are very similar, both as to maximum and minimum monthly

values. Note that the latitudes are almost the same. The July temperature of

Volvograd is about 5° higher, and so has a higher annual range.

(3) Compare the precipitation regimes of Volvograd and Pueblo. Explain the important difference in the cycles.

Whereas the monthly precipitation values for Volvograd are more-or-less uniform throughout the year (except for a small peak in June), Pueblo shows a well-developed annual cycle with maximum in the summer months. The Pueblo cycle is explained by the summer influx of mT airmasses from the Gulf of Mexico and the Pacific Ocean off Baja California. Volgograd is effectively cut off from such air masses by great land areas to the south.

(e) Inch'on, Korea

Inch'on, a west-coast port city for South Korea's capital, Seoul, provides access to the shallow Yellow Sea, but is an industrial city in its own right. During the Korean War, Inch'on was the site of a daring landing of U.S. troops in September, 1950, executed under the command of General Douglas MacArthur to relieve enemy pressure on beleagured UN troops within the Pusan perimeter far to the south.

(1) Fans of TV's indestructable saga, MASH, have experienced the intense continentality of Korea's climate, including vivid episodes of summer heat wave and winter deep-freeze. Compare the climographs of Inch'on and Madison, Wisconsin, text Figure 9.26, as to the annual temperature cycle. Explain the important differences.

The cycles are quite similar in form and annual range, but that of Inch'on is displaced a few degrees higher. The cold-month values differ by about 10°, explained by the marine effect of the Yellow Sea. The warmer winter values

(2) Veteran MASH viewers notice that rain is almost never shown, perhaps because the filming was done only on dry days in southern California's Santa Monica Mountains. Compare the precipitation cycles of Inch'on and Madison. Explain the significant difference.

Not only does Inch'on have a greater annual total rainfall, but it is strongly concentrated in July and August because of the Asiatic monsoon effect. Totals for July and August are impressively large. The monsoon effect is comparatively weak in the cycle for Madison. Both stations have moderately small amounts in the winter months.

(f) Vardo, Norway

Vardo is a small port town on the Arctic Ocean lying 4 degrees of latitude north of the arctic circle, equivalent to about 280 miles. Despite this arctic location, the port is ice-free, thanks to the mild waters of the North Atlantic drift than sweep eastward around the North Cape. History has it that in 1320 the northernmost fortress in the world was built at Vardo, which thereafter enjoyed brisk trade with Russia and Finland. The explorer Fridtjof Nansen used Vardo as his base of operation for arctic expeditions in the 1890s.

(1) Placing Vardo's climate in one of the standard types described in your textbook poses some problems. Careful analysis of the temperature data is needed, along with comparisons with both the boreal forest climate (11) illustrated by the climograph for Ft. Vermilion, Alberta (text Figure 9.29), and the tundra climate (12) shown in the climograph for Upernivik, Greenland (Figure 9.34) and that for Point Barrow in this exercise. Make these comparisons in terms of (a) annual temperature range and (b) mean annual temperature. Explain the similarities and differences.

(a) Annual range for Vardo ($26\frac{1}{2}°F$) is much less than for Ft. Vermilion (74°) and also substantially less than that for Upernivik (51°). This suggests a greater affinity for the tundra climate, but with added potency of the marine influence of the warm ocean current. (b) Mean annual temperature of Vardo (34F°) higher than that of Ft. Vermilion (25°) and much above that of both Upernivik (17°) and Point Barrow (10°). This suggests closer affinity with the boreal forest climate (11). Perhaps a unique type of climate is indicated for Vardo. (This climate has been identified as the "marine subarctic climate.")

(2) Make the same climograph comparisons with respect to total annual precipitation. Draw conclusions as to the meaning of the data.

Annual precipitation totals (inches): Vardo, 23.5; Fort Vermilion, 12; Upernivik 9; Point Barrow, 4.1. Vardo' precipitation, almost double that of Fort Vermilion and much greater than the tundra stations, suggests it doesn't belong in either the boreal forest or tundra climate. It appears to be a unique climate. (Exceptionally high annual precipitation is a feature of the marine subarctic climate.)

(3) Explain the exceptionally high precipitation of Vardo as compared with the other stations examined above.

Exceptionally warm surface water of the North Atlantic drift produces an unusually moist maritime polar air mass for this high latitude. Moreover, the air mass tends to be unstable. These factors explain the comparatively high precipitation in all months.

(g) Point Barrow, Alaska

Point Barrow, on the shores of the Arctic Ocean, is the most northerly point of Alaska; it lies at latitude 71°N, which is about 5° north of the arctic circle (equivalent to about 330 mi). Nearby is the small city of Barrow, once a rather insignificant but ancient Eskimo settlement. After World War II, the U.S. Navy placed an arctic research station in Barrow, and the U.S. Air Force set up a radar station close by as part of the the Distant Early Warning Line. Wiley Post, a pioneer American aviator, used Point Barrow as a stopover on his arctic round-the- world flight in 1931, but later he and a distinguished passenger, humorist Will Rogers died in a plane crash near Point Barrow.

(1) Barrow's temperature cycle was noted in the previous exercise as having a much greater annual range than Vardo, although both are coastal stations on the Arctic Ocean. How do you explain this striking difference?

Note that the July-August tempertures are almost the same for both stations. The winter months, however, are much colder at Barrow--more than 40° colder at Barrow in February. Not only is the warm ocean current absent at Barrow, but the sea is solidly frozen over off this coast in winter, so that the marine effect is eliminated and the snow-covered ice has a strongly negative radiation balance, as does the land surface.

(2) Explain the sharp increase in precipitation in July and August at Barrow.

In July and August, the adjacent sea surface is largely free of ice, while the land surface is free of snow cover and is water-saturated. These sources of water vapor increase precipitation from the summer air masses.

Name _____   Date _____

_____   _____

Exercise 9-D   Soil-Water Budgets of Midlatitude and High-Latitude Climates
[Text p. 190-93, 196-97, 200-203, 207-8, 211-13; Figures 9.2, 9.5, 9.11, 9.17, 9.21, 9.27, 9.30, 9.35.]*

In this exercise, you continue to build on Exercise 7-D, in which you learned to plot the annual soil-water budget. Refer back to that exercise for instructions as to how the data are to be entered on the blank graphs.

As in Exercise 8-D, the unit used is the centimeter. In the data tables given you, four lines are left blank. You will need to calculate the missing figures and enter them into the table. Instructions for carrying out these calculations are given in detail below.

The data given consist of monthly values of precipitation (P), water need (Ep), and water use (Ea). From these data, calculate monthly values of the following terms and enter the figures in the blank spaces provided in the tables:

D   Soil-water shortage (Subtract Ea from Ep.)

-G   Storage withdrawal (Subtract P from Ea, where Ea is the larger quantity.)

+G   Storage recharge (Subtract Ep from P, where P is the larger quantity.)

R   Water surplus (Subtract Ep from P, but only for those months during and following completion of the soil-water recharge.)

Credit: Data used in this exercise are from C.W. Thornthwaite Associates, <u>Average Climatic Water Balance Data of the Continents</u>, Laboratory of Climatology, Publications in Climatology, Centerton, N.J., Vols. 15-18, 1962-1965.

*Modern Physical Geography, 3rd Ed., p. 180-85, Figures 10.21-10.30.

Station Data

(a) Sevilla, Spain   37½°N 6°W

	J	F	M	A	M	J	J	A	S	O	N	D	Year
P	4.4	6.5	7.0	5.0	3.6	2.7	0.2	0.2	2.7	7.0	9.9	6.7	55.9
Ep	1.7	2.2	3.9	5.7	9.8	14.0	18.0	17.6	12.1	7.1	3.3	1.9	97.3
Ea	1.7	2.2	3.9	5.5	8.1	8.9	6.1	3.5	3.8	7.0	3.3	1.9	55.9
D	—	—	—	0.2	1.7	5.1	11.9	14.1	8.3	0.1	—	—	41.4
-G	—	—	—	0.5	4.5	6.2	5.9	3.3	1.1	0.0	—	—	21.5
+G	2.7	4.3	3.1	—	—	—	—	—	—	—	6.6	4.8	21.5
R	—	—	—	—	—	—	—	—	—	—	—	—	—

(b) New Plymouth, New Zealand   39°S 174°E

	J	F	M	A	M	J	J	A	S	O	N	D	Year
P	11.2	10.4	9.1	11.9	15.2	15.5	16.0	14.2	12.7	14.0	12.2	11.2	153.6
Ep	9.3	8.1	7.6	5.9	4.2	3.0	2.8	3.3	4.2	5.9	7.0	8.7	70.0
Ea	9.3	8.1	7.6	5.9	4.2	3.0	2.8	3.3	4.2	5.9	7.0	8.7	70.0
D	—	—	—	—	—	—	—	—	—	—	—	—	—
-G	—	—	—	—	—	—	—	—	—	—	—	—	—
+G	—	—	—	—	—	—	—	—	—	—	—	—	—
R	1.9	2.3	1.5	6.0	11.0	12.5	13.2	10.9	8.5	8.1	5.2	2.5	83.6

(c) Xi'an (Sian), China   34½°N 109°E

	J	F	M	A	M	J	J	A	S	O	N	D	Year
P	0.4	0.8	1.7	4.0	5.2	5.6	9.1	10.6	10.5	5.9	1.5	0.5	55.8
Ep	0.0	0.3	2.4	5.7	11.2	15.6	17.6	15.1	8.9	5.0	1.4	0.1	83.3
Ea	0.0	0.3	1.9	4.3	6.2	6.9	9.9	10.9	8.9	5.0	1.4	0.1	55.8
D	—	—	0.5	1.4	5.0	8.7	7.7	4.2	—	—	—	—	27.5
-G	—	—	0.2	0.3	1.0	1.3	0.8	0.3	—	—	—	—	3.9
+G	0.4	0.5	—	—	—	—	—	—	1.6	0.9	0.1	0.4	3.9
R	—	—	—	—	—	—	—	—	—	—	—	—	0.0

(d) Leipzig, East Germany   51½°N 12½°E

	J	F	M	A	M	J	J	A	S	O	N	D	Year
P	4.0	3.3	4.1	4.7	6.0	6.7	8.5	6.7	4.9	5.0	4.0	4.2	62.1
Ep	0.0	0.2	1.8	4.5	8.9	11.4	12.6	10.5	7.2	3.9	1.4	0.4	62.8
Ea	0.0	0.2	1.8	4.5	8.8	10.7	11.5	9.1	6.2	3.9	1.4	0.4	58.5
D					0.1	0.7	1.1	1.4	1.0				4.3
-G					2.8	4.0	3.0	2.4	1.3				13.5
+G	4.0	2.0								1.1	2.6	3.8	13.5
R		1.1	2.3	0.2									3.6

(e) Irkutsk, U.S.S.R.   52½°N 104°E

	J	F	M	A	M	J	J	A	S	O	N	D	Year
P	1.0	0.9	0.7	1.6	3.1	5.8	7.7	6.8	4.4	1.7	1.6	0.5	36.9
Ep	0.0	0.0	0.0	0.5	6.5	10.9	12.9	10.4	5.2	0.0	0.0	0.0	46.4
Ea	0.0	0.0	0.0	0.5	5.1	8.5	9.9	8.2	4.7	0.0	0.0	0.0	36.9
D					1.4	2.4	3.0	2.2	0.5				9.5
-G					2.0	2.7	2.2	1.4	0.3				8.6
+G	1.0	0.9	0.7	1.1						1.7	1.6	1.6	8.6
R													0.0

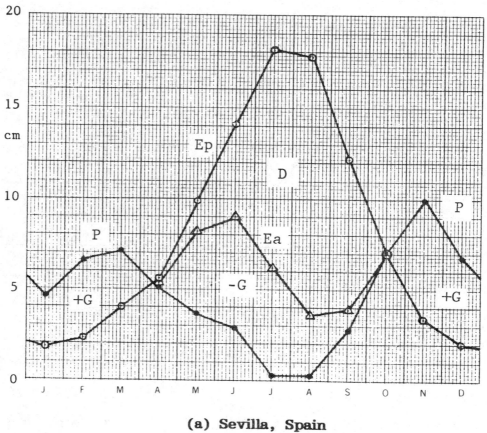

(a) Sevilla, Spain

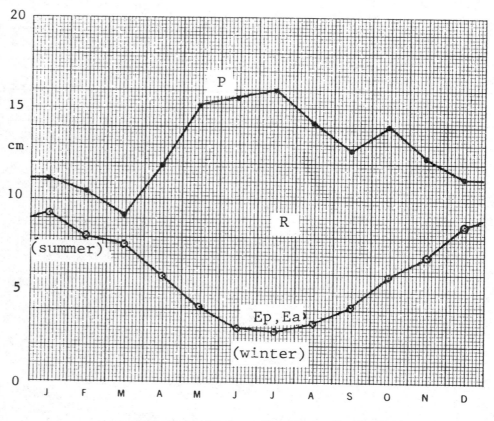

(b) New Plymouth, New Zealand

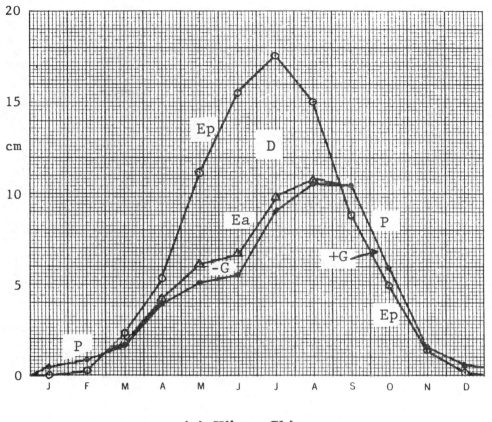

(c) Xi'an, China

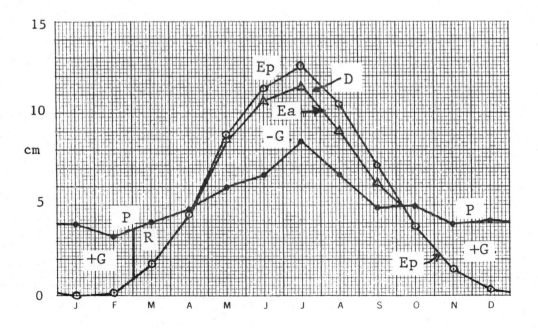

(d) Leipzig, East Germany

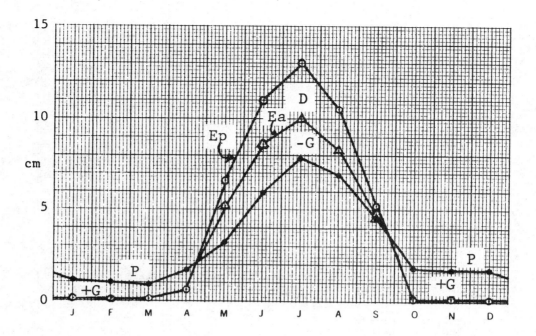

(e) Irkutsk, U.S.S.R.

QUESTIONS

(a) Sevilla, Spain

Sevilla--"Seville" to Americans who know it as the bullfighting capital of the world--is the leading city of the province of Andalusia, in southwestern Spain. It lies in the broad, flat lower valley of the Guadalquivir River, which has served for centuries as access for ocean-going ships. It became the prosperous chief port of trade with Spanish colonies of the New World. Often cited as one of the world's most beautiful cities, Sevilla retains the architectural flavor gained in five centuries of Moorish occupation.

(1) What climate does this soil-water budget illustrate? (Name and number.)

Mediterranean climate, semiarid subtype (7s)

(2) Describe the outstanding features of this soil-water budget.

The annual cycles of precipitation, P, and water need, Ep, are almost exactly out-of- phase, so that a large summer soil-water shortage, D, results. Storage recharge, +G, produces no water surplus, R.

(3) Compare the soil-water budget of Sevilla with that of Los Angeles, text Figure 9.11.

The cycles are generally quite similar, but the water need, Ep, of Sevilla peaks much more strongly in the summer, while the winter drought is shorter. This effect can be explained by the inland location of Sevilla, with more continental climate in the absence of a cool marine air layer typically present at Los Angeles.

(4) What forms of natural vegetation would you expect to find in the region surrounding Sevilla?

Native plants must be able to survive a long dry summer with high air and soil temperatures; they would probably be mostly grasses or shrubs. Shrubs and trees would bear thick leaves (sclerophylls) or spines. The growing season would be in the winter.

(5) Would perennial streams originate within this region? Explain.

With no water surplus, perennial streams would not originate within this area. However, channels would be expected to carry water on brief occasions during rains in the moist winter season.

(6) Would irrigation be required for cultivated crops in this region?

Some crops, such as wheat and hay, would flourish on winter rainfall, but irrigation would be needed for crops, such as vegetables, grown during the summer.

(b) New Plymouth, New Zealand

New Plymouth is a small port city located on the west coast of the North Island of New Zealand. Here, a strong oceanward bulge in the coastline surrounds a beautifully symmetrical dormant volcano, Mount Egmont, its snow-clad summit rising to 8,300 ft (2,500 m). If New Plymouth were to designate a "sister city," it might well be Valdivia, Chile, also located on the Pacific Ocean and at the same latitude. The vast expanse of South Pacific waters and a scarcity of landmasses gives to such locations a distinctive climate flavor not found in the northern hemisphere.

(1) What climate type and subtype is illustrated by the budget for New Plymouth?

Marine west-coast climate, humid subtype (8h) [World climate map does not show this small coastal area as 8p.]

(2) Describe the outstanding features of this soil-water budget.

The precipitation cycle shows a winter maximum, as typical of the marine west-coast climate. Precipitation, P, is large in every month, and always greater than water need, Ep. As a result there is a very large water surplus, R, and no soil-water shortage, D.

(3) Compare the budget of New Plymouth with that of Cork, Ireland, text Figure 9.17. Explain the important differences.

For Cork, the summer drop in precipitation carries P to values below those of water need, Ep, so that a small soil-water shortage occurs in the summer. Total annual precipitation for New Plymouth is about 50% larger than for Cork, perhaps because the latitude of New Plymouth is considerably lower and the impinging maritime air masses hold larger quantities of water vapor.

(4) What feature does the soil-water budget of New Plymouth share in common with the Mediterranean budget shown by Sevilla and Los Angeles?

Both climate types have a well-marked winter maximum of precipitation and a tendency to dryness (or less wetness) in summer. In both climates the cycles of water need, Ep, and precipitation, P, are exactly out of phase.

(5) What forms of natural vegetation would you expect to find in the vicinity of New Plymouth?

Natural vegetation would be some form of forest. Whereas in the northern hemisphere, evergreen needleleaf forest would be found, in New Zealand and Chile, rainforest of broadleaf evergreen trees would be most likely. (See Figure 20.17.)

(6) What conditions of stream flow would you expect in this region?

Even small streams would be of the perennial type, flowing throughout the year. In winter, stream flow would be very substantial, with possible flooding.

(c) Xi'an (Sian), China

Xi'an, capital of Shanxi Province, is a city of over 2 million inhabitants in the semiarid interior of China. It lies close to the great Huanghe River (also Hwang-Ho, or Yellow River) which is famed for its enormous load of silt, picked up in deeply eroded hills of windblown dust, called loess. (See text Figure 17.50 for a photograph of the loess area near Xi'an; accompanying descriptive text on p. 399.) The city has numerous pagodas of the T'ang Dynasty (618-906), but standing in sharp contrast are a modern university of science and technology, a medical college, and several technical institutes.

(1) What climate type and subtype is represented by the soil-water budget of Xi'an? (Note: On the world climate map, Figure 7.9, the orange area in this part of China, representing 9sd, is erroneously labeled "3d.")

Dry midlatitude climate (9), semidesert subtype (sd) (Accept subtype s.)

(2) Describe the outstanding features of this soil-water budget.

Both water need, Ep, and precipitation, P, follow a strong annual cycle, though not precisely in phase. The Asiatic summer monsoon effect is delayed here by about two months. The large soil-water shortage, D, is conspicuous, and no water surplus, R, is generated.

(3) Compare the budget of Xi'an with that of Medicine Hat, Alberta, text Figure 9.21.

An important difference is that the peak of precipitation at Medicine Hat comes in June, whereas at Xi'an it is displaced to August and September. Another important difference is that because Medicine Hat is much farther north, it shows five consecutive months (Nov.-Mar.) with zero Ep, when the soil water is frozen, whereas at Xi'an, only two months are zero (or almost zero). Total Ep is also substantially less, for the same reason.

(4) What forms of natural vegetation would you expect to find in this region?

Because of the aridity, grasses would be the dominant form of vegetation, and these would be short grasses, typical of steppe. Shrubs and small trees would be expected on moister, shaded slopes.

(5) What conditions of stream flow would you expect in the region?

Stream channels would be dry most of the year, but torrential flooding would be expected locally from June through September.

(6) What kinds of agricultural productivity would you expect from this region?

Without irrigation, field crops would not grow here, with the possible exception of some varieties of wheat. Grasslands might support grazing animals--sheep, goats, and cattle--in limited numbers.

d) Leipzig, East Germany

Leipzig is East Germany's second largest city and one of its chief industrial centers. Its rich cultural history is studded with the names of famous persons who made it a home at one time or another--Leibnitz, J.S. Bach, Schiller, Goethe, Mendelssohn, Schumann, Wagner. The Leipzig Music Conservatory, founded by Mendelssohn, became world-renowned, while the city also gained the reputation of foremost center of music publishing. It was at Leipzig in 1813 that Napoleon suffered a crushing defeat at the hands of Austrian, Russian, and Prussian armies.

(1) What climate type and subtype is illustrated by the soil-water budget for Leipzig?

Moist continental climate, humid subtype (10h)

(2) Describe the outstanding features of the soil-water budget of Leipzig.

Water need, Ep, shows a strong annual cycle with strong summer maximum, whereas precipitation, P, shows a weak annual cycle with only a moderate summer peak. As a result, the soil-water shortage, D, is small. In early spring, a small water surplus, R, is produced.

(3) Compare the soil-water budget of Leipzig with that of Pittsburgh, text Figure 9.27.

The graphs are strikingly similar in form, but differ in important ways. First, Pittsburgh's precipitation is greater in all months and the annual total is substantially larger. As a result, Pittsburgh has a much greater water surplus. Pittsburgh's summer water need, Ep, is less than for Leipzig, explained by the much lower latitude of Pittsburgh.

(4) Explain why water need, Ep, is extremely small or zero during December, January, and February at both Leipzig and Pittsburgh.

Because air temperatures average close to freeezing and soil water is frozen for long periods, plants are dormant. Under these cold conditions there is little transpiration from foliage (which may be absent) or direct evaporation from the soil surface.

(5) What kind of natural vegetation would have been found in this region prior to deforestation?

Forest would have been the natural vegetation. It would have consisted of deciduous trees, needleleaf evergreen trees, or mixed stands of both.

(e) Irkutsk, U.S.S.R.

Irkutsk is an industrial city in southern Siberia, not far north of the Mongolian border. It became infamous as a place of exile as early as the 18th century; later it became strategically located as major stop on the Trans-Siberian Railroad. Irkutsk also occupies a key position near the southern end of Lake Baikal, the largest fresh-water lake in Eurasia and the world's deepest lake. It is fed by numerous streams that drain surrounding mountain ranges. Voluminous flow at the lake's outlet has been harnessed at Irkutsk as a major source of hydroelectric power.

(1) What climate type and subtype are represented by the data for Irkutsk?

Boreal forest climate, subhumid subtype, 11sh.

(2) Describe the outstanding features of the soil-water budget for Irkutsk.

Both water need, Ep, and precipitation, P, are very low (or zero) in the long winter season, then both rise sharply to strong summer peaks. However, Ep is substantially greater than P during the summer and a moderate soil-water shortage, D, is developed. There is no water surplus. Note the six consecutive months of zero, or near-zero Ep, when soil-water is frozen.

(3) Compare the soil-water budget of Irkutsk with that of Trout Lake, Ontario, text Figure 9.30. Explain the major differences.

Water need, Ep, is about the same for both stations, but Trout Lake has much greater precipitation, P, both for the year and throughout the winter months (as snow). The result is that Trout Lake has a large water surplus, R, whereas Irkutsk has none. For this reason, Trout Lake is classifified as 11h, whereas Irkutsk is 11sh.

(4) Which of the above two stations would have the most winter snow?

Trout Lake would build up a heavy snow cover, that at Irkutsk would be much lighter.

(5) Which of the above two localities would have the denser forest cover?

Trout Lake, with its large water surplus, would have denser needleleaf forest. At Irkutsk vegetation would be comparatively sparse (perhaps dominantly open larch forest).

Exercise 10-A   The Annual Cycle of Rise and Fall of the Water Table
[Text p. 158-62, Figures 7.10-7.13; p. 222-23, Figure 10.5.]*

The Cape Cod peninsula juts into the Atlantic Ocean, its shape like a flexed Yankee arm with closed fist, boldly challenging any foreign invaders who might attempt a landing. Today, however, the Cape's threat is by Yankee invasion from the mainland, as each year developers build more houses to fill the demand for both year-around residence and summer use. The Cape's resident population in 1988 is up more than 500 percent from 60 years ago, but the quantity of fresh ground water available is just about the same. There is no other viable source of fresh water. What's worse, the existing supplies are prone to pollution from sewage and the leaching of toxic substances from town refuse dumps. Environmental planning to preserve this priceless resource requires scientific knowledge of the Cape's ground water system.

In this exercise we make use of the principles of the soil-water balance and the annual soil-water cycle, covered in Chapter 7 of your textbook. We apply this information to the data of observation wells on Cape Cod. An observation well is one that is set aside to monitor the natural rise and fall of the ground-water table; it is located some distance away from other wells that are being pumped. The record of rise and fall of water level is what we interpret in this exercise.

Part 1   The Soil-Water Budget of Cape Cod.

The table below gives data for the soil-water budget of East Wareham, Massachusetts, based on the 22-year record 1931-1952. Following the method you used in Exercise 7-D, plot the data on the blank graph, Graph A. (Note that we are using inches instead of centimeters for this example.) Label the graph fully.

East Wareham is located only a few miles west of Cape Cod. (See map, Figure A.) Its climate is very similar in all respects to that of Cape stations such as Provincetown and Hyannis, but it has a longer record.

(1) Briefly describe the soil-water budget of East Wareham, comparing it with the budgets for Pittsburgh, PA, (text Figure 9.27) and Leipzig, East Germany (Exercise 9-D).

All three budgets are similar in the annual cycles of P and Ep, since all three

are in the moist continental climate. East Wareham has somewhat greater

precipitation than Pittsburgh, and considerably more than Leipzig. Water deficit

(D) is small for all three stations. Surplus (R) of both East Wareham and

Pittsburgh is large. _____

_____

_____

*Modern Physical Geography, 3rd Ed., p. 169-73, Figures 10.6, 10.10;
   p. 186-87, Figure 11.2.

Data Table

	J	F	M	A	M	J	J	A	S	O	N	D	Year
P	3.98	3.31	3.31	3.19	2.72	2.91	3.00	3.70	4.88	3.82	3.90	4.09	42.81
Ep	0.00	0.00	0.35	1.26	2.91	4.25	5.23	4.72	3.27	1.89	0.79	0.00	24.67
Ea	0.00	0.00	0.35	1.26	2.91	4.17	4.80	4.41	3.27	1.89	0.79	0.00	23.85
D	0.00	0.00	0.00	0.00	0.00	0.08	0.43	0.31	0.00	0.00	0.00	0.00	0.82
-G	0.00	0.00	0 00	0.00	0.19	1.26	1.80	0.71	0.00	0.00	0.00	0.00	3.96
+G	0.00	0.00	0.00	0.00	0.00	0.00	0.00	0.00	1.61	1.93	0.42	0.00	3.96
R	3.98	3.31	2.96	1.93	0.00	0.00	0.00	0.00	0.00	0.00	2.69	4.09	18.96

(2) Cape Cod is largely underlain by a thick layer of sand and gravel of glacial origin. The surplus water readily sinks into this porous substratum to become part of the water table. Flowing surface streams are generally absent. Thus for practical purposes, we can assume that all of the water surplus (R) goes to recharge of the ground water system. What is the total annual recharge for East Wareham? Does the water table rise in elevation by an equivalent amount each year? Explain.

Total annual recharge will be about 19 inches. During recharge, the water table will tend to rise in elevation, but at the same time, outflow of the ground water to the ocean will tend to lower the water table. On the average, year after year, the water table will remain approximately at the same elevation, because average outflow will balance recharge (asssuming no important climate change.)

(3) How does the level of the water table respond to the annual cycle of alternate winter water surplus (R) and summer water deficit (D and -G combined)?

By early winter (Nov.-Dec.) the onset of a large recharge (R), amounting to 3 to 4 in. of water per month, will cause the water table to rise. The rise continues until late spring (May), when Ep exceeds P; after which recharge ceases and the water table begins to fall. The decline will continue through the summer and fall.

(4) On the blank graph above the soil-water budget, sketch in the approximate curve of annual rise and fall of the water table, illustrating your description given in answer to Question 3.

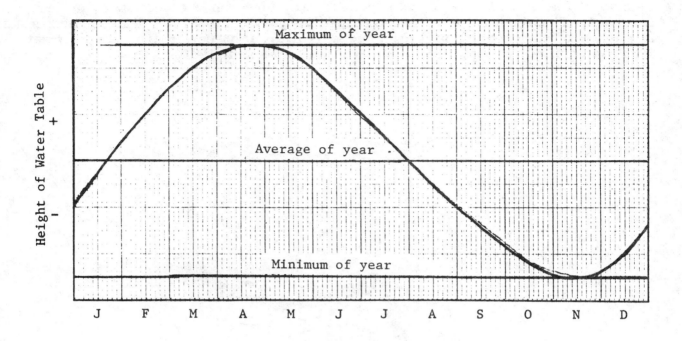

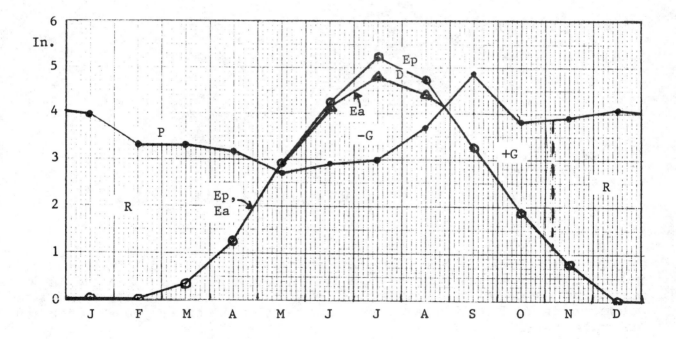

Graph A

Part 2 Fluctuating Water-Table Levels in Cape Cod Observation Wells

Graph B shows the records of four observation wells maintained by the U.S. Geological Survey on Cape Cod. Locations are shown on the accompanying map, Figure A. The record spans nine consecutive years, 1962-1970. The vertical scale is marked off in one-foot units. Included is a drought period that began in late 1963, intensified through 1965 and 1966, and ended in 1967.

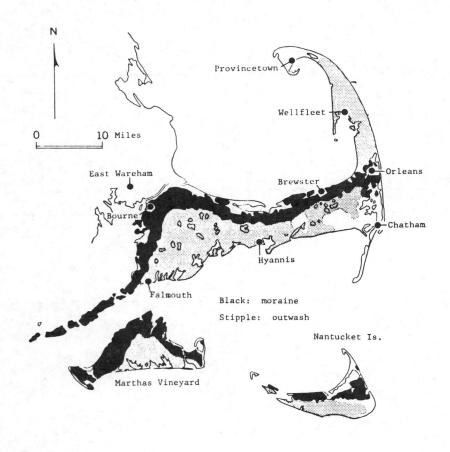

Figure A Cape Cod, Massachusetts. (A.N. Strahler.)

(5) Using those parts of the graphs for the years 1962-63 and 1968-70 (before and after the drought), designate the typical month in which the annual peak occurs and the month of lowest level. How does this schedule relate to the soil-water budget for East Wareham? Does it agree with the curve you sketched above the water-budget graph?

<u>The peak level typically occurs in May or June (sometimes earlier); the lowest</u>

<u>level, typically in November or December. This schedule is in general agreement</u>

<u>with that predicted in the answer to Question 3 and shown on the curve sketched</u>

<u>on the graph.</u>

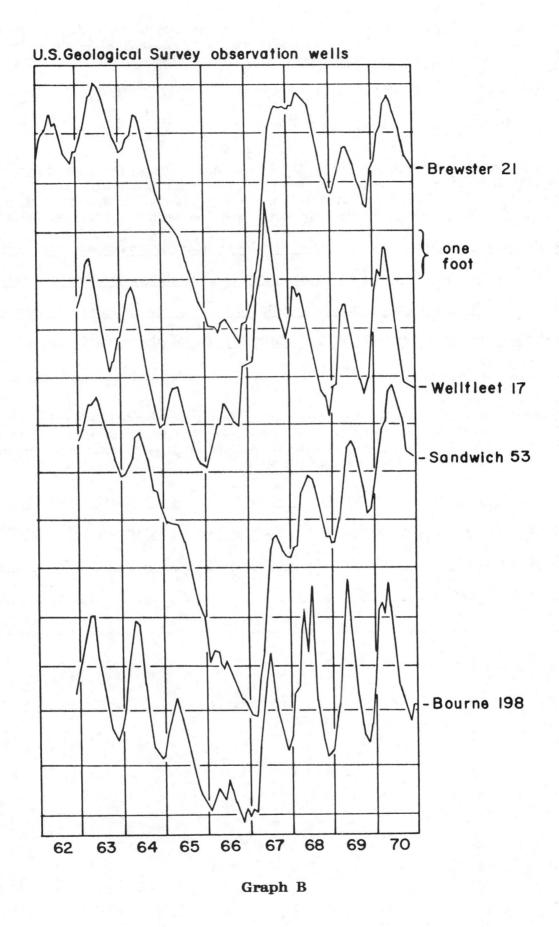

U.S. Geological Survey observation wells

– Brewster 21

one foot

– Wellfleet 17

– Sandwich 53

– Bourne 198

62 63 64 65 66 67 68 69 70

Graph B

(6) During the drought period, what was the approximate total decline in water table level at each of the four stations? (Measure from 1963-64 minimum to minimum of lowest year of drought cycle.)

Brewster: __4__ ft   Wellfleet: __2__ ft

Sandwich: __5__ ft   Bourne: __1½__ ft

(7) In the period 1968-70, the records of the four wells show some very marked differences in pattern. Bourne had three similar years; Sandwich rose in each successive year; Wellfleet and Brewster showed a dip in the maximum of the middle year. Can you suggest one or more possible causes for these differences?

Three different precipitation patterns may have been experienced, but this seems unlikely in a region of such small size. Another possibility is that the wells were affected by water pumping in nearby wells, and that the effects of these differences were superimposed on the natural cycle of the observation wells.

(8) Suppose that in the area surrounding an observation well, many new housing developments were added, each unit using a new well and disposing of its sewage in a septic tank with leaching field. What change might you expect to observe in the average level of the water table? Explain.

The water table would fall noticeably and reach a new equilibrium level. Although the ground water is being recycled through sewage disposal, a substantial loss occurs through evapotranspiration as irrigation is applied to lawns and gardens in dry summer periods.

(9) Eventually, as new construction of homes continues, raw sewage returned to the ground water threatens to pollute that water source. To forestall this disastrous event, municipalities install sanitary sewer systems carrying the raw sewage to the coastline, where it is discharged into the ocean. What effect on the ground water table can you predict as a result of this outfall installation? Explain.

The water table will decline drastically in level, threatening the local water supply. Deepening of household wells would be required. Importation of fresh water from distance sources may also be required. Diversion to ocean outfall reduces the regular annual replenishment of ground water that would occur in the natural soil-water budget.

(10) In what other way might the scenario you predict in Question 9 cause the ground water to become unpotable (not drinkable)?

Drastic lowering of the water table permits salt ground water to move landward, eventually entering the supply wells. When this happens, the water is unfit for drinking.

Name _____   Date _____

_____   _____

Exercise 10-B   Stream Gauging and Stream Discharge
[Text p. 226-27, Figure 10.10.]*

Measuring the rate of flow of our nation's rivers is an important task carried out daily by a special branch of the U.S. Geological Survey--the Water Resources Division. While driving cross-country you may have spotted one of their river-gauging stations near a highway bridge. What catches your eye is a steel cable spanning the river and suspended on it a cagelike cable car that allows observers to traverse the river in short steps, stopping to lower a current meter with which to measure the speed of flow (velocity) of the current threads at various depths below the water surface.

Figure A is an idealized diagram showing a typical stream gauging installation. The elevation of the water surface is continuously recorded by the recording gauge in a stilling well located on the river bank. The cross section of the river is imagined to be divided up into a grid of squares. Proceeding from one bank across to the other, the observer reads the current speed at the center of each grid square. The <u>area</u> (A) of the cross section is obtained by counting the number of squares. Totalling the individual velocity readings and dividing by the number of squares yields a statistic called the <u>mean velocity</u> (V) for the river cross section as a whole. The stream discharge (Q) is obtained by using the equation: Q = A • V

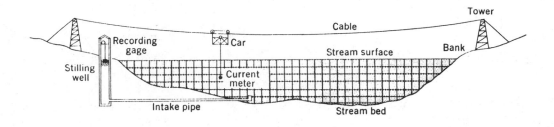

Figure A

As you can imagine, the stream gauging procedure takes a lot of time. There must be an easier way--and there is. It works best at a cross section where the river channel is carved in hard rock (or confined in a concrete channel) and does not continually change in form, as would a channel floored by gravel, soft sand, or mud. The idea is quite simple. Using past records of the discharge at a wide range of water levels, you plot on a graph each water level--the <u>stage</u>--against the corresponding computed discharge.

Graph A shows how a curve has been drawn through several plotted points and how the curve is used to read discharge from a particular gauge height. Let's plot a typical <u>stage-discharge</u> curve, using the data in Table A, and the blank graph, Graph B.

---

*Modern Physical Geography, 3rd Ed., p. 192-93, Figures 11.15, 11.16.

231

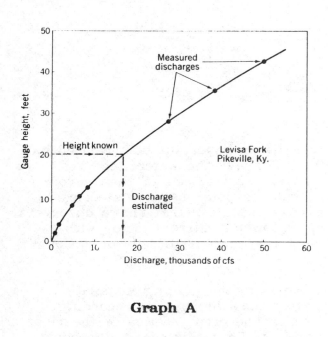

Graph A

Table A

Little Ossipee River, Maine

Gauge height (ft)	Discharge (cu ft/sec)
0.38	11
0.80	55
1.5	220
2.2	650
3.0	1400
3.8	2100
4.4	2800

(1) For each height value, draw a horizontal line to intercept the curve and a vertical line down to intercept the discharge scale. Read off the discharge for the following gauge heights:

0.5 ft: ___20___ cfs   1.0 ft: ___75___ cfs

2.6 ft: ___975___ cfs   4.1 ft: ___2450___ cfs

(2) Describe the shape of the plotted curve. Is it similar in shape to that of Figure A. Can you offer an explanation of this curve form in terms of the factors that control stream flow?

Both curves start out steep, then become more gentle and also becoming straighter. Part of this relationship, the upward slope, is because with greater depth, average speed of flow (V) increases. The curvature itself (convex-up) may be caused by the typical U-shaped cross section of the channel.

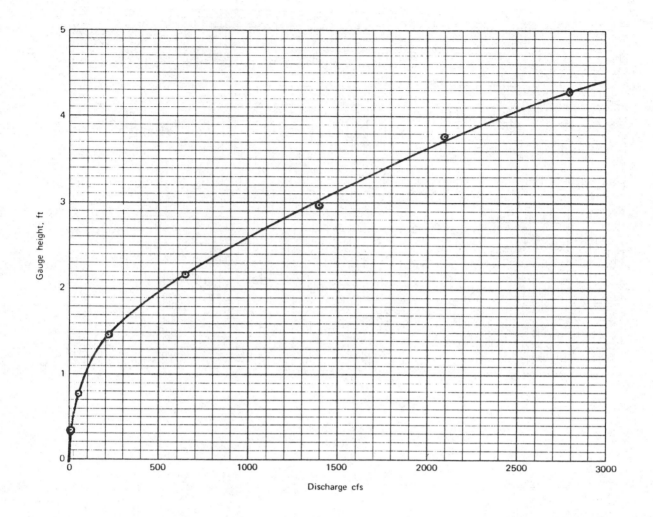

Gauge height, ft

Discharge cfs

Graph B

Special-credit problem for mathematicians:

Graph B you have used for Question 1 has arithmetic (linear) scales on both abscissa and ordinate. The curvature of the fitted line suggests that the mathematical relationship is not an arithmetic-linear one (Y = a + bX), but may instead be of exponential or logarithmic form.

Blank Graph C has logarithmic scales on both abscissa and ordinate; i.e., it is a "log-log" graph. Plot the data of the Little Ossipee River on the log-log graph. Fit the points with a straight line.

(3) Is the straight-line fit a good one with respect to the points?

Yes, the fit is excellent.

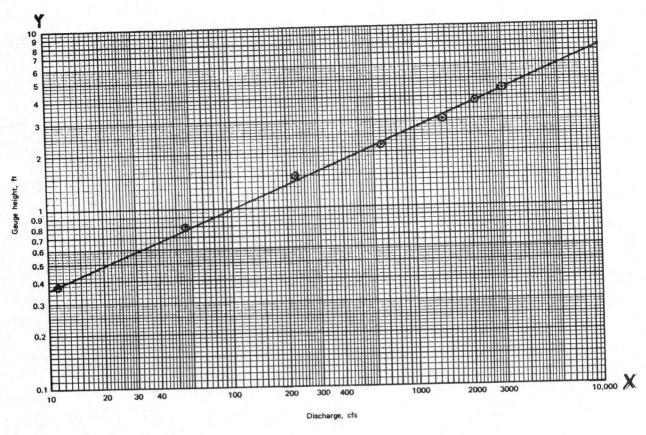

Graph C

(4) What is the significance of the straight-line fit on the log-log graph? Gauge height is related to discharge by a power function of the form $Y = aX^b$.

(5) In very rough terms, what exponent can be assigned to the log equation? The exponent is close to 0.5, which is to say that gauge height varies about as the square root of discharge.

(6) Which of the two rating curves you have prepared is the easier to use for low discharges? Explain.

The curve on log-log paper is much easier to use for low values because the scale is greatly expanded in that direction.

Name _____ Date _____

_____ _____

Exercise 10-C   Drawing Isohyets of a Rainstorm
               [Text p. 228-29.]*

Hydrologists study stream floods in order to be able to predict the maximum height, or crest, a particular flood will reach, and when that crest will arrive at various downstream points. To do this, they need to understand the relationship between rainfall and runoff. How much rainfall produces how much runoff?

Rainfall is measured routinely at hundreds of rain gauges scattered over the nation. Many are manned by volunteer observers, who daily measure the amount of water caught by the rain gauge. We will use such data in our exercise. The accompanying map, Figure A, shows a large area of the lower Ohio River valley. Each number represents a rain gauge and tells the total depth of water received during a four-day rainstorm in August. It must have been a whopper of a storm to drop more than 10 inches of rain in that period. Flooding occurred in many streams and small rivers in Illinois, Indiana, Ohio, Kentucky, Arkansas, and Tennessee. In response, the Ohio River rose to flood stage, and the rise was later felt in the Mississippi River below Cairo.

To show the distribution of rainfall over a given area, hydrologists draw isopleths of a kind known as <u>isohyets</u> (text p. 27 and Table 1.1). On the exercise map, isohyets at one-inch intervals are drawn for a part of the map. Your assignment is to complete the drawing of the isohyets for the remaining area of the map. Label the 5-inch and 10-inch isohyets, drawing them in bold lines.

(1) What weather situation (fronts, air masses) may have been responsible for this pattern of heavy rainfall? (Consult text Figure 6.8.)

<u>Quite possibly a slowly moving cold front or occluded front lay over this region</u>

<u>during the storm. Perhaps a squall line was the cause, running ahead of the</u>

<u>front. The center of concentration of the thunderstorms may have moved slowly</u>

<u>northeastward across the area.</u>

*Modern Physical Geography, 3rd Ed., p. 193-94.

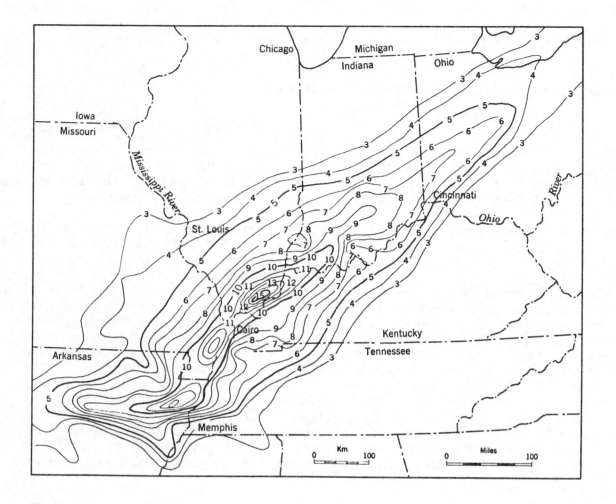

Figure A

(2) How would you go about estimating the total volume of rain that fell during the storm within the area enclosed by the 3-inch isohyet?

It would be necessary to measure the area enclosed by each isohyet. This can be done with an instrument known as a planimeter, but most operations of this kind are now carried out with a computer-graphics scanner. Area of each "layer" is multiplied by the value of the enclosing isohyet, giving volume. Volumes are summed for all isohyets from 3 through 15.

Name _____   Date _____

_____   _____

Exercise 10-D   Precipitation and Runoff
[Text p. 228-29, Figure 10.14.]*

This exercise builds on Exercise 10-C by plotting the quantity of runoff (as stream flow) produced by a single rainstorm. The graph will appear similar to that in text Figure 10.14, but here the data are for a much larger watershed. Whereas the watershed of Sugar Creek, OH, has an area of about 300 sq mi, that of the Delaware River watershed above the Port Jervis, NY, gauge covers more than 3,000 sq mi--larger by a factor of about ten.

(1) Table A gives the depth of rainfall in inches for four time periods. Note that the periods are of different duration. On Graph A, plot the precipitation amounts as a bar graph, using the scale on the right. Runoff, given in cfs, should be plotted as a point centered on the number of the calendar day. Connect the points with a smooth curve. Note that three of the runoff values are designated as "base flow." Estimate the missing base flow values from September 30 through October 8 and connect them by a smooth curve.

(2) Estimate the position of the centers of precipitation (CMP) and runoff (CMR). Enter a black dot for each and label it. Estimate the lag time between precipitation and runoff. (Subtract CMP from CMR.)

Lag time: <u>1.75</u>   days;   <u>42</u>   hrs

Table A   Delaware River Watershed. Port Jervis, N. Y.

Rainfall	Inches
6 P.M., Sept. 28 to 6 A.M., Sept. 29:	0.1
6 A.M., Sept. 29 to 6 P.M., Sept. 29:	0.9
6 P.M., Sept. 29 to 6 P.M., Sept. 30:	3.7
6 P.M., to midnight, Sept. 30:	0.1
Total	4.8

Runoff	cfs
Sept. 28	1000 (base flow)
Sept. 29	1000 (base flow)
Sept. 30	12,000
Oct. 1	74,000
Oct. 2	47,000
Oct. 3	21,000
Oct. 4	13,000
Oct. 5	9000
Oct. 6	8000
Oct. 7	7000
Oct. 8	6000
Oct. 9	5000 (base flow)

*Modern Physical Geography, 3rd Ed., p. 193-95, Figure 11.17.

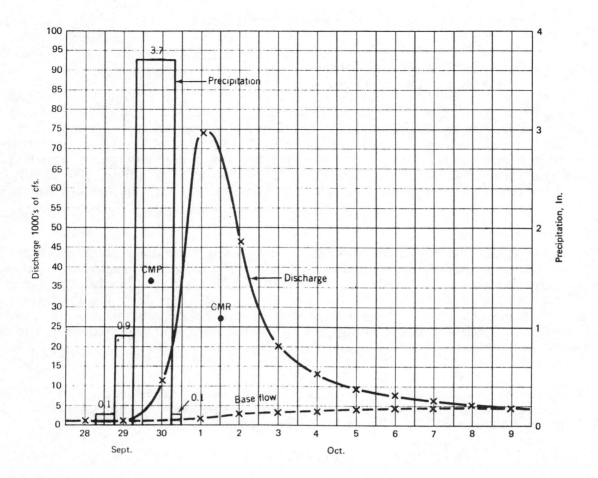

Graph A

(3) Compare the lag time obtained above with that for Sugar Creek in text Figure 10.14. Explain.

That for Sugar Creek is about half the time for Port Jervis. The shorter lag time is expected for a smaller basin.

Extra credit investigation:

From the data already given, it is possible to determine approximately what proportion of the rainfall falling on the Port Jervis drainage basin actually passed through the gauging station. First, the daily base flow (estimated by you) is subtracted from the daily discharge, yielding the net discharge. Next, the net discharge is converted to daily volume in cu ft, by multiplying by the number of seconds in one day (86,400). The daily volumes are totaled to obtain the total storm runoff. Use the blank form below to carry out these calculations.

Day	Daily Discharge, cfs	Base Flow, cfs	Net Discharge, cfs	Daily Volume, cu ft
Sept. 30	12,000	1,000	11,000	950,400,000
Oct. 1	74,000	2,000	72,000	6,220,800,000
Oct. 2	47,000	3,000	44,000	3,801,600,000
Oct. 3	21,000	3,500	17,500	1,512,000,000
Oct. 4	13,000	4,000	9,000	777,600,000
Oct. 5	9,000	4,500	4,500	388,800,000
Oct. 6	8,000	5,000	3,000	258,200,000
Oct. 7	7,000	5,000	2,000	172,800,000
Oct. 8	6,000	5,000	1,000	86,400,000

Total storm volume   14,168,600,000 cu ft

(4) Calculate the mean depth in feet of the total storm volume of surface runoff. (Watershed area is 3076 sq mi.)

Watershed area in sq ft: (3076 X $(5280)^2$ =   85,754,000,000

Next, divide storm runoff volume by watershed area:

Mean depth of storm runoff: 0.165   ft

Converted to depth in inches: 1.98   in.

(5) Compare the depth obtained in (4) with total depth of rainfall of the storm. Explain your results.

Rainfall depth: 4.8   in.

The rainfall depth is about $2\frac{1}{2}$ times greater than runoff depth. The difference is accounted for by infiltration of rain into the ground and by evaporation back to the atmosphere.

Exercise 10-E   Flood Regimes and Climate
         [Text p. 231-32, Figure 10.18.*]

In the summer of 1988, the mighty Mississippi River suffered a great humiliation as it dwindled to record low stages. Commercial barge traffic came to a halt in some reaches of the river, seriously disrupting industrial activities dependent on cheap transport of bulk commodities. So, too little water in the channel of a large river can cause problems, though not comparable with the devastation of a major flood. A distinguished authority on the Mississippi River, Gerard Matthes, tells us that the Muddy Mississippi is a comparatively clear stream. He says that in late September, 1936, during a low stage, "the Misississppi flowed water as blue as that of the Danube (which, incidentally, also is blue only during low-water periods." In our modern industrial era, chemical pollution has taken away the romance of blue rivers. Raw sewage and industrial wastes, dumped into the river, increase in concentration as the river stage lowers.

In this exercise we examine four strange-looking diagrams (Graph A), each telling the average discharge regime of an American river throughout the annual cycle. The measured quantity is stage (gauge height) in feet, which is directly related to discharge, as we found in Exercise 10-B. Like the exam grades of a college class, the frequency data are treated on a percentile basis. This form of presentation plays down the effect of differences in river magnitudes, allowing us to compare them in terms of response to the seasonal cycles of precipitation and runoff. Your knowledge of midlatitude climates and their soil-water cycles will be put to use in analyzing these graphs.

The accompanying map of the United States, Figure A, shows the major rivers and their organization into 14 major regions or areas, according to the U.S. Geological Survey, as needed for reporting flow data in its Water-Supply Papers. With the aid of an atlas, locate the Lower Mississippi, Sacramento, Colorado (Texas), and Connecticut rivers. Use a color pencil or pen to highlight these rivers. Make a dot at the location of each gauging station and label with the name of the station.

Mississippi River at Vicksburg, Mississippi

(1) Describe in general terms the annual cycle of stages (gauge heights) at Vicksburg.

The annual cycle of rise and decline is simple in form, as in a sine wave, and

the percentage divisions (quartiles) tend to be fairly uniform (especially the two

intermediate quartiles). The peak occurs in late spring (April-May); the minimum

in late fall (October).

_____

*Modern Physical Geography, 3rd Ed., p. 196-97, Figure 11.22.

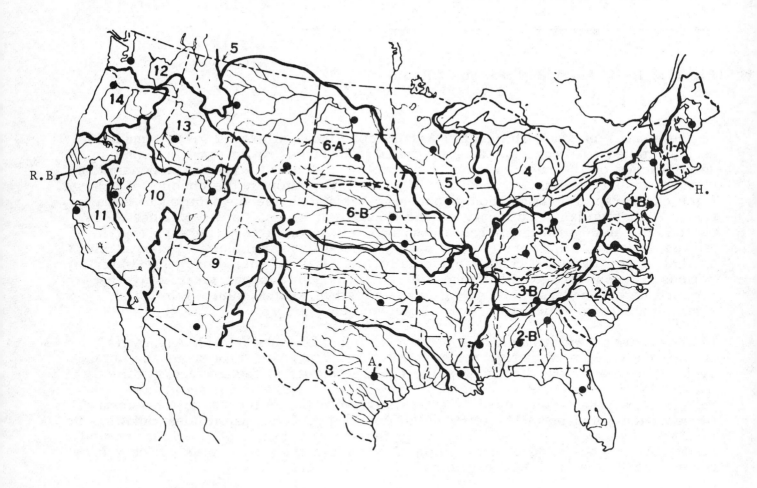

Figure A

(2) Study the the entire Mississippi River system as shown in Figure A. The areas included are numbered 3, 5, 6, and 7. In the table below, enter the name of the principal river (or rivers) in each area or sub-area. Highlight these rivers in color on the map and label them. Identify and label those states, all or parts of which lie in these drainage areas.

Area	Principal River(s)
3-A	Ohio
3-B	Cumberland, Tennessee
5	Upper Mississippi, Red
6-A	Missouri
6-B	Missouri, Platte
7	Arkansas, Red (Canadian)

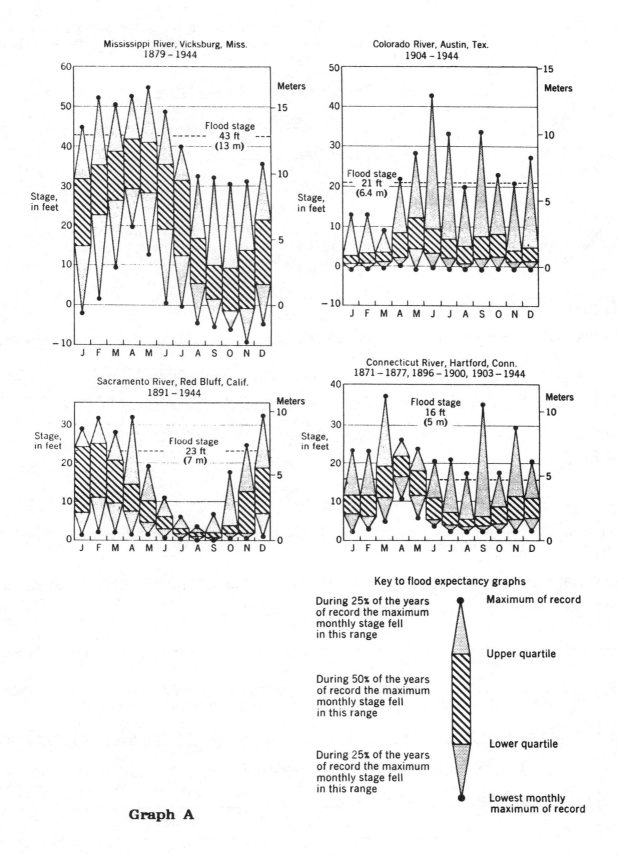

Graph A

Key to flood expectancy graphs

During 25% of the years of record the maximum monthly stage fell in this range — Maximum of record

Upper quartile

During 50% of the years of record the maximum monthly stage fell in this range

Lower quartile

During 25% of the years of record the maximum monthly stage fell in this range — Lowest monthly maximum of record

(3) On the map, Figure A, follow up the above rivers to their extreme upper limits. Using an atlas, identify and name the mountain ranges in which each of the following river systems have their headwaters:

Area 3: <u>Central and Southern Appalachians</u>

Area 6-A: <u>Northern and Middle Rockies</u>

Area 6-B: <u>Southern Rockies</u>

Area 7: <u>Southern Rockies</u>

(4) We can expect the release of runoff by snowmelt in the above mountain systems to have a major controlling effect on spring and summer stages of the lower Misssissippi. Anticipate the order in which these runoff events would occur. Explain.

<u>Snowmelt would occur first in the Appalachians, which are relatively low in elevation. This contribution would act through the early spring (February-March) rise of the Ohio River. Snowmelt of the higher Rocky Mountains, feeding the Missouri system, would be delayed until early summer, continuing into midsummer. This program accounts for the broadly-rounded form of the spring/summer maximum shown on the graph.</u>

(5) How do you account for the low stages in the middle and lower quartiles in September and October?

<u>These low stages coincide with the deficit period of the soil-water budgets, and with the soil-water recharge periods that immediately follow.</u>

(6) Looking at the upper quartile and maximum of record, how do you explain the extreme lengthening of this quartile in September and October?

<u>These maxima may represent invasions of tropical storms (hurricanes) passing over the Southern Appalachians. Very heavy rainfall from a slowly moving cyclone would cause flooding in area 3.</u>

Sacramento River at Red Bluff, California

Using a good relief map of the western United States, examine the topographic features of northern California. Notice that Red Bluff lies near the northern end of the Great Valley and that tributaries to the Sacramento River extend north and east into the Sierra Nevada and the southern end of the Cascade Mountains.

(7) Describe the general features of the annual cycle for this river. How and why does it differ significantly from that for the Missisippi River?

The most obvious feature is the collapsing of the expectancy bar to very short lengths, starting in June and reaching the minimum in August. This feature is explained by the soil-water budget for the Meditereranean climate, with its nearly rainless summer and large summer soil-water deficit.

(8) Examine the cycle of lowest monthly maximum of record. Describe and explain this cycle.

Most obvious is the small amplitude of the cycle, with all months showing a maximum lowest stage of 2 ft or less, and five consecutive months of zero or near-zero values. This suggests that in some years winters are exceptionally dry, with little snow on the mountains.

Colorado River at Austin, Texas

Using an atlas, examine in detail the topographic features of the drainage basin of the Colorado River in Texas. Notice that the headwater drainage area is limited to the Llano Estacado, an elevated but flat part of the Great Plains. This high plain is separated from the Southern Rockies by the valley of the Pecos River.

(9) What are the most striking features of this graph? Explain.

The strong compression (shortening) of the middle and lower quartiles contrasts greatly with the elongation of the upper quartile and the extreme flood peaks of record. The lowest monthly maxima, shown as lying at or below zero stage, indicates that river flow has in every month dropped at least once to zero.

(10) How do you explain the extremely large maximum values occurring May through December?

These extremes represent occurences of intense rainstorms of several days' duration, such as occur when an intense cyclone moves inland and stalls over the watershed. Maritime tropical air masses supply large quantities of moisture. In the August-December period, these cyclones are tropical storms or hurricanes. Those of earlier occurrence are associated with fronts of strong interaction between polar and tropical air masses.

Connecticut River, Hartford, Connecticut

Use an atlas to examine the long, narrow watershed of the Connecticut River, extending almost due north to the Canada border. Mountainous and hilly terrain, including the Taconic and Green Mountains, is typical of this region.

(11) Describe the annual cycle of the graph. Explain the salient features.

From June through February, the middle and lower quartiles show a rather uniform pattern, with only a small summer dip, whereas a sharp peak occurs in March, April, and May. This spring maximum can be attributed to the melting of frozen soil water and snow pack. Because the water surplus is typically large, ground-water recharge maintains base flow through the entire year.

(12) Suggest an explanation for the extremely high value of the maximum of record in September. Would a similar explanation hold for the exceptional maximum in March?

This September value might represent a single great storm, such as a hurricane. [This event was the great hurricane of September 1938, in which torrential rains fell on soils surfaces already fully saturated by earlier rains. The city of Hartford experienced a great overbank flood during this hurricane.] The maximum of March would not likely have been caused by a tropical cyclone, as this is well out of season of such storms. An intense middle-latitude cyclone beneath a stationary upper-air low might explain the flood.

Name _____   Date _____

_____   _____

Exercise 10-F   Evaporation from American Lakes
[Text p. 232-35.]*

The need for great increases in supplies of fresh water is nowhere greater than in the arid climates, where irrigation agriculture is highly developed and urban centers are growing--that goes without saying. The catch is that loss of water by evaporation from lakes behind irrigation dams increases in proportion to their size and number. Engineers can easily design a dam so high, and with an area of water impoundment so vast, that it will never fill with enough water to supply downstream irrigation systems or to run the turbines installed in the dam to generate electric power. To say "never" may be an overstatement because, in due time, the blocked valley behind the dam will fill with sediment, reducing the water-storage volume sufficiently to allow the dam to fill, but that might take a half-century or more to come about. Small wonder the dam builders have been called "damn builders" by conservationists!

In this exercise, we study rates of evaporation from free water surfaces of lakes or ponds. Since evaporation is difficult to measure directly from such water bodies, hydrologists make use of the <u>evaporating</u> <u>pan</u>, which is simply a circular container 1.2 to 1.8 m (4 to 6 ft) in diameter and 25 cm (10 in.) or more in depth (Figure A). Water is added as required and the amount of surface lowering due to evaporation is measured with a gauge. Evaporation data are collected by the National Weather Service at many observing stations. Evaporation from the pan is generally greater from that of a lake or reservoir at the same location, so the pan readings require correction by a reduction factor ranging from 60 to 80 percent.

Figure A An evaporating pan with anemometer at side. Notice a thermometer immersed in the water (upper left). The cylindrical device at the near edge of the pan is a hook gauge to measure the height of the water level. (U.S. National Weather Service.)

---

*Modern Physical Geography, 3rd Ed., p. 198-99.

EX. 10-F

Table A gives total annual evaporation, E, in inches for some actual reservoir surfaces . Also given are mean annual air temperature, T (°F), and mean annual relative humidity, RH (%). (Wind speed is another important factor, but we will not deal with it in this exercise.)

Table A

Code		E (in.)	T (°F)	RH (%)	Climate symbol	Köppen symbol
IT	Ithaca, NY (42½°N)	23	47	78	10h	___
FA	Fallon, NV (39½°N)	57	51	52	9s	___
BI	Birmingham, AL (33½°N)	43	63	72	6h	___
EL	El Paso, TX (32°N)	71	64	36	5d	___
SA	San Juan, PR (18½°N)	55	78	78	2	___
GA	Gatun, CZ (9°N)	48	80	84	1	___

(Data of National Weather Service.)

The stations are arranged in order of decreasing latitude. The first two make a northerly U.S. latitude pair; the second two a southerly U.S. pair. The last two, San Juan and Gatun, represent the tropical and equatorial latitude zones, respectively.

(1) Using the world climate map, text Figure 7.9, find these six stations and determine the climate of each. Enter the climate symbol in Table A. Space is also given for the Köppen climate symbol (see text Figure 7.17).

(2) The accompanying U.S. map, Figure A, shows isopleths of annual evaporation from shallow lakes. Locate the U.S. stations on the map, marking each with a dot and labeling it with the station code.

(3) What meteorological factors are responsible for the low value of E for Ithaca?

Three factors can be assigned: low air temperature, moderately high RH, and the fact that lakes and reservoirs will be frozen over during the winter months.

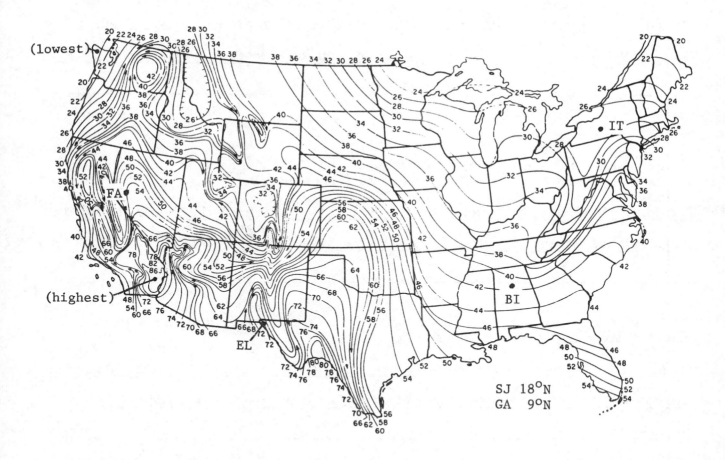

Figure B Average annual evaporation (inches) from shallow lakes. (U.S. National Weather Service.)

(4) Fallon has about 2.5 times the evaporation of Ithaca. Explain why this should be so.

The lower RH of Fallon could account for most of the difference. Also, at Fallon winter air temperatures are not nearly as cold as at Ithaca, so that no ice cover would be expected. (Fallon lies in the dry midlatitude climate (9s) of the Great Basin.)

(5) Why is Birmingham's evaporation about double that of Ithaca?

Birmingham's warmer temperature and lack of ice cover would largely explain its greater E.

(6) Why is evaporation higher at El Paso than at Birmingham?

The low RH at El Paso would largely explain the difference.

(7) San Juan, although warmer than any of the U.S. mainland stations, has an intermediate value of E, about the same as for Fallon. Explain.

The high RH at San Juan would act to reduce E throughout the year, despite the prevailing higher temperatures in this tropical zone.

(8) Gatun's evaporation is even less than that of San Juan, and not much different from that of Birmingham. Explain.

Gatun, in the wet equatorial climate, has high RH throughout the year, along with almost uniform 80-degree temperature. The effect would to suppress evaporation.

(9) Where in the U.S. is evaporation least? Give both location and climate type.

E value of 20 or less occurs in (a) northern Maine (11h) and (b) westernmost Washington (8h).

(10) What is the highest value of E shown on the U.S map? With what geographical region is this high associated?

Values over 86 occur in southeastern California, along the border with Arizona. This is the valley of the Lower Colorado River, in the Sonoran Desert.

Exercise 11-A   Crustal Elements and Life on Earth
[Text, p. 241, Figure 11.1; p. 464-66, Table 20.2.]*

The Book of Genesis (2:7,9) reveals that "the Lord God formed man of the dust of the ground" and "out of the ground made the Lord God to grow every tree that is pleasant to the sight." Taking our clue from these legends of the ancient Hebrews, we ask whether there is any clear and strong relationship between the element chemistry of the biosphere and that of the lithosphere, hydrosphere, and atmosphere.

The first indications of life on our planet are found in rock about $3\frac{1}{2}$ billion years old; these are spheroidal microscopic structures resembling living procaryotic cells. Geologists find it reasonable to suppose that about a half-billion years earlier (i.e., about 4 billion years ago) the earth had a rock crust with elevated continents and with extensive oceans occupying deep crustal depressions. A primitive atmosphere was already in action recycling the ocean water through storms that precipitated rain on the exposed continental surfaces. Exposed rock was being chemically altered and the weathering products were being carried by streams to the oceans.

A scientific hypothesis favored today is that the first living cells were synthesized from available chemical elements in the form of ions and molecules contained in shallow seas or inland waters. One way you can investigate this hypothesis is to compare the abundances of elements in the modern biosphere with those of crustal rock, seawater and stream water, and the atmosphere. What you find will also be helpful in understanding today's terrestrial ecosystems and their role in recycling nutrient elements (Chapter 20). That knowledge will, in turn, help you understand the global patterns of soils and natural vegetation.

We start with the biospheric data in text Table 20.2, p. 465, reproduced below as Table A. Data for abundances of crustal elements, shown in text Figure 11.1 are also reproduced here (Table B), but with the list extended to include the first 15 elements. A third table gives data for seawater (Table C). Elements of the atmosphere are shown in Table D.

(1) You are asked first to plot the data of Tables A, B, C, and D on Graph A. The elements are arranged from left to right in the same order of abundance as in Table A. The vertical scale is marked off in powers of ten; i.e., it is a logarithmic scale. Use of this kind of scale allows us to show very small quantities more easily than on a uniform (arithmetic) scale, but at the same time, values are more difficult to plot. To assist you in plotting, we show in detail in Graph B the full scale of values within two log cycles. Estimate as best you can the position of the value you are plotting as a point on the graph. Then, using a felt-tipped pen, make a bold vertical bar for each element, as shown for the selected values plotted for Table A. Where the quantity of an element is smaller than 0.001 and cannot be shown on the graph, enter the word "trace."

---

*Modern Physical Geography, 3rd Ed., p. 202, Figure 12.1; p. 423-27, Table 24.2.

Table A Abundances of Elements Comprising Global Living Matter
(Percentage of atoms of total atoms of the 15 most abundant elements.)

Rank	Name & symbol	Percent
1	Hydrogen (H)	49.74
2	Carbon (C)	24.90
3	Oxygen (O)	24.83
4	Nitrogen (N)	0.272
5	Calcium (Ca)	0.072
6	Potassium (K)	0.044
7	Silicon (Si)	0.033
8	Magnesium (Mg)	0.031
9	Sulfur (S)	0.017
10	Aluminum (Al)	0.016
11	Phosphorus (P)	0.013
12	Chlorine (Cl)	0.011
13	Sodium (Na)	0.006
14	Iron (Fe)	0.005
15	Manganese (Mn)	0.003

(Data of E.S. Deevey, Jr., 1970, Scientific American, vol. 223.)

Table B Abundances of Elements in Average Crustal Rock
(Percent of atoms.)

Rank	Name	Percent
1	Oxygen (O)	60.39
2	Silicon (Si)	20.46
3	Aluminum (Al)	6.24
4	Hydrogen (H)	2.87
5	Sodium (Na)	2.70
6	Calcium (Ca)	1.88
7	Iron (Fe)	1.86
8	Magnesium (Mg)	1.78
9	Potassium (K)	1.37
10	Titanium (Ti)	0.19
11	Phosphorus (P)	0.07
12	Fluorine (F)	0.07
13	Barium (Ba)	0.06
14	Manganese (Mn)	0.04
15	Strontium (Sr)	0.01

(Data of B. Mason, 1966, Principles of Geochemistry, 3rd. ed., John Wiley & Sons, Table 3.3.)

(2) For tables B, C, and D, the ranking order of the elements is not the same as for Table A. For tables B, C, and D, enter on Graph A the rank number of the element above the top of each bar.

(3) For tables B, C, and D, several of the elements are not among the 15 elements listed in Table A. Enter the names and rank numbers of these extra elements in the blank space at the right of each table in Graph A.

Table C   Abundances of Elements in Seawater
(Percent of atoms.)

Rank	Name	Percent
1	Hydrogen (H)	65.90
2	Oxygen (O)	33.40
3	Chlorine (Cl)	0.331
4	Sodium (Na)	0.284
5	Magnesium (Mg)	0.032
6	Sulfur (S)	0.017
7	Carbon (C)	0.014
8	Calcium (Ca)	0.0061
9	Potassium (K)	0.0060
10	Bromium (Br)	0.0005
11	Boron (B)	0.0002
12	Silicon (S)	(trace)
13	Strontium (Sr)	(trace)
14	Fluorine (F)	(trace)
15	Lithium (Li)	(trace)

(Data of K. Turekian, 1972, _Chemistry of the Earth_, Holt, Rinehart and Winston, Table 6-3.)

Table D   Abundances of Elements in the Atmosphere
(Percentage by volume.)

Rank	Name	Percent
1	Nitrogen (N)	78.084
2	Oxygen (O)	20.946
3	Argon (A)	0.934
4	Carbon (as CO_2)	0.035
5	Neon (Ne)	0.00182
6	Helium (He)	0.00053
7	Krypton (Kr)	0.00012
8	Xenon (Xe)	0.00009
9	Hydrogen (as H_2)	0.00005

(Data of American Meteorological Society.)

(4) Hydrogen and oxygen rank 1 and 3 respectively in global living matter (Table A). What are the most obvious abundant sources of these elements? In what chemical forms are these two elements available?

Both H and O are abundantly available in the molecular form of water (H_2O) in the hydrosphere, represented here by seawater in Table C. Oxygen is abundantly available in the atmosphere as free oxygen in molecular form, O_2.

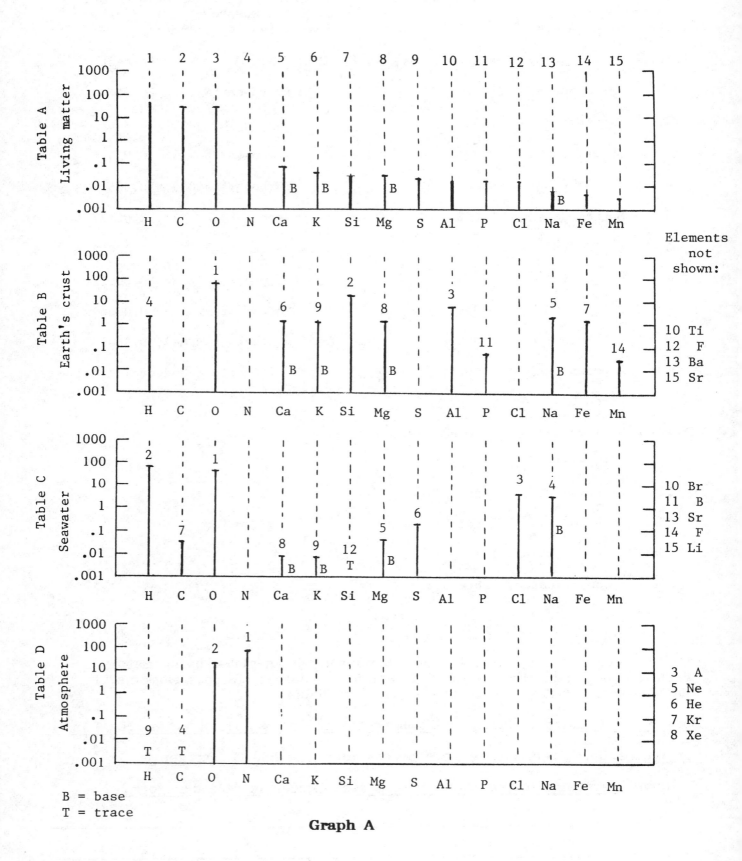

B = base
T = trace

Graph A

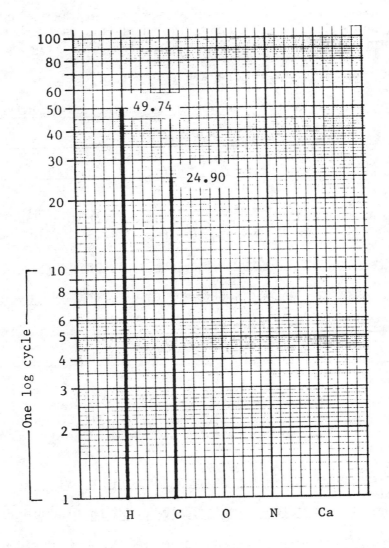

Graph B

(5) A large amount of oxygen resides in the lithosphere, as shown in Table B. Do living cells make use of this oxygen source? Explain.

Oxygen in the crust is locked up in silicate and oxide minerals and cannot be

directly taken up by living cells.

(6) From what source do organisms derive carbon, the second most abundant element in living matter? In what chemical form is this carbon available to plants to carry out photosynthesis? (Consult text p. 459-60 and Figure 20.6.)

Both the hydrosphere and atmosphere are sources of carbon. It occurs in

available form as gaseous carbon-dioxide (CO_2)in the atmosphere and dissolved in

water of the oceans, lakes, streams, and soils.

(7) Nitrogen is a major element in living matter. From what basic source is it obtained by plants? In what chemical form does it exist in this source?

Abundant nitrogen is present in the atmosphere in the gaseous form of molecular

nitrogen, N_2. Nitrogen gas is also dissolved in all water of the hydrosphere and

is available to land plants in soil water.

(8) A special question for the chemists: The element values given in Table B do not agree with those shown in text Figure 11.1, p. 241. Not only are the numbers different, but the ranking is also different. Can you explain this seeming discrepancy?

Figure 11.1 gives percentage of the element by weight, whereas Table B gives

percentage of atoms of the given element in relation to total atoms. Elements in

this list differ greatly among themselves in terms of atomic weight (atomic mass),

so that hydrogen--lightest of all elements--appears in rank 4 in Table B, but is

not among the first 8 elements in Figure 11.1. Iron (Fe), a comparatively heavy

element, ranks 4 by weight-percentage in Figure 11.1, but drops to rank 7 in

percent of atoms.

(9) In text Figure 11.1, four elements are labeled "bases." Name these elements. Find them on the graph for Table A and write the word "base" beside the bar. Are these four bases directly available to living matter from minerals of the crust? Explain. (Consult text p. 428-29.) To assist you further, consult the table of composition of salts in seawater, given in Exercise 11-B.

The bases are Ca, K, Mg, and Na. They are abundant in crustal rock, but not directly available to organisms in the mineral forms in which they occur. The processes of chemical alteration release these elements as ions in solution in soil water, stream water, and seawater. As ions, they are readily taken up by plant cells.

(10) Another question for the chemists: The elements silicon, aluminum, phosphorus, iron, and manganese present us with a difficult problem. They are on the list for living matter (Table A), also well represented in crustal rock (Table B), but almost completely unrepresented in seawater (Table C). Can you make a suggestion as to why these elements are not found abundantly in seawater?

[Few, if any students will supply a rational answer to this question, but it can serve as a challenge.] Although these four elements are released by chemical weathering and carried by streams into the ocean, their residence time in the ocean is very short. Silicon, combined with oxygen as silica (SiO_2) is quickly precipitated, as are oxides of Al, P, Fe, and Mn. These oxides go into storage as mineral matter of seafloor sediments and eventually, by tectonic processes become ingredients of the continental crust.

We leave unsolved another puzzle. Two important elements in living matter, sulfur and chlorine, are not listed in crustal rock, but they both appear in important amounts in seawater. Where do these two mystery elements come from? This question comes up next in Exercise 11-B.

Name _____ Date _____

_____   _____

Exercise 11-B   Volcanoes and Volatiles
[Text, p. 244-45, 249, 252, Figure 11.16.]*

Geology as a science is much, much more than collecting mineral and rock specimens, identifying them by name, and arranging them neatly in drawers. The same can be said of biology. A distinguished evolutionary biologist, Ernst Mayr, put it in a nutshell when he said: "Biology is not postage stamp collecting." Nor is geology. Our study of the lithosphere, hydrosphere, atmosphere, and biosphere emphasizes the concept of flow systems of matter and energy. Planetary materials --chemical elements, that is--flow from each of these spheres into another.

In Exercise 11-A we began to look for these material flow paths in and out of the biosphere. Here, we continue that quest in an attempt to clear up the mystery of two elements essential to life: sulfur and chlorine. The problem we identified was that sulfur and chlorine, ranking 9 and 12, respectively on the abundance list for living matter (Table A), don't appear in the earth's crust (Table B), but are prominent in seawater (Table C) where they rank 6 and 3, respectively. Now, if seawater gets its elements from chemical weathering of minerals in the crust exposed on the continents, why don't we find sulfur and chlorine on the list for crustal rock? Actually, if the list were continued, it would show sulfur ranking 16 and chlorine 20, but the amounts are very small. Clearly, some part of the material flow system remains to be identified.

One clue may lie in the occurrences of two other elements. Carbon ranks second in the biosphere, is missing from the crustal list, and appears in a small quantity in the atmosphere. Nitrogen, missing on our lists for both crust and seawater, ranks first in the atmosphere. Chemists tell us that sulfur, chlorine, carbon, and nitrogen belong to a class of inorganic substances known as <u>volatiles</u>. In everyday life we think of volatiles as substances that evaporate easily at room temperature; for example, alcohol, nail-polish remover, and naphtha. If so, the commonest of all volatiles must be water itself--the compound H_2O--and indeed it is.

Table A of this exercise gives a list of the volatiles of the earth's atmosphere and hydrosphere.

Note: You might wonder why we don't list oxygen as a volatile. After all, it is a component of the water molecule and it is present as O_2 in the atmosphere in the amount of 21 percent. Oxygen is a very special element, extremely active chemically, and equally at home in solids, liquids and gases. Only because plants continually give off oxygen gas do we find it in the free, gaseous state. If all plants on earth died off, that free oxygen would quickly link up with other elements. On the other hand, hydrogen is classed as a volatile, largely in the form of water.

---

*Modern Physical Geography, 3rd Ed., p. 204, Table 12.1; p. 213-14, Figure 12.14.

Table A Volatiles of the Earth's Hydrosphere and Atmosphere
(Percent by weight.)

			Rank in Table B
1	Water, H_2O	92.8	1
2	Carbon, as CO_2	5.1	2
3	Chlorine, Cl_2	1.7	6
4	Nitrogen, N_2	0.24	4
5	Sulfur, S_2	0.13	3
6	Hydrogen, H_2	0.07	7
	Fluorine, F	(trace)	
	Neon, Ne	(trace)	
	Argon, A	(trace)	5
	Krypton, Kr	(trace)	
	Xenon, Xe	(trace)	

(Data of W. W. Rubey, 1952;
K. Turekian, 1972.)

Table B Volcanic Gases Contained in Basaltic Lavas of Mauna Loa and Kilauea
(Percent by weight.)

1	Water, H_2O	57.8
2	Carbon, as CO_2	23.5
3	Sulfur, S_2	12.6
4	Nitrogen, N_2	5.7
5	Argon, A	0.3
6	Chlorine, Cl_2	0.1
7	Hydrogen, H_2	0.04

(Same data source as Table A.)

(1) What general statements can you safely make as to the overall chemical composition of the earth's hydrosphere and atmosphere, taken together?

(a) Both hydrosphere and atmosphere are composed of volatiles. (b) The overwhelming bulk of this matter (nearly 93%) consists of water.

_____

(2) Name in order the next five volatiles on the list, following water.

Carbon as CO_2 (2), chlorine (3), nitrogen (4), sulfur (5), hydrogen (6).

_____

Table C Principal Salts in Seawater

Name of salt	Chemical formula	Grams of salt per 1000 g of water
Sodium chloride	NaCl	23
Magnesium chloride	$MgCl_2$	5
Sodium sulfate	Na_2SO_4	4
Calcium chloride	$CaCl_2$	1
Potassium chloride	KCl	0.7
With other minor ingredients, to total		34.5

(3) Of the above five, which are important in the hydrosphere, and which in the atmosphere? (Refer to tables in Exercise 11-A.)

Hydrosphere: hydrogen (in H_2O), chlorine, sulfur, carbon as CO_2, fluorine

(trace). Atmosphere: nitrogen, plus traces of the gases hydrogen, argon, neon,

helium, krypton, and xenon.

Let's now search for the origin of these volatiles. Our hypothesis will be that they emerged from deep within the earth throughout geologic time. If so, they will be found today in the gases emanating from active volcanoes. This geologic process is called <u>outgassing</u>. Table B gives the volcanic gases found emerging along with erupting basaltic lavas of the active Hawaiian volcanoes, Mauna Loa and Kilauea.

(4) On Table A, enter the rank numbers of the ranked volatiles of Table B. Make a general statement describing the ranking differences.

Water and carbon hold their leading ranks. Sulfur is much more abundant in the

volcanic gases, chlorine is less.

(5) Does the great predominance of water in both hydrosphere and volcanic gases suggest how the oceans of our planet came into being? Speculate on this question.

The oceans came into existence by the condensation of water vapor emitted by volcanoes through geologic time. Very likely there was little, if any, free water in our planet's atmosphere to start with, about 4 billion years ago.

Table C shows the composition of the dissolved solids in seawater. Study the elements present in chemical formulas of these five principal salts.

(6) Identify and list the bases present in seawater. What is the primary source of the bases? Name the volatiles present. What is their original source?

Bases: Ca, Mg, K, Na. These elements are derived by chemical weathering from the igneous rocks of the crust. Volatiles: Cl, S. These were derived by outgassing, followed by precipitation in rainwater.

(7) The composition of sea salts is thought to have remained nearly constant over recorded geologic time. How could this condition be sustained if these elements are continually being brought into the oceans? (Consult text p. 249.)

Salts of the oceans have been precipitated into thick layers of evaporites (rock salt) in basins in continental interiors and margins, present at various periods of geologic time. This process has removed the dissolved salts about as fast as they are added to the oceans. This is a recycling process.

(8) From the list of elements in living matter (Table A in Exercise 11-A) we know that two volatiles, sulfur and chlorine, are essential ingredients (nutrients). Land plants also require these volatiles, but have no access to salts of the ocean. How do you suppose land plants get these elements? Speculate about this problem.

Three possibilities need to be considered. First, sea salts that were incorporated into marine strata are now exposed on land; these may supply S and Cl directly to soils. Second, wind may act as an agent in two ways. Sea salts lifted into the air are carried over the lands and deposited there. Winds also lift mineral particles from desert surfaces and transport these to distant lands. Third, S and Cl emitted by erupting volcanoes are brought down to the land surfaces by washout and fallout.

(9) The plasma (watery solution) of human blood is a saline solution quite similar in chemical composition to seawater. A saline solution is injected into persons who have lost blood during trauma or surgery. Speculate on the significance of these facts in tracing the origin of land animals back through millions of years of geologic time.

This evidence suggests that terrestrial animal life originated in the oceans, and that as these organisms evolved, they "came ashore" to become air breathers and occupy the terrestrial environment. Nevertheless, they retained the original blood salinity. [The transition from fishes to amphibians to reptiles is well documented in the geologic record of the early Paleozoic Era (Silurian through Carboniferous periods).]

Exercise 11-C The Igneous Rocks and Their Minerals
[Text p. 242-45, Figures 11.2. 11.3.]*

Geologists keep looking for the oldest rocks in the earth's crust, and every few years a new world's record is claimed. The ancient shield of western Greenland and eastern Labrador has been one of the best hunting grounds. In a mountainous region called Isua, close to the edge of the Greenland Ice Sheet, are some strongly deformed metamorphic rocks that were once ordinary extrusive volcanic and sedimentary rocks. Their age: a record 3.8 billion years! But they aren't the oldest rocks of our planet, because the marine-type sediments in them must have been derived from even older rock.

Individual grains of the mineral zircon within some ancient Australian metamorphic rocks (quartzites) have yielded ages between 4.1 and 4.2 billion years. Zircon, a silicate of the element zirconium, is present as a secondary mineral in most igneous rocks, and especially common in the granitic (felsic) group. Zircon is also extremely durable, and grains of the mineral can travel long distances in sand and gravel of streams. That's a good reason to infer that the first crustal rocks were igneous types, among them granite, intruded over 4 billion years ago.

The Granite-Gabbro Series

Information in text Figure 11.2 provides good basic material on igneous rocks and their silicate minerals. The information in that diagram is descriptive, using only words and pictures. We will now go a step further and make this information quantitative. How much of each of the minerals shown is found in each of the important igneous rock varieties? This information is shown in Figure A. Study it carefully to understand how it is constructed.

(1) What is the meaning of the horizontal lines drawn across the graph in Figure A? What units are used to express the quantity of each important mineral in the rock?

Each horizontal line is drawn to represent a particular rock variety. Numbers on

the line give the percentage by volume of each important mineral in the rock.

The numbers total to 100% on each line. Any number of horizontal lines can be

drawn on the graph to show the compositions of intermediate subvarieties of

rocks.

_____

_____

*Modern Physical Geography, 3rd Ed., p. 203-206, Figure 12.2.

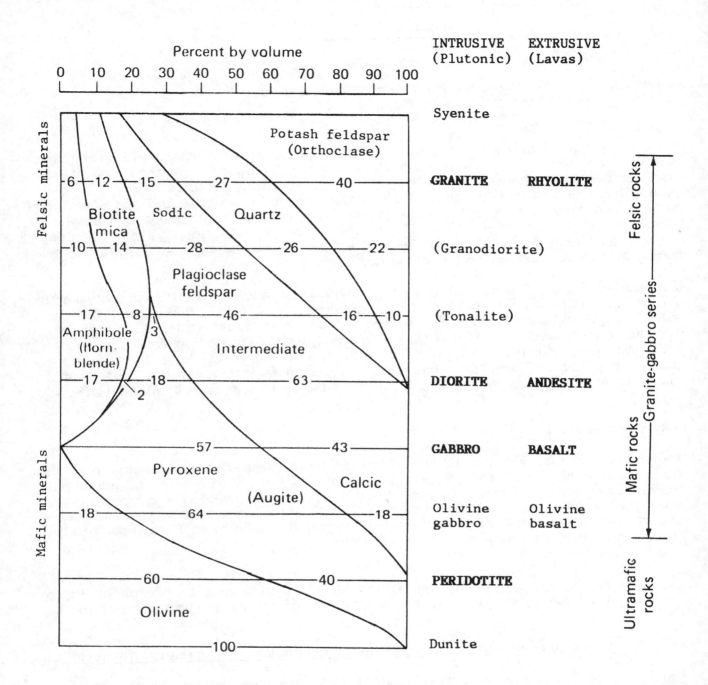

Figure A

(2) Explain the curved lines that cut across the horizontal lines on the graph.

The curved lines mark off fields occupied by a single silicate mineral variety.

More about the feldspars: Text Figure 11.2 describes two major kinds or groups of feldspars: One is potash feldspar, a silicate of aluminum rich in potassium. A common variety of potash feldspar is orthoclase. The other is plagioclase feldspar, which is actually a group of several distinctly different minerals. All are silicates of aluminum; all have some of both calcium and sodium in their composition. To understand our graph, we need to elaborate on the plagioclase group. It consists of three kinds or classes:

Sodic plagioclase: rich in sodium; poor in calcium. A common variety is albite.

Intermediate plagioclase: about equal parts of calcium and sodium. A common variety is labradorite.

Calcic plagioclase: rich in calcium; poor in sodium. A common variety is anorthite.

We will find that each of these three groups has a different association with the igneous rock varieties.

(3) A bracket at the right of the graph (Figure A) designates the granite-gabbro series. Name the intrusive igneous rocks in this series in order from top to bottom and name the two most abundant minerals in each. Give percentages of each mineral. Follow with the ultramafic rocks. Include only those rocks in bold capitals.

Name	Mineral 1	Mineral 2
Granite	Orthoclase (40)	Quartz (27)
Diorite	Intermediate plagioclase (61)	Pyroxene (18)
Gabbro	Pyroxene (57)	Calcic plagioclase (43)
Peridotite	Olivine (60)	Pyroxene (40)

Five other igneous rocks are named on the graph for the sake of completeness, but you may ignore them so as to concentrate on the four major kinds described in your textbook. Knowledge of the four major kinds and their extrusive equivalents (lavas) will be put to good use in Chapter 12 to explain the kinds of crust and their origin through processes of plate tectonics.

Rock Densities

Perhaps the most important message you can read from Figure A is the typical density of each of the four major igneous rock types. The rock density can be estimated from the mineral percentages given in the graph. Each percentage must be multiplied by its corresponding mineral density and the average density calculated as in the following example for granite. Use density values given in Table A.

(4) Complete the following table to find the density of each igneous rock listed.

	Granite	Diorite	Gabbro	Peridotite
Quartz	0.27 x 2.65 = 0.716	(none)	(none)	(none)
Potash feldspar	0.40 x 2.57 = 1.028	(none)	(none)	(none)
Plagioclase feldspar	0.15 x 2.62 = 0.393	0.63 x 2.71 = 1.707	0.43 x 2.76 = 1.187	(none)
Biotite mica	0.12 x 3.00 = 0.360	0.02 x 3.0 = 0.06	(none)	(none)
Amphibole (hornbende)	0.06 x 3.20 = 0.192	0.17 x 3.20 = 0.544	(none)	(none)
Pyroxene (augite)	(none)	0.18 x 3.30 = 0.594	0.57 x 3.30 = 1.881	0.40 x 3.30 = 1.320
Olivine	(none)	(none)	(none)	0.60 x 3.40 = 2.040
Rock density: (rounded off)	2.689 2.69	2.905 2.91	3.008 3.01	3.350 3.35

Table A Mineral densities
(g/cc)

Quartz	2.65
Potash feldspar (orthoclase)	2.57
Plagioclase feldspars:	
Sodic (albite)	2.62
Intermediate (labradorite)	2.71
Calcic (anorthite)	2.76
Biotite mica (average)	3.00
Amphibole (hornblende)	3.20
Pyroxene (augite)	3.30
Olivine (common)	3.40

(5) Offer a general statement of the relationship between mineral varieties present in the rock and its density.

Rock density increases persistently as the minerals forming the rock include fewer of the felsic type and more of the mafic type.

Igneous Rocks under the Microscope

Figure B is a drawing of the appearance of mineral grains of two different kinds of igneous rocks seen under the microscope. Enlargement is about 10 times actual size. Very thin slices of the rock are mounted on a slide and illuminated from below by polarized light. Notice that the patterns of the minerals resemble those shown in text Figure 11.2. You can identify the minerals in the slides by comparing texture patterns with those in Figure 11.2.

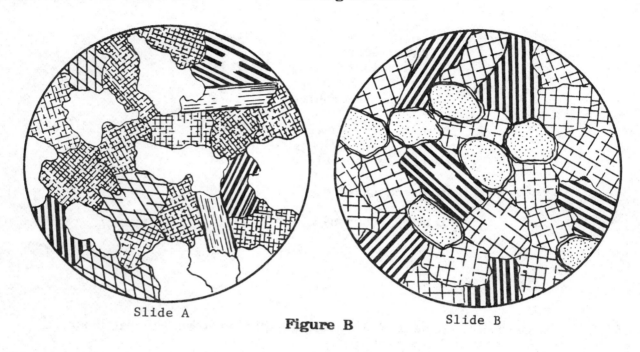

Slide A   **Figure B**   Slide B

(6) Using the code letters for seven minerals given below, list the minerals found in each slide. On the slide illustration write the code letter on at least one example of a grain of each mineral variety present.

Code	Mineral name	Slide A	Slide B
Q	Quartz	Q	___
K	Potash feldspar	K	___
P	Plagioclase feldspar	P	P
B	Biotite mica	B	___
H	Hornblende (amphibole group)	H	___
A	Augite (pyroxene group)	___	A
O	Olivine	___	O

(7) Identify the varieties of igneous rock shown in the slides. Give the names below:

Slide A: ___granite___   Slide B: ___olivine gabbro___

Name _____   Date _____

_____   _____

Exercise 11-D   Size Grades of Sediment Particles
[Text p. 248-49, Figure 11.11; p. 427, Figure 19.3.]*

Mineral particles that make up clastic sedimentary rock were at one time in a loose state, spread in layers over continental surfaces in direct contact with the atmosphere or submerged under water on the floors of lakes, streams, coastal estuaries and lagoons, or oceans. Just how these particles are transported is a question we deal with at several points in later chapters. Running water, wind, waves and currents, and glacial ice are prime movers of sediment, each having its own unique form of action and leaving a distinctive form of sediment layers.

To prepare for what is to come in later chapters, we concentrate here on the vast range in sizes of the individual particles of sediment. Scientists not only describe discrete objects in words, but find it necessary to measure objects in terms of length, area, and volume. The study of sediments--sedimentology, that is--is a quantitative form of science. Classes of sizes are set up to make this task easier and more meaningful.

Particle Size and Surface Area

To get acclimated to the world of particles, we start with a geometrical concept that is of great importance in both the geology of sediments and the science of agricultural soils. The mechanical disintegration of rock to form sediment (Chapter 14) is one of "making little ones out of big ones." As this process proceeds, an important effect is that the surface area of all the particles contained in a cube of given size, such as 1 cm on a side, increases very greatly as the particles are made smaller. Chemical alteration of minerals takes place on the exposed mineral surfaces, so the smaller the particles, the greater is the speed with which alteration can proceed.

Assume that we start with a single mineral cube 1 cm in height, width, and length. Imagine that we can slice this cube into perfect cubes of smaller and smaller dimensions in steps of powers of ten. What then will be (a) the increase in number of particles and (b) the increase in total surface area of all the particles?

---

*Modern Physical Geography, 3rd Ed., p. 207, Table 12.2.

(1) Fill in the missing figures in the following table:

Surface Area Resulting from the Subdivision of a One-Centimeter Cube

Cube dimension (length, cm)	Reciprocal of length	Particle name	Number of particles	Total surface area (cm^2)
1	1	Pebble	1	6
0.1	10	Coarse sand	10^3	60
0.01	100	Silt	10^6	600
0.000,1	10,000	Fine clay	10^{12}	60,000
0.000,01	100,000	Colloidal clay	10^{15}	600,000

(2) Describe below the rate of increase in number of particles and in total surface area with respect to the cube dimension. (The answer requires knowledge of powers of ten.)

> The number of particles increases as the ___third power___ of the ___reciprocal___ of the cube dimension.

> The total surface area is equal to ___six___ times the ___reciprocal___ of the cube dimension.

Note: Colloidal particles of clay are chemically active because they bear electrical charges, by means of which they can hold and exchange chemical ions. This important property of colloids is emphasized in text Chapter 19 (p. 428-29, Figures 19.6 and 19.7).

The Wentworth Scale of Particle Grades

Grades of mineral particles are most commonly described in terms of the Wentworth scale (named after a prominent geologist who did pioneering research on sediments). This scale is reproduced here as Table A, with the names of the grades, their subdivisions, and the diameter in millimeters that marks the boundary between each grade and subgrade. Equivalents in inches are given for the coarser grades; in microns for the finer grades.

(3) Study the progression of diameters downward from boulders, through cobbles, pebbles, and sand. Describe this progression.

Each value is half that of the one above it and twice that of the one below it.

Table A The Wentworth Scale of Size Grades

Grade Name		mm	in.
		4096	160
Boulders	Very large		
		2048	80
	Large		
		1024	40
	Medium		
		512	20
	Small		
		256	10
Cobbles	Large		
		128	5
	Small		
		64	2.5
Pebbles	Very coarse		
		32	1.3
	Coarse		
		16	0.6
	Medium		
		8	0.3
	Fine		
		4	0.16
	Very fine		
		2	0.08
			Microns
Sand	Very coarse		
		1	1000
	Coarse		
		0.5	500
	Medium		
		0.25	250
	Fine		
		0.125	125
	Very fine		
		0.0625	62
Silt	Coarse		
		0.0312	31
	Medium		
		0.016	16
	Fine		
		0.008	8
	Very fine		
		0.004	4
Clay	Coarse		
		0.002	2
	Medium		
		0.001	1
	Fine		
		0.0005	0.5
	Very fine		
		0.00024	0.24
	(Colloids down to 0.001 microns)		

For the particle grades in the range from coarse sand down through silt, a natural sample containing a mixture of grades is separated out into grades by passing the sample through a series of wire sieves. The mesh opening of each sieve corresponds to one of the boundary values betweeen grades. Particles larger than the sieve mesh diameter are caught in the sieve, while those that are smaller pass on through to the next sieve.

(4) Figure A shows the outlines of several mineral grains, enlarged ten times. A length scale is included. Measure the maximum diameter of each grain. In the spaces below, enter the grain diameter and the name of the grade, according to the Wentworth scale.

Code letter on grain	Max. diameter (mm)	Wentworth grade name
A	45	Very Coarse Pebbles
B	24	Coarse Pebbles
C	6	Fine Pebbles
D	1.5	Very Coarse Sand
E	0.38	Medium Sand
F	0.18	Fine Sand
G	0.10	Very Fine Sand
H	0.046	Coarse Silt
I	0.025	Medium Silt
J	0.003	Coarse Clay

(5) Examine text Figure 19.3 (p. 427) showing size grades as defined by the U.S. Department of Agriculture for soil texture descriptions. Compare the size boundaries it uses with those of the Wentworth scale. Be specific.

The USDA upper limit for sand is 1.0 mm, whereas it is 2.0 in W scale. In USDA, larger grains are given the name "gravel;" but "pebbles" in W scale. The USDA silt/sand boundary is 0.05, rather than 0.0625. The silt/clay boundary is 0.002, corresponding with the medium silt/fine silt boundary in the W scale. Colloids in the USDA are shown to begin below 0.001 microns, whereas in the W scale they start at 0.24 microns.

(6) Using horizontal lines, show the positions of the U.S.D.A. boundaries on the Wenthworth scale diagram, Table A. Label the diameter values. Write the names of the USDA classes along the side of the diagram.

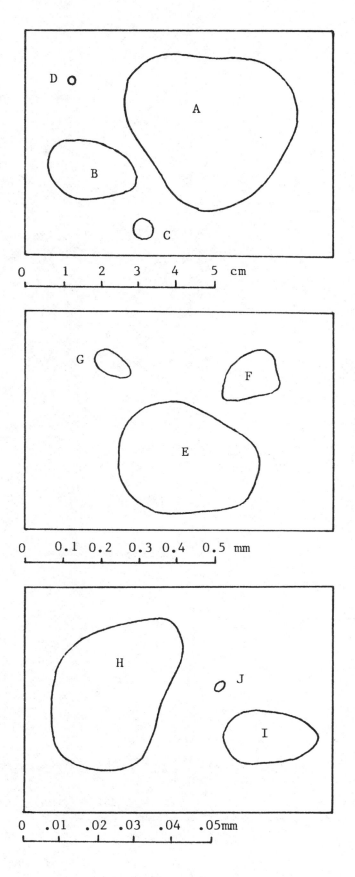

Figure A

_____   _____

Exercise 11-E   Age Relationships of Geologic Units--A Puzzle
           [Text p. 244-45, Figure 11.6.]*

A large portion of scientific research carried out by geologists consists of making specialized maps of the earth's surface. These geologic maps show the areal distribution of different kinds and/or ages of the bed rock that is exposed at the surface or lies beneath a thin cover of soil, weathered rock materials, or other shallow kinds of overburden. In a real sense, this map-making process is a geographical activity--showing the areal distribution of some property.

Geologic mapping requires (in most cases) direct examination of the surface rock, including tests performed directly on the rock, and often the taking of small rock samples for laboratory analysis. Modern techniques of remote sensing, ranging from the study of stereoscopic photos to the analysis and interpretation of satellite images, provide valuable data for reconnaissance geologic maps or maps of special rock and mineral properties.

Traditional geology has required maps of two kinds of information: (a) the kind or variety of rock present (lithology), and (b) the age of the rock in terms of the name of the age (period, epoch, stage, etc.). In classical geology, fossils have been the basis for determining relative rock age. Today, radiometric methods can establish actual age in terms of years-before-present.

A Geologic Map and Structure Section

Figure A contains a geologic map of a small area. Below it is a structure section, which is a side view of an imaginary vertical slice through the map along a selected line of traverse. Structure sections are partly inferred from the surface arrangements of the rock units, but are often supplemented by information from rock cores obtained by drilling into the crust.

(1) What two major classes of rock are included in this geologic map? Name the rock varieties in each class. Use the legend at right of map to identify the rock units.

(a) Igneous rock: granite. (b) sedimentary rock: shale, conglomerate, limestone,

sandstone.

*Modern Physical Geography, 3rd Ed., p. 244, Figure 14.2.

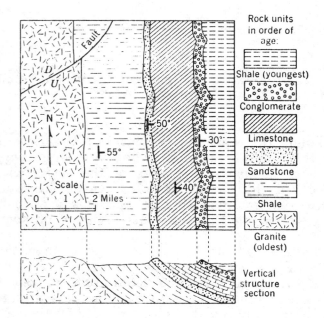

Figure A

(2) Does the structure section seem to confirm the correctness of the sequence of ages given in the map legend? Explain. (Refer to text p. 247 and Figures 11.9 and 11.10.)

The ages agree with the layering succession; youngest at the top and progressively older downward to the bottom layer. [This is the principle of stratigraphic succession.] The oldest layer of sedimentary rock was deposited on a surface eroded on older granite.

(3) The strata are shown as strongly inclined layers. Why are the edges of the layers abruptly ended at the ground surface? Explain. (Refer to text p. 350-52 and Figure 16.2.)

Strata are deposited (in most cases) in more-or-less horizontal layers, because the sediment settled out from suspension in either air or water. Since being deposited, the strata have been deformed and subsequently partly eroded to appear truncated.

(4) What is the meaning of the T-shaped map symbol and the numeral beside it? (Refer to text p. 352 and Figure 16.3.)

<u>The symbol shows the strike and dip of the strata. Strike is the horizontal</u>

<u>direction or trend of the bedding lines; dip is the angle in degrees measured</u>

<u>from the horizontal.</u>

Age Relationships Among Contiguous Rock Units

Figure B is a cross-sectional diagram that has been synthesized and idealized from typical relationships of one rock body to another. You might think of it as a drawing of the steep wall of a canyon, such as the Grand Canyon in Arizona.

Letters on the diagram are placed on either a rock unit or the line of contact between two rock units. Here is the key to the letters, in alphabetical order:

A   Modern land surface

B   Batholith of granite

D   Dike of mafic magma

F   Feeder pipe to volcano

G   Gneiss

L   Lava flow

M   Marine strata (laid down on sea floor)

P   Line separating M from rocks below. (Do not label.)

S   Sill of mafic rock

T   Terrestrial strata (laid down on land surface)

U   Line separating T from B and G (Do not label.)

V   Volcano

(5) On the diagram, Figure B, label the rock bodies (formations) in the above list. You may also color each of the rock units with a different color of your choosing.

Before you start to unravel the events of geologic history revealed in this diagram, the following concepts need to be understood:

<u>Principle of superposition</u>: Within a series of strata whose attitude is approximately horizontal or moderately inclined, each layer (stratum) is younger than the bed beneath but older than the bed above. The same principle applies to lava flows.

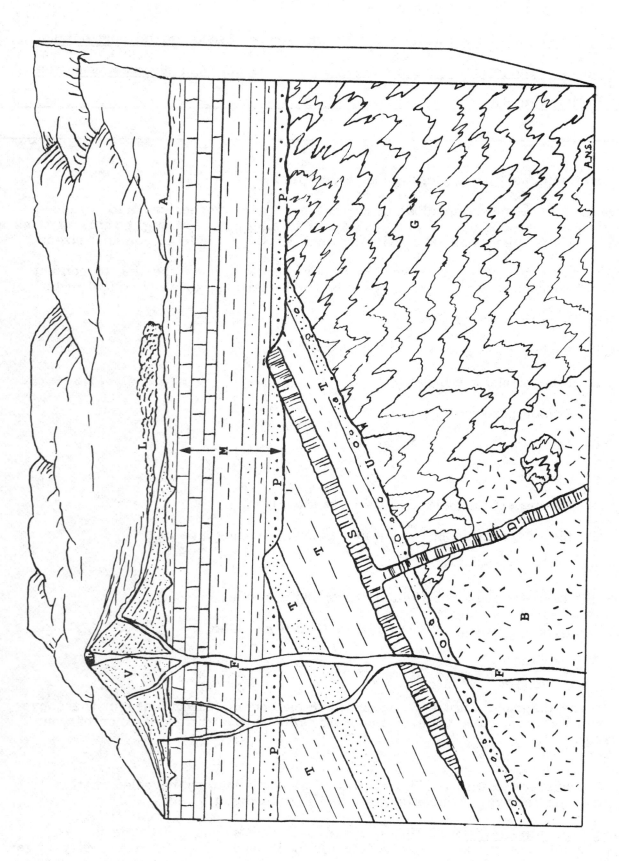

Figure B

Principle of unconformities: Long-continued erosion of large crustal masses of igneous, metamorphic, and deformed (tilted or folded) strata produces an erosion surface of low relief (a peneplain) beveling the older rock structures. Subsequent deposition of strata on that erosion surface produces an unconformity, seen in cross section as a line of separation between the younger and older rocks. Examples are shown in Figure C, illustrating two forms of unconformities.

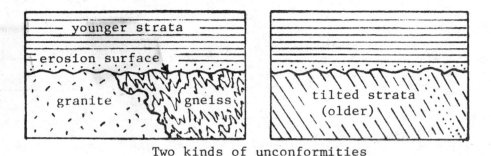

Two kinds of unconformities

Figure C

Principle of crosscutting relationships: A body of intrusive igneous rock (a pluton), such as a dike or the feeder pipe of a volcano (shown in text Figure 11.6), cuts across the structures of older rocks, showing that the intruding igneous rock is younger than the rock body which it penetrates. Another form of crosscutting is produced by the intrusion of a batholith, replacing the older rock (also shown in text Figure 11.6). Note that an unconformity is also a kind of crosscutting relationship.

(6) Your assignment is to arrange the letters on the diagram, Figure B, in order of a sequence of historical events. Enter the appropriate letters in the blanks below. For each, write a brief note naming the feature it represents and explaining the associated geologic event.

Code letter:	Rock, geologic feature, or event:
G	Ancient banded gneiss produced by metamorphism; the oldest rock unit.
B	Granite batholith intruded into gneiss.

U   Unconformity between gneiss/granite basement and basal conglomerate of the series of overlying strata, T. An erosion surface (peneplain) was first produced, then buried.

T   Terrestrial strata, deposited on continental surfaces by streams or wind.

D   Basalt dike cutting across the granite and gneiss and feeding the sill, S.

S   Sill of mafic rock, forced between units of the terrestrial strata.

P   Unconformity between tilted strata and younger sequence of marine sediments, M. Tilting occurred during a mountain-making event, followed by much erosion.

M   Marine strata, consisting of sandstone, shale, and limestone. These began to accumulate on the peneplain as it became submerged.

A   Present land surface, produced by erosion following uplift and emergence of youngest marine strata.

<u>F</u>   <u>Feeder pipe to volcano neck in tube through which lava</u>
<u>reached the surface. It crosscuts all older rocks.</u>
_____

<u>V</u>   <u>Volcano formed of extrusive lavas and ash.</u>
_____
_____

<u>L</u>   <u>Lava flow; the most recent feature of the landscape.</u>
_____
_____

Name _____   Date _____

_____   _____

Exercise 12-A   The Lithospheric Plates
          [Text p. 270-73, Table 12.2, Figures 12.14, 12.15.]*

Plate tectonics has been hailed as a major revolution in the geosciences. The idea of scientific revolutions was put forward in 1962 by science philosopher Thomas S. Kuhn. He applied the term <u>paradigm</u> to whatever major scientific theory happens to dominate a given science field at a given period of history. The word was strange to most scientists, but they quickly adopted it as a buzz-word and looked around for examples. Kuhn thought that a paradigm gradually became unsatisfactory because it increasingly failed to cope with new information. Its loyal followers held fast to it as long as they could, but finally revolted, discarding the old paradigm and replacing it with a new one. The king is dead, long live the king! Kuhn's ideas soon came under fire, and today they look a bit simplistic, to say the least, and perhaps even unrealistic. A more moderate and perhaps more accurate view is that major theories of science evolve in sequence, but that much of the scientific evidence gathered in support of a discarded theory remains valid and is incorporated to newer theories.

Plate tectonics--the notion that gigantic lithospheric plates move, separate or collide, grow or disappear--was indeed a new paradigm, but did it bring the executioner's axe to bear on an earlier paradigm? Not really. Geology, though long-established as a science, never had even a good working hypothesis of how continents grow and ocean basins come into existence, or how mountain ranges rise. Now geologists have their very own first-born paradigm to love and cherish. True, it arrived a century later than the biologists' paradigm of evolution, but better late then never.

All of our exercises for this chapter relate to plate tectonics, so that you can become better acquainted with this strange newcomer to the world of science. As geographers looking always for global patterns and relationships, we begin with a geographical survey of the lithospheric plates.

Using the three maps of text Figure 12.14 as your information source, enter from text Table 12.2 the name of the plate or subplate that satisfies each of the following descriptions.

(1) It consists almost entirely of oceanic lithosphere. Its eastern boundary is largely a spreading boundary; its western boundary is largely a converging boundary.

                                        ___Pacific___

(2) Formed entirely of oceanic lithosphere, its eastern boundary is a converging boundary; its northern, western, and southern boundaries are spreading boundaries.

                                        ___Nazca___

*Modern Physical Geography, 3rd Ed., p. 226-28, Figures 13.14, 13.15.

(3) Formed almost entirely of oceanic lithosphere and volcanic island arcs, this plate is bounded almost exclusively by converging boundaries.

_____Philippine_____

(4) It consists of a central mass of continental lithosphere completely surrounded by a broad zone of oceanic lithosphere.

_____Antarctic_____

(5) It consists of two widely separated masses of continental lithosphere, one of which is almost completely surrounded by oceanic lithosphere; the other bounded on the north by a converging boundary.

_____Austral-Indian_____

(6) It consists largely of continental lithosphere. Oceanic lithosphere forms its western western and northern border zone. Converging boundaries in the form of island arcs comprise much of its eastern and southern boundary.

_____Eurasian_____

(7) Its outline and boundaries exemplify the plate model shown in text Figure 12.13, described as the "sunroof" model.

_____Arabian_____

(8) It is about equally divided into continental lithosphere and oceanic lithosphere. Oceanic lithosphere forms the eastern part; continental, the western part.

_____American_____

(9) Composed entirely of oceanic lithosphere, its western and southern boundaries are spreading boundaries; its northern boundary is a converging boundary.

_____Cocos_____

(10) Elongate in the east-west direction, its eastern boundary is a converging boundary, and both its northern and southern boundaries are transform boundaries. (Two answers)

_____Caribbean_____   _____Scotia_____

(11) A small, narrow plate sandwiched between two great plates, one of which is formed of oceanic lithosphere, the other of continental lithosphere.

_____Juan de Fuca_____

(12) It consists of a large central core of continental lithosphere, surrounded on the west, south, and southeast by oceanic lithosphere. Its northern boundary is partly a converging boundary. Its northeastern boundary is a spreading boundary.

_____African_____

Exercise 12-B   Kinds of Plate Junctions and Their Meaning
[Text p. 270-73, Figure 12.14.]*

Lithospheric plates may seem to be highly varied in their sizes and outlines and in the kinds of plate boundaries they possess, but actually there are a few simple observations that apply to all of them. The common point of meeting of three plates is called a <u>triple junction</u>. Although in abstract theory it would quite permissible for four plates to share a common point--like the Four Corners common point of Utah, Arizona, New Mexico and Colorado--the physics of rupture of a thin brittle plate makes a quadruple junction highly improbable.

Figure A shows six common kinds of plate junctions consisting of the three kinds of plate boundaries: converging, spreading, and transform. Note that the legend of the diagram corresponds with the map legend of text Figure 12.14. (The continental suture can be included within the converging boundary class.) What is added on the accompanying diagram is a system of broad, open arrows to show relative plate motions. ("S" means the plate is stationary relative to the other two.)

Code letters are applied to each of the three boundary symbols:

   R   for "rift"   codes for a spreading boundary,

   T   for "trench"   codes for a converging (subduction) boundary, and

   F   for "fault"   codes for a transform boundary.

Thus, each kind of triple junction can be expressed as a three-letter code. The six kinds we will refer to have the codes RRR, TTT, TTF, TFF, RTF, and RFF.

(1) Your assignment is to find on the world map of plates (text Figure 12.13), a clear representative of each type and enter the information on Figure B. Write the letter code beneath the drawing of the junction. Write in the names of the three plates involved in each. Draw arrows to show relative plate motions. One of the three plates can be designated as stationary (S), if that seems reasonable.

---

*Modern Physical Geography, 3rd Ed., p. 225-26, Figure 13.14.

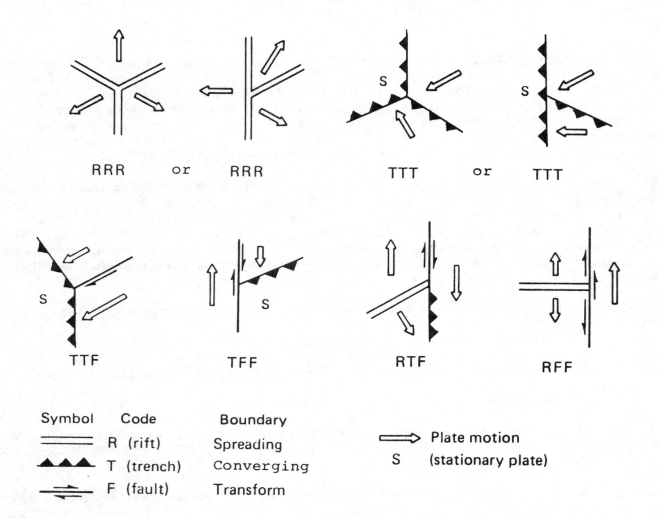

Figure A

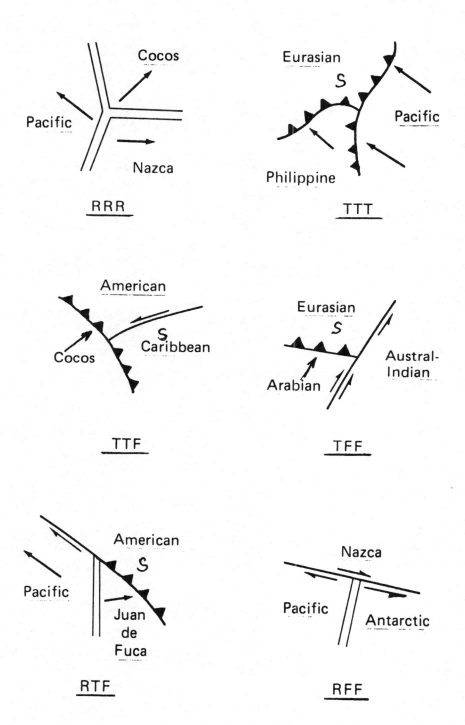

Figure B

(2) How many triple junctions does each of the following plates have? The meeting point of three colors on the world map, text Figure 12.14, will define a triple junction. (Disregard the Caroline and Bismark plates.)

Pacific	9
Antarctic	7
African	5
Austral-Indian	5
Nazca	5
Cocos	4
Caribbean	3
Arabian	3
Scotia	2
Philippine	2
Juan de Fuca	2

Exercise 12-C   Those Magnetic Stripes on the Ocean Floor
[Text p. 268-70, Figure 12.11; p. 280-81, Figure 12.22.]*

Figure 12.22 in your textbook shows how continents can rift apart, forming a new ocean basin floored with young basaltic crust. These diagrams and Figure 12.11 show that basaltic lava rises to the ocean surface along the widening axial rift. The basalt congeals in the rift, adding new oceanic crust to the separating plates. How did this explanation come to be known to geologists and geophysicists? What evidence could they produce to support the hypothesis?

By about 1960, it was firmly established that at certain periods in the geologic past, the earth's magnetic field reversed its polarity, so that the north and south magnetic poles switched (or flipped) direction by 180 degrees. In those reversed polarity intervals, the compass needle that now points toward the north geographic pole would have pointed toward the south geographic pole. Polarity reversal is observed and measured in samples of basalt. Those samples can, in turn, be dated in years-before-present by radiometric methods. Thus a polarity reversal timetable has been established and its time-units given names.

Figure A sums up the information and applies it to a spreading plate boundary on the ocean floor. As polarity reversals occur, paired bands of the same polarity are generated in a mirror-image relationship. A second diagram, Figure B, gives you the names of the major magnetic polarity epochs. Those of polarity the same as today are called normal epochs; those of opposed polarity are reversed epochs. Several lesser reversals are recognized as "events," but we don't need to refer to them. (The epochs are named after scientists noted for their work in geomagnetism.) Figure B also gives you an age scale for the reversals, back to 4 million years before present.

Look next at the world map of lithospheric plates, text Figure 12.14. Notice that the spreading plate boundaries of the ocean floors are cut across by faults (black lines). Each fault is expressed by an offset of the plate boundary. Our objective in this exercise is to explain how this offsetting occurs and what it means.

Let's approach our objective by considering two alternative hypotheses, illustrated in Figure C. The first hypothesis (left) proposes a transcurrent fault. (Refer to p. 297-78 of the textbook and Figure 13.21 for an explanation.) The great San Andreas fault of California is an excellent example of a transcurrent fault. The second hypothesis (right) proposes that a special variety of transcurrent fault, called a transform fault, is present. It is found almost exclusively in oceanic lithosphere along spreading plate boundaries.

---

*Modern Physical Geography, 3rd Ed., p. 224-25, Figure 13.11; p. 234-36, Figure 13.25.

Magnetic polarity epochs and events

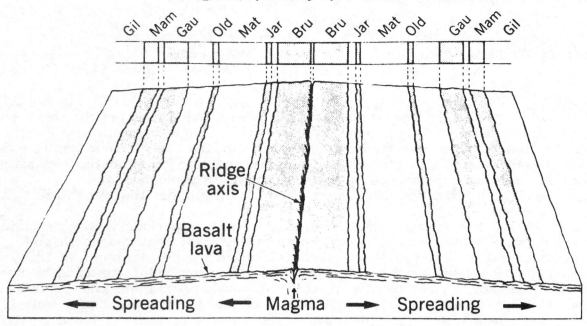

Figure A

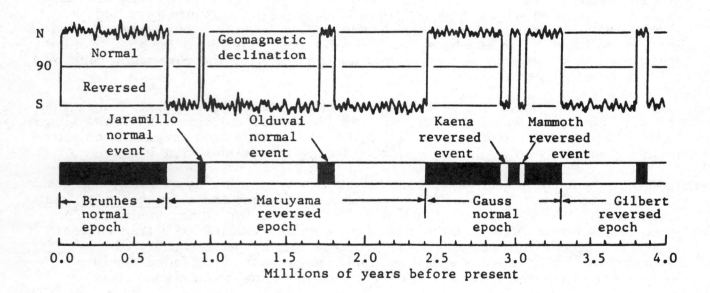

Figure B

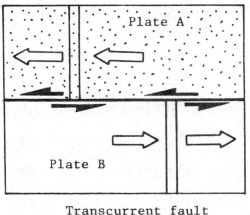

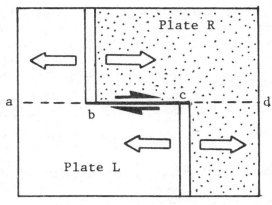

Transcurrent fault   Transform fault

Figure C

(1) In what way or ways do the two schematic maps of Figure C differ in the interpretation of the same data? Describe each in precise terms.

The transcurrent fault extends continuously across the map, and presumably far beyond in both directions. The transform fault exists only as a line segment between the offset ends of the spreading plate boundary.

(2) Using a color pencil or pen, apply one color to the entire area of Plate A in the left figure and to the entire area of Plate R in the left figure. Analyze the difference in the two colored areas. Does the interpretation of a transcurrent fault make good sense? Explain.

The transcurrent-fault hypothesis makes no sense, because the spreading boundaries now lie within a single plate, meaning that the plates are not spreading apart along those boundaries. Instead, the entire boundary between plates A and B is the transcurrent fault.

(3) Describe the relative motion of the transform fault in the righthand map of Figure C. Does the fault change in length with the passage of time as plate separation continues? Explain.

The transform fault exists only between the offset ends of the spreading boundary. Its relative motion is opposite to that shown for the transcurent fault. There is no reason for the length of the fault line to change as time passes.

(4) What is the meaning of the dashed lines a-b and c-d in the righthand diagram? Give a name to this class of feature and write it on the diagram. (Hint: Examine text Figure 12.11 closely; then examine Figure 12.5.)

As the plates spread, what was formerly an active transform fault becomes a trace, or scar, within each plate. On the ocean floor, these traces take the form of low ridges or escarpments (cliffs). These are called "transform scars" (or "healed transforms"). [Note: Long before plate tectonics came along, these features were named "fracture zones," a term that continues in use today in the geographical names of these features.]

(5) Assuming that your answer to Question 4 is correct, predict for each hypothesis (transcurrent vs. transform) where earthquakes should be generated, and where not. (Consult text p. 303 for important information.)

If transcurrent, earthquakes should occur along the entire fault as far as it extends across the map. If transform, earthquakes should be generated only in the active fault segment between the offset plate boundaries.

Special Workshop Project

A paper model of magnetic stripes and transform faulting.

Carry further the concepts of the formation of magnetic stripes and how they relate to transform faults. Figure D shows four stages in the formation of paired magnetic stripes going back four million years. Figure E is a cut-out version of the same diagram. Two separate parts--one for Plate A and one for Plate B--are inserted in slots cut in a cardboard base. As the two strips of paper are pulled apart at the same speed, the new magnetic anomaly stripes appear to be simultaneously generated along the two slots.

[Reference: Edward C. Stoever and Ronald E. Armstrong, 1978, Sea-Floor Spreading and Transform Faults, Journal of Geological Education, vol. 26, p. 19-21.]

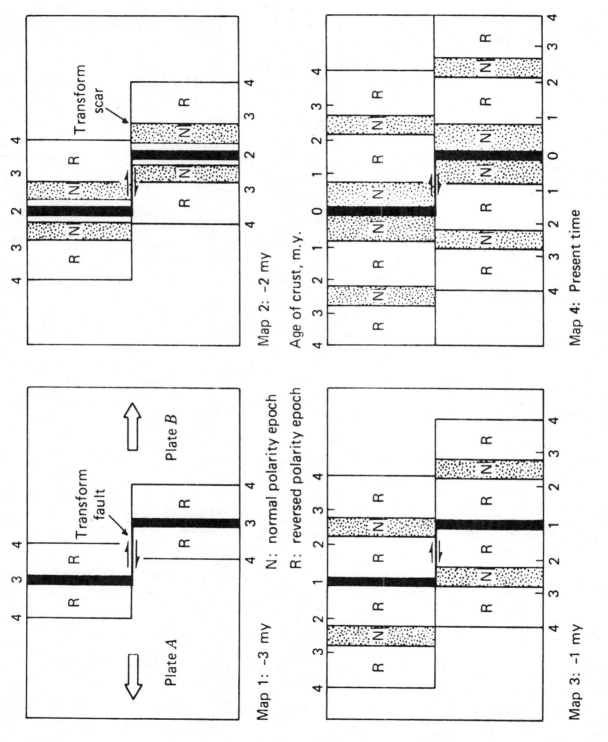

Figure D

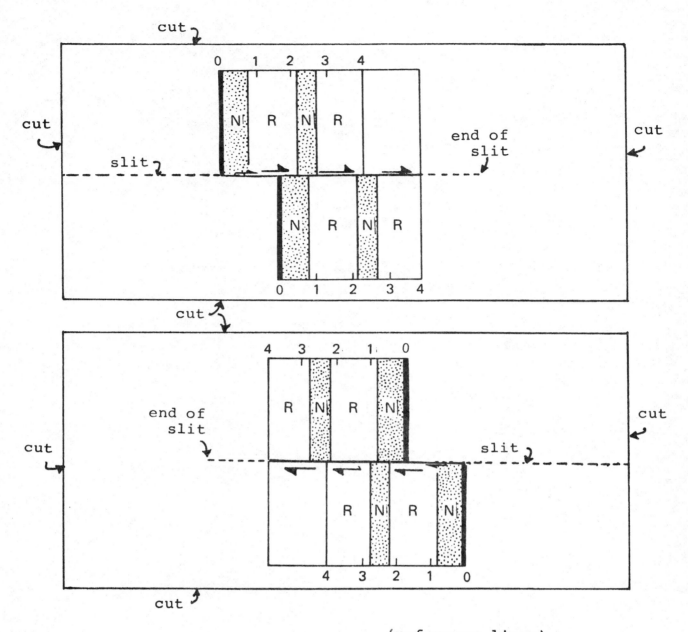

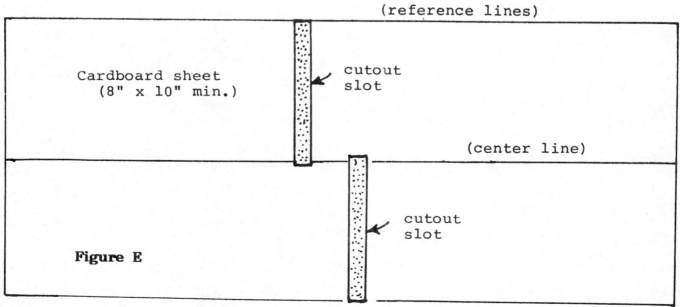

Figure E

Exercise 12-D   How Fast Do Lithospheric Plates Move?
   [Text p. 281-84, Figure 12.25.]*

That the great lithospheric plates move--one relative to another--seems well enough established by the evidence of the magnetic stripes we examined in Exercise 12-C. All one needs to do to get firm figures is to look at a map of the magnetic stripes of a part of an ocean basin, measure the distance in kilometers across a series of stripes, and divide by the elapsed geologic time. You might, for example, find that the separation rate (rate of separation of reference points on the two sides of the mid-oceanic rift) amounts to 5 cm/yr. During the lifetime of a person 20 years of age, the rift would have widened by 100 cm, or one meter. (This on the order of magnitude of the rate of lengthening of human hair or fingernails.)

Estimating the rate of closing of an active (converging) subduction boundary would be a lot more difficult, if not impossible, with no magnetic stripes to measure. For a transform plate boundary within a continent, such as the San Andreas Fault between the Pacific plate and the North American plate, some excellent evidence can be obtained by using the number of meters of offsetting during a single great earthquake and the average number of years elapsing between those events. With that evidence, geologists obtain a figure of about 5 cm/yr. Geologic evidence going back through the past 150 million years gives a total horizontal slip displacement along the San Andreas Fault of about 560 km, for an average of nearly 4 cm/yr.

Our exercise uses the data of the magnetic stripes to calculate rates of relative plate movement. Be sure you have thoroughly mastered Exercise 12-C and understand the principle of the formation of magnetic polarity zones as spreading takes place.

The accompanying map (Figure A) shows bands of spreading stripes in two parts of the floor of the Atlantic Ocean. The upper map is for the North Atlantic between 25° and 30°N latitude; the lower map is for the South Atlantic at a comparable latitude. Instead of magnetic polarity reversal stripes (as in Exercise 12-C) the stripes represent epochs and periods of geologic time. The geologic ages assigned are for the basaltic crustal rock itself, based on magnetic data. (Consult text Table 12.1, p. 257, for the names of the epochs of the Cenozoic Era and the Mesozoic periods.)

Even a brief study of the two maps shows immediately that the stripes are narrower for equivalent epochs in the North Atlantic than in the South Atlantic. Our plan is to measure the widths of the stripes and derive actual spreading rates.

Figure B is a graph showing how we can manipulate the measurements of stripe widths to give us estimates of spreading rates. The vertical scale gives distance in kilometers away from the spreading plate boundary (spreading axis). The horizontal scale gives time in millions years before present. Vertical lines divide up the graph into the geologic epochs and periods. Using the maps of the geologic stripes on the sea floor, we can measure distances to the several geologic time boundaries and plot these distances as points on the graph.

---

*Modern Physical Geography, 3rd Ed., p. 236-38, Figure 13.27.

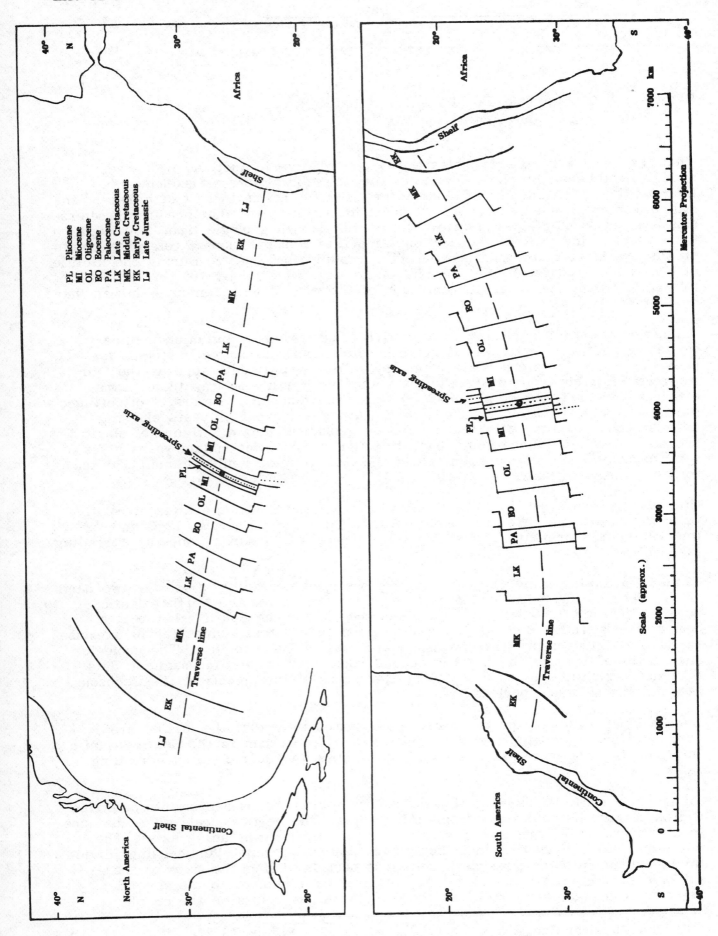

Key:

PL	Pliocene
MI	Miocene
OL	Oligocene
EO	Eocene
PA	Paleocene
LK	Late Cretaceous
MK	Middle Cretaceous
EK	Early Cretaceous
LJ	Late Jurassic

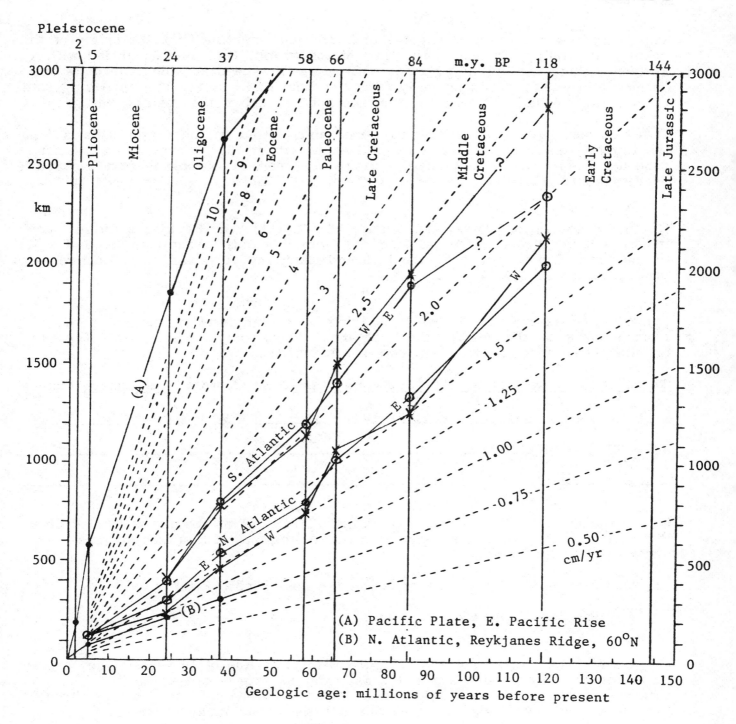

Figure B

Figure A (facing page)

[Data source: R.L. Larson, W.C. Pitman, _et al_, 1985, The Bedrock Geology of the World, W.H. Freeman & Company, New York. Copyright © 1985 by R.L. Larson and W.C. Pitman III. Used by permission. Map details simplified for student use.]

The graph has a system of diagonal lines (dotted) starting from the origin of the graph and fanning out across the field of the graph. Each line is labeled with a number, telling the <u>spreading</u> <u>rate</u>, defined as the speed at which the plate one on side of the rift is moving away from the axial line. (Note: The spreading rate is equal to one-half of the separation rate, mentioned in the opening paragraph.)

Two examples have already been measured and plotted on the graph. One is from the North Atlantic basin just south of Iceland, at lat. 60°N. Here the axial rift zone has been named the Reykjanes Ridge. The second example is from the South Pacific basin, on the western side of the East Pacific Rise, about at lat. 30°S.

On the world map of plates, text Figure 12.14, locate the Reykjanes Ridge. Mark it with a short straight line crossing the spreading axis at right angles and write in the name. Then turn to Figure 12.8 and carry out this same identification procedure.

(1) Refer to the graph, Figure B. Judging by the angle of inclination of the plotted points on the line for the Reykjanes Ridge, what spreading rate (cm/yr) is indicated? Label the estimated rate on the graph.

<u>Three points were plotted. The line connecting them shows a rate greater than</u>

<u>0.75 cm/yr, but less than 1.0. An estimate of 0.85 cm/yr is close to the value</u>

<u>obtained by computation.</u>

Locate the East Pacific Rise on text Figure 12.14B. (This feature is labeled on Figure 12.5.) Draw and label a short straight line from the spreading axis westward into the Pacific Plate.

(2) What two plates are separating along the East Pacific Rise?

    <u> Pacific </u> plate and <u> Nazca </u> plate

(3) Examine the plotted points on Figure B for this line of traverse. They are connected by straight-line segments. Estimate the rate of spreading by comparison with the labeled guide lines. Was the spreading rate constant through the entire time period shown? When was it more rapid? When less rapid?

<u>Rate for Miocene and Oligocene epochs averaged about 10 cm/yr. Through the</u>

<u>Pliocene, it was much faster. In the Eocene, it was much less--between 2.5 and 3</u>

<u>cm/yr.</u>

Turn next to the map, Figure A. Measure the distance from the marked point on the axis of spreading to each of the geologic boundaries. Use a compass to transfer each distance to the scale of kilometers. (The edge of a piece of blank paper can also be used for this purpose.) Record each measurement as a dot on the graph, Figure B. Connect the dots by straight line segments. Do this first in the eastward direction on the map, then repeat for the westward direction. Label the two lines "E" and "W." (Note: For the South Atlantic, do not include Early Cretaceous in your measurements.)

Find each of the two map traverses on the world map of plates, text Figure 12.14. Mark and label the two lines.

(4) Name the two plates that meet along the Mid-Atlantic Ridge:

    __American__ plate and __African__ plate

(5) From the plotted data on the graph, Figure B, estimate the overall average spreading rate for each of the two pairs of traverses. (Fit a single straight line to each region and estimate its slope.)

    North Atlantic: __1.6__ cm/yr

    South Atlantic: __2.2__ cm/yr

(6) Make a careful comparison of the sets of points for the North and South Atlantic traverses. Note how the rate changes for each successive geologic epoch. Describe these changes in rate. Do they show a correspondence between the two areas? Offer an explanation for your findings.

Within each region, the two plots (E and W) show a rather close correspondence in changes of trends from slower to faster and vice versa. Also, the two regions show very similar patterns of rate change. The rate became faster for both in the Oligocene, slower in the Eocene, faster in the Paleocene, slower in the Late Cretaceous. (Changes do not correspond in the Middle Cretaceous, but the data for the South Atlantic are questionable.)

(7) Why should spreading rate seem to change abruptly from one geologic epoch to the next? Could this kind of change appearing on the graph result from some cause or causes other than actual change in rate of plate movement?

If the geological ages assigned to the time boundaries between epochs are

incorrect, the spreading rates will seem to change abruptly.

(8) Is it realistic to connect the plotted points by straight line segments, rather than by a smoothly curving line fitted to the points?

If the plotted change actually corresponds to true change in speed of motion, a

smoothly curving line fitted to the points would perhaps be more realistic in

principle. It is most unlikely that an enormous lithospheric plate could suddenly

change speed.

(9) (Special bonus question.) Both sets of data are from the same two lithopspheric plates along their common boundary. How then can you explain the faster rate of spreading in the S. Atlantic than that in the N. Atlantic? In the space below, sketch a very simple map to illustrate your hypothesis.

The paths of plate motion may lie on a set of concentric small circles having

a common pole of rotation. The pole would lie to the north of both traverses.

With the same angular rotation, the separation rate would be greater in the

more southerly zone. (Note: the very slow spreading rate of the Reykjanes Ridge

traverse is consistent with this explanation.)

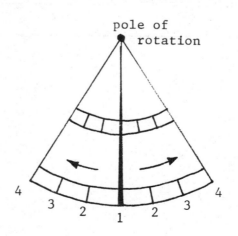

pole of
rotation

(10) (Special bonus question.) Of the two traverses, which includes the oldest geologic time units? Name the oldest units in each. Are they found at both ends of each traverse? What is the significance of your observation in terms of the geologic history of the Atlantic ocean basin? (Refer to text Figure 12.25 and include information contained on the set of maps in that figure.)

The oldest geologic unit in the S. Atlantic traverse is Early Cretaceous; that of the N. Atlantic is Late Jurassic. These units are matched at both ends of the traverses. Significance: Opening up of the North Atlantic basin began earlier (in Jurassic time), than the opening of the South Atlantic, which was delayed until Early Cretaceous time. Maps B and C of Figure 12.25 clearly show this sequence of opening to have begun first between North America and Europe, while South America and Africa were still joined.

Exercise 12-E   Quasars Tell How Fast the Plates Move
[Text p. 270-73, Figure 12.14.]*

During the 1980s, the science of geodesy (from the Greek, "to measure the earth") took a giant stride ahead, thanks to special orbiting earth satellites that can be used to measure with great precision the surface distance between two points on the earth. A satellite known as LAGEOS was used to bounce back laser beams with such precision as to yield actual rates at which points on two different plates are either separating or converging. For a time, the measurement errors were so large, compared with the spreading or converging rates the data were yielding, that the results were in question. Then, in 1982, the measurement errors had been reduced to the point that plate movement rates were virtually a certainty within acceptable limits of error. Best of all, the observed motions were in the same directions predicted by plate tectonics and the rates were in the same ball park, at least.

Then came Very Long Baseline Interferometry (VLBI), making use of radio signals received from distant astronomical objects, such as quasars. To use these data, NASA had already set up an organization called the Crustal Dynamics Project, with participating observing stations in North America, South America, Europe, Australia, and Japan. By 1986, excellent VLBI results had been obtained, in close agreement with plate movement rates derived by geophysicists from the evidence of plate tectonics--paleomagnetic data, particularly. The data we will use are given in Table A. A plus sign means separation; a minus sign convergence (closing).

Table A   VLBI Data of Relative Plate Motions

Stations	VLBI Rate cm/yr	Geophysical rate, cm/yr
Westford, CT, and Onsala, Sweden	+1.7	+1.7
Kauai, HI, and Fairbanks, AL	-3.9	-5.0
Kauai, HI, and Tokyo, Japan	-8.3	---
Perth, Australia, and California (N. American plate)	-7	---
Perth, Australia, and Nazca, Peru	+2 to +3	---
Vandenberg A.F.B., CA, and Fairbanks, AL	-7.9	-5.2

[Data sources: W.E. Carter and D.S. Robertson, 1986, Scientific American, vol. 255, no. 5, p. 46-54; Research News, Science, 1987, vol. 236, p. 1425-26.]

---

*Modern Physical Geography, 3rd Ed., p. 236-38, Figure 13.27.

(1) You are asked to plot the data of the Table A on the world map of lithospheric plates, text Figure 12.14A. You should use a sheet of tracing paper taped to the textbook page. First, copy the rectangular outline of the map. After the data are plotted, transfer the tracing sheet to the blank rectangle provided. Locate and label the following items of information:

Observing stations at the ends of each line.

Connecting straight line between stations.

VLBI rate of relative motion (beside line).

Geophysical rate, in parentheses, near VLBI rate.

(2) Westford, CT, and Onsala, Sweden. (Use Göteborg as the equivalent location of Onsala.) Sweden is not shown on Map A, but you can draw the line to graze the northern end of Ireland. What two plates are in relative motion on this line? Does this rate compare favorably with the long-term geologic rates you obtained for the North Atlantic spreading rift in Exercise 12-D?

The North American and Eurasian plates are moving apart along this line. The rate is close to that at 30° lat. in Exercise 12-D (about +1.5 cm/yr), but more than for the Reykjanes Ridge at 60°N (+0.75 cm/yr).

(3) Kauai, HI, and Fairbanks, AL. What plates are involved in this motion? What kind of plate boundary separates these plates? What relative motion and rate are observed on this line?

The Pacific and North American plates are in relative motion, separated by an active (converging) subduction boundary in which the Pacific plate is plunging down beneath the North American plate. The closing rate is -3.9 cm/yr.

(4) Judging from the geographical position of the northern Pacific plate boundary in the Aleutian region from Alaska to Kamchatka, does the Pacific plate motion follow the straight line connecting Kauai and Fairbanks? Draw a few arrows to show your interpretation of the motion of the Pacific plate at this northern boundary.

No, the Pacific plate appears to be moving in a general northwesterly direction. Measured in this direction, the relative motion with the North American plate will actually be faster than on the VLBI line. [Note: A plate may move at an acute angle with respect to a plate boundary. Where the boundary forms an arc a single direction of motion should be indicated.]

(5) Kauai, HI, and Tokyo, Japan. Is it reasonable that the rate of motion on this line is much higher (-8.3 cm/yr) than on the line to Fairbanks (-3.9 cm/yr)? Can you explain this discrepancy?

The Kauai-Tokyo line may be more nearly in the true compass direction of the motion of the entire Pacific plate. [Other measurements, taken along a northwesterly line from Kwajalein to Tokyo, show a rate of -11 cm/yr, which is probably close to the true direction and rate of plate motion.]

(6) Perth, Australia, and California. The California station is located on the North American plate, inland from the San Andreas Fault. The convergence measured is thus between the Austral-Indian plate and the North American plate. Is the converging rate of -7 cm/yr adequately explained by the plates and boundaries lying between the two stations?

Yes, the converging (subduction) boundary between the Pacific and Austral-Indian plates explains the negative sign. The transform plate boundary of the San Andreas Fault crosses the observation line at about right angles, so has no effect here.

(7) Perth, Australia, and Nazca, Peru. Nazca is located at lat. 14°S, near the Pacific coast. (Aim for the sharp bend in the coastline.) Perth and Nazca are obviously moving part, and this is opposite to the closing motion of Perth and California. Study the plate boundaries along the track and offer an explanation of the observed separating rate of between +2 and +3 cm/yr. Refer back to Table A of Exercise 12-D for helpful information.

Opposite sides of the Pacific plate are on the traverse line. That on the west is a converging (subduction) boundary; on the east, a spreading boundary. We know from Exercise 12-D, Table A, that the latter boundary has a very high spreading rate (much over +10 cm/yr in the past 5 million years). The converging rate on the west may be substantially less (say -7 to -8 cm/yr). What we read is a balance in favor of separation of the two stations.

Special note: NASA scientists who have analyzed the VLBI data plot the connecting lines on a Mercator projection, just as you have done. Straight lines on the Mercator grid are true rhumb lines (lines of constant compass direction), as explained in Exercise 1-G. Evidently the scientists assume the rate of separation or closing between any pair of stations applies pretty well to the rhumb line, and not to the great circle, which is the shortest surface distance between the two stations. The great circle route between Perth and Nazca follows a course passing over the edge of the Ross Ice Shelf on Antarctica. As you see from Map C of Figure 12.14, the great circle path would completely miss the Pacific plate and would cut across the Antarctic plate. Using that course in an attempt to explain the observed separation rate of Perth and Nazca would end in

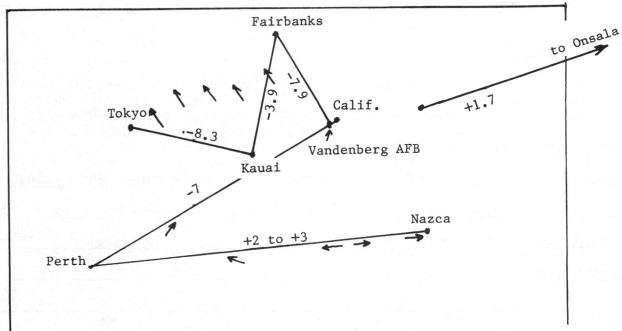

frustration. The point is that relative plate motions are referred to small circles relative to the common pole of rotation of two plates. For relatively short arcs in middle and low latitudes the small circle and rhumb line are not greatly different.

Name _____   Date _____

_____   _____

Exercise 13-A   Topographic Contour Maps
              [Text p. 26-28, Figures 1.18, 1.23.]*

The topographic contour map is a special kind of isopleth map, as explained in text Chap. 1, p. 27, and illustrated in Figure 1.18. The U.S. Geological Survey publishes topographic contour maps on various scales covering the United States. These maps are bought in large numbers by hikers and campers, who learn to use them as an aid in following roads and trails in recreational areas and for taking off across country in wilderness areas lacking in any trails or guideposts.

In physical geography, the topographic contour map is used to display scientific information in quantitative form. That information depicts landforms, which can be studied and compared as a means of identifying and classifying them, and for developing and testing hypotheses about the processes that shape landforms.

In the next several chapters, we use topographic contour maps to represent many kinds of landforms, along with photographs and remote-sensing images that help you to visualize a landform from the contours alone. First, however, you need to develop your skills in drawing contours and learning how they provide information on the configuration of the land surface.

The illustrations on the following page will help you understand the meaning and interpretation of topographic contour maps.

*Modern Physical Geography, 3rd Ed., p. 480-83, Figures A.III.2, A.III.10.

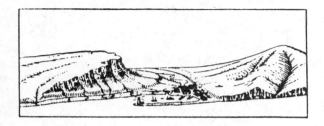

Perspective drawing
of a landscape.

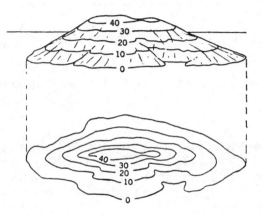

Contour map of a small island.
The contour interval is 10 feet.

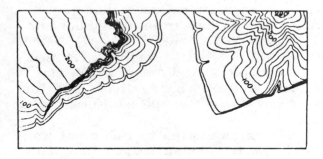

Contour lines only.

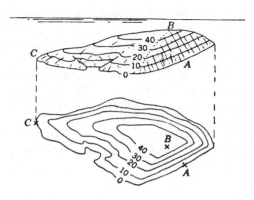

Contour map of a small island
with a steep slope (cliff)
at one end.

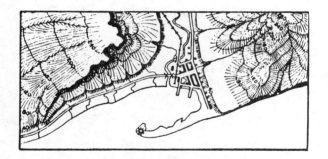

Contour lines and slope lines
(hachures) combined.

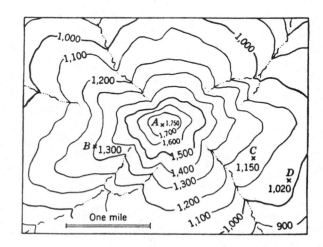

Contour map of a hill with stream
valleys carved into its sides.
Contour inverval is 100 feet.
Estimated elevations are given
for points A, B, C, and D.

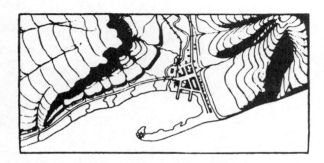

Contour lines with shading
to bring out slope forms.

Topographic Contour Maps

A Contour Map of an Imaginary Island

In the blank space provided, draw a contour map of a semicircular island, the straight side of which has a very steep slope as compared with the remainder of the island. The island is about 5 mi wide and 6 to 7 mi long. Use a map scale of one inch to one mile. The summit point of the island is 105 feet above sea level. The contour interval is 10 ft. Two stream valleys extend down the gently sloping sides of the island.

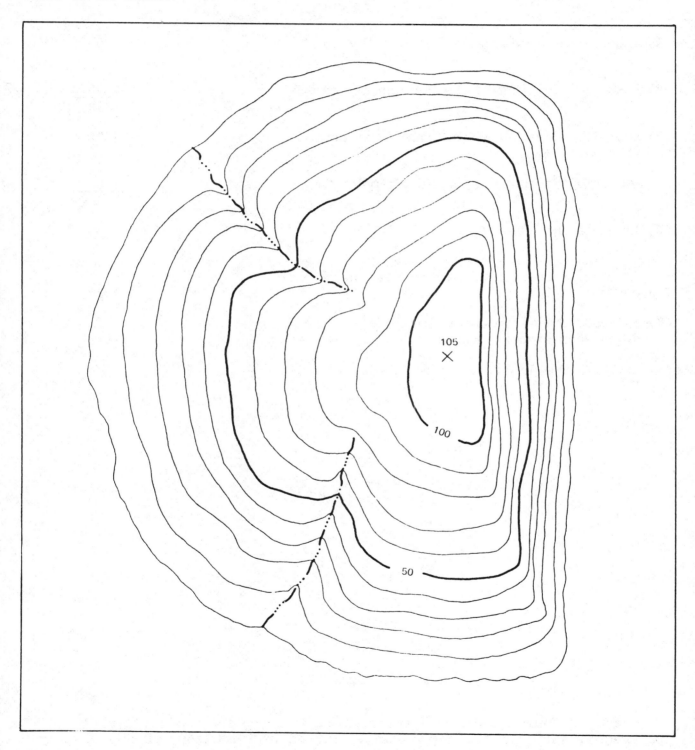

Contour Map of an Imaginary Island

Determining Elevations from a Contour Map

Figure A is a contour map constructed to give you practice in estimating the surface elevation of points in various locations. Note that in two places contours that form a closed loop are distinguished by short, straight lines at right angles to the contour line; these are <u>hachured contours</u> and indicate <u>closed depressions</u>, which are hollows in the surface. Figure B gives further information on these depression contours.

(1) What contour interval is used on this map? 25 ft

(2) Give the elevation of the contour at point A. 1050 ft

(3) Give the elevation of the contour at point B. 1050 ft

(4) Estimate the elevation of summit point C. 1180 ft

(5) Estimate the elevation at point D. 1035 ft

(6) Give the elevation of the hachured depression contour at E. 1025 ft

(7) Estimate the elevation of the bottom of the depression at point F. 1020 ft

(8) Estimate the depth of the depression at F below the lowest point of the rim that encloses it. 40 ft

(1060 - 1020 = 40)

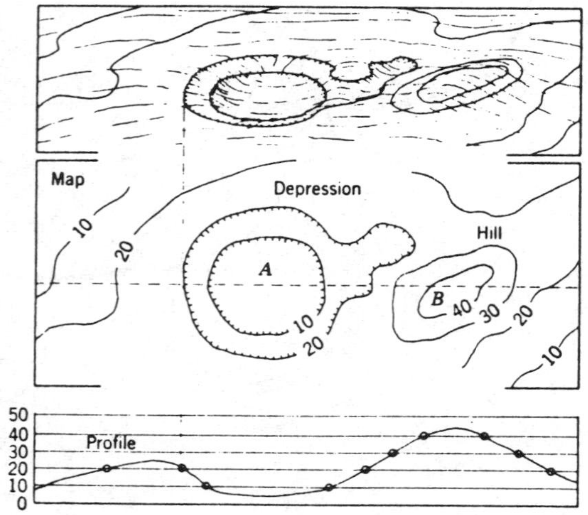

Figure B Hachured contours forming closed loops on the map (center). A perspective sketch of the depression, compared to a hill (top). Profile through the depression and hill (bottom).

Figure A

Drawing Topographic Contours

On Figure C, complete the drawing of contours at intervals of 100 ft. Use a soft pencil and draw lightly at first. Then go over the lines in heavy pencil or ink.

Figure C

(9) How high is the cliff at Point C in the southeastern part of the map?

_____245_____ ft

(10) Determine the average slope of the ground, in feet per mile, along the profile line between the 200 ft and 500 ft contours in the northeastern part of the map. (Two bold dots show the profile segment to be measured.)

_____1170_____ ft per mi

Drawing a Topographic Profile

Construct a topographic profile along the line A-B of the contour map, Figure C. Use the graph provided (Graph A). Fold under the left side of the page along the dashed line, and temporarily attach the folded edge to the profile line A-B on the map. Use a vertical exaggeration of five times the horizontal scale. Label elevations at 100-ft intervals on the vertical scale. Figure D gives you the necessary information on the construction of a topographic profile and the calculation of its vertical exaggeration.

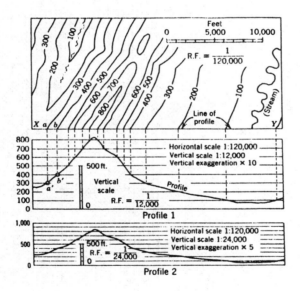

Figure D

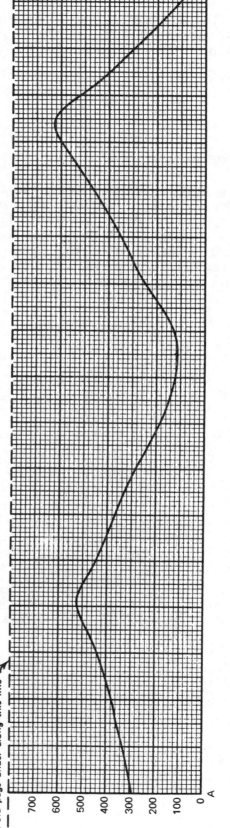

Graph A

_____   _____

Exercise 13-B   Mount Shasta--A Composite Volcano
[Text p. 288-290, Figures 13.3, 13.4.]*

Volcanoes of the Cascade Range of northern California, Oregon, Washington, and southern British Columbia provide splendid examples of composite cones, two of which have delivered major eruptions in the this century. Greater and more devastating volcanic eruptions have occurred elsewhere in the same period, but the eruptions of Mount Lassen (1914-1915) and Mount St. Helens (1980) provide good examples of intense volcanic eruptions and their environmental consequences. Included in the Cascade chain is a great caldera, Crater Lake, relict of a former lofty composite volcano that has been named Mount Mazama. In this exercise, we study the great Cascade volcano, Mount Shasta, in northern California. It is a towering and strikingly beautiful landform.

Our Grid System

In this and following exercises on landforms, our data sources consist not only of photographs but also of topographic contour maps, most of which are selected portions of topographic quadrangles published by the U.S. Geological Survey. To assist you to find and refer to designated features on these maps, we use a special grid system with 1000-yard units, scaled along the left and bottom margins of each map. We will give the locations of features in grid coordinates, as illustrated in Figure A. The lower left-hand corner has the grid coordinates 0.0-0.0. The first number gives the distance to the right, the second number the distance upward. Keep this in mind by using the slogan: "Read right up." For example, Point A has the grid coordinates 0.7-0.8, meaning "Right 700 yards, up 800 yards." Many of the maps also show a graphic scale of miles. Contour elevations are given in feet and it is up to you to determine the contour interval.

Nearly all topographic maps currently sold by the U.S. Geological Survey give contour intervals and elevations in feet. Why don't we use meters and kilometers as our units of grid measurement and elevation? Most Americans continue to use the English units of length and distance measurement with which they are thoroughly familiar. These allow us to relate the dimensions of landforms to real-life experience. Consider that the radio messages of U.S. Navy fighter aircraft pilots, warning of the headon approach of hostile fighter planes, state the closing distance in miles! Need we say more?

Figure B is a photo of Mount Shasta, taken from a ground viewpoint on the southwest side of the peak, along a line from 0.10-0.0. Using infrared film, it shows distant features in great detail. The blue sky appears black, clouds and snow are white. The main part of the volcano is dissected (carved up) by streams and glaciers and no longer shows a crater. Several small glaciers, shown in gray overprint on the map, Figure C, remain on the mountain as relict features of the most recent (Wisconsinan) glacial epoch. A subsidiary volcanic cone, named Shastina, is seen on the western slope of the main cone. Shastina is a relatively young feature and still shows a crater rim. Ski lifts and trails are located on the slope between Shastina and the main peak.

_____

*Modern Physical Geography, 3rd Ed., p. 244-46, Figures 14.3-14.5.

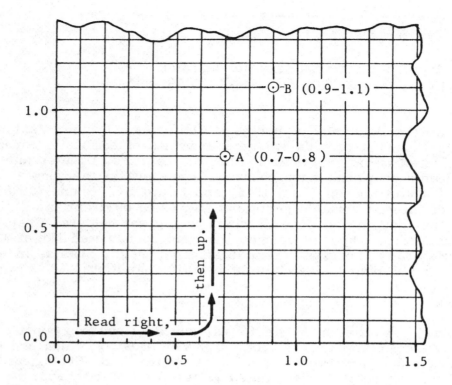

Figure A

(1) Give the grid coordinates of the summit of Mount Shasta. Estimate the summit elevation of this highest point.

      Grid coordinates: <u>16.5-17.0</u>

      Summit elevation: <u>14,000</u> ft, plus <u>100 to 200</u> ft

(2) Estimate the summit elevation of Shastina, 13.5-17.7.

      Elevation: <u>12,000</u> ft. plus <u>200 to 300</u> ft

(3) How wide is the volcano at its base, assuming the 5000-ft contour to represent the base? (Measure width on a line from 11-27 to 23-7.)

      Width: <u>about 13</u> mi

(4) Calculate the angle of slope of the main volcano between the 10,000 and 12,000 foot contours. Take your measurement on the southwest side. On Graph A, plot the vertical and horizontal distances. Draw the hypotenuse of a right triangle to represent the average surface and measure its angle with a protractor.

      Slope angle: <u>26</u> degrees

Figure B Mount Shasta. (Infrared photograph by Eliot Blackwelder.)

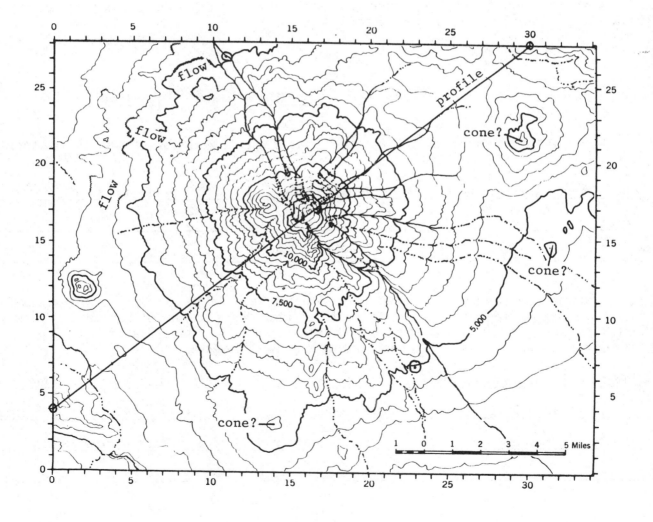

Figure C

(5) Refer to text Figure 13.3, p. 288. Estimate the angle of slope of the uppermost (steepest) part of the slope of Mount Mayon. With a straightedge, define straight lines matched to the profile on both left and right. Measure the angle with a protractor. How does this angle compare with that you measured for Mount Shasta. Offer an explanation for the difference, if any.

Slope angle is about 37° (left) and 34° (right), which is substantially steeper than for Shasta (26°). An explanation may that Mayon is a young, active volcano, rapidly accumulating ash (tephra) on its slopes. Shasta has been subject to erosion for thousands of years, a process that reduces the slope angle of a dormant or extinct cone.

(6) In the map area of 9.0-26.0 are special "sawtooth" contours, called <u>serrate contours</u>. They are symbolic of rough, blocky lava flows of recent date. Judging by the configuration of these contours, can you locate a likely point source for this area of fresh lava? (Give coordinates.) Find two additional areas of fresh lava. Label the three areas as "lava."

Source of lava flow: __10.0-24.2__

Two other areas of lava flows: __5.0-22.0__ and __4.0-18.0__

(7) Describe the pattern of the streams that drain the slopes of Shasta. (Consult text p. 367 and Figure 16.35.)

The streams form a radial pattern. It resembles the pattern of spokes of a wagon wheel.

(8) Examine the round hill at 2.0-12.0. What is the origin of this feature? Look for it on the photograph. Are other similar features present in the eastern and southern areas of the map? Give coordinates. Look for these features on the photograph. Label these features on the map as "cone?"

The round hill is probably a cinder cone. A similar, but larger volcanic cone is located at 30.0-22.0. Possible small cones are at 31.5-15.0 and 14.0-3.0. [Another possible answer is a plug dome, which would not have a summit depression.]

Special project:

(9) On blank Graph B, construct a topographic profile across Mount Shasta from 0.0-4.0 to 30.0-28.0. Calculate and label the vertical exaggeration of the profile.

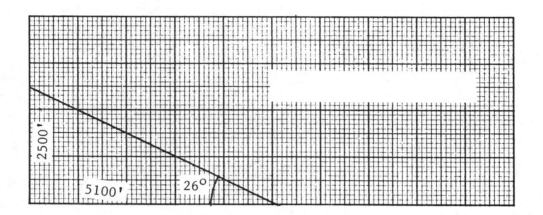

Graph A

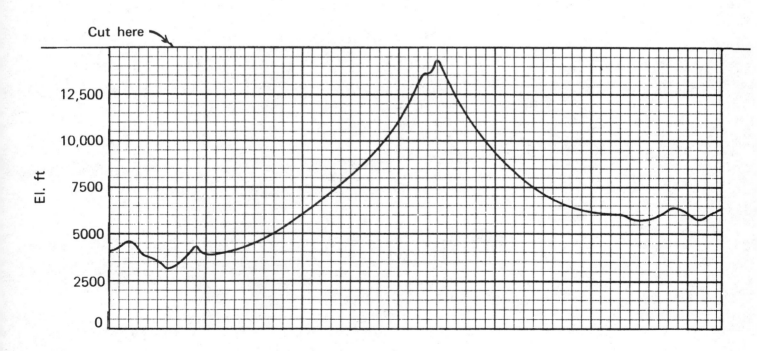

Graph B

_____   _____

Exercise 13-C   Crater Lake--A Huge Explosion Caldera
[Text p. 289-90, Figure 13.5.]*

The great volcanic explosion that accompanies the birth of a caldera has often been described as a volcano "blowing its top," but this is only part of the story. We need to add "and falling through the floor." In the midst of the spewing out of great quantities of volcanic ash and other tephra, to the accompaniment of local earthquakes and seismic sea waves, the collapse of a great mass of rock to leave a basinlike depression defies observation. We see it only when "the smoke clears." The paramount example in recent human history was the explosion of the Indonesian volcano Krakatoa in 1883, mentioned in your textbook.

The example of caldera formation we focus on in this exercise occurred long before European settlers arrived in the American West. What was going on in the Old World in 4,400 B.C., when the great ancestral Cascade volcano, Mount Mazama, blew its top and fell through the floor? Civilization was doing nicely in Mesopotamia and Egypt, but writing and the wheel were yet to be invented. Peoples from Asia had already crossed the Bering Strait and spread south through North America, so we can imagine the inhabitants of this region witnessing in terror the catastrophic demise of Mount Mazama.

Figure A shows three schematic cross sections of inferred stages in the formation of a caldera such as Crater Lake. Diagram 1 shows the final stage in the construction of the composite cone, Mount Mazama. Diagram 2 shows the eruption of large quantities of hot gases driving ash and pumice from the magma chamber, while at the same time the magma is draining down to greater depth, leaving a cavity. Collapse follows (diagram 3) and the caldera floor is covered by a layer of pumice.

We present Crater Lake for your study as a landform depicted through a photograph, a topographic contour map, and a perspective drawing (a block diagram). The block diagram, Figure C, was painstakingly created from map data by a distinguished geographer, Erwin Raisz. He was both a skilled cartographer and a free-hand landscape artist; his art shows in the softness and delicacy of his rendition of block diagrams, but allows no comprise with accuracy. We shall encounter more of his diagrams in later exercises, and in later chapters of your textbook.

---

*Modern Physical Geography, 3rd Ed., p. 245-46, Figure 14.6.

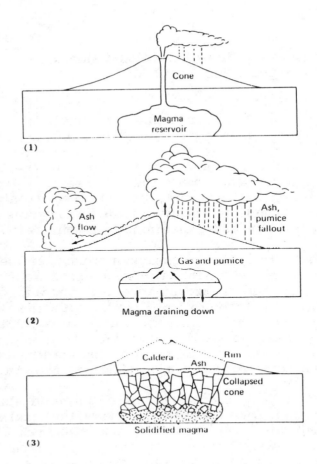

Figure A

(1) On the topographic map, Figure B, measure the diameter of the Crater Lake caldera from rim to rim, both on an east-west line and a north-south line. Use the top of the cliff as the reference point. Give approximate figures in yards, miles, and kilometers.

	Yards	Miles	Kilometers
East-west:	10,000	6	9.5
North-south	9,500	5½	8.5

(2) Figure E is a map showing depths of the lake floor below the mean water level. The bottom contours are in fathoms. (One fathom equals 6 ft.) What is the difference in elevation of the deepest point in the lake and the highest point on the rim? Give in feet and meters. First, find and name the highest rim point bearing a bench mark elevation (BM or VABM).

Name: __Hillman Peak__   Elevation: __8,156__ ft; __2,486__ m

Elevation of deepest point on lake floor: __4,244__ ft; __1,294__ m

Elevation difference: __3,912__ ft; __1,192__ m

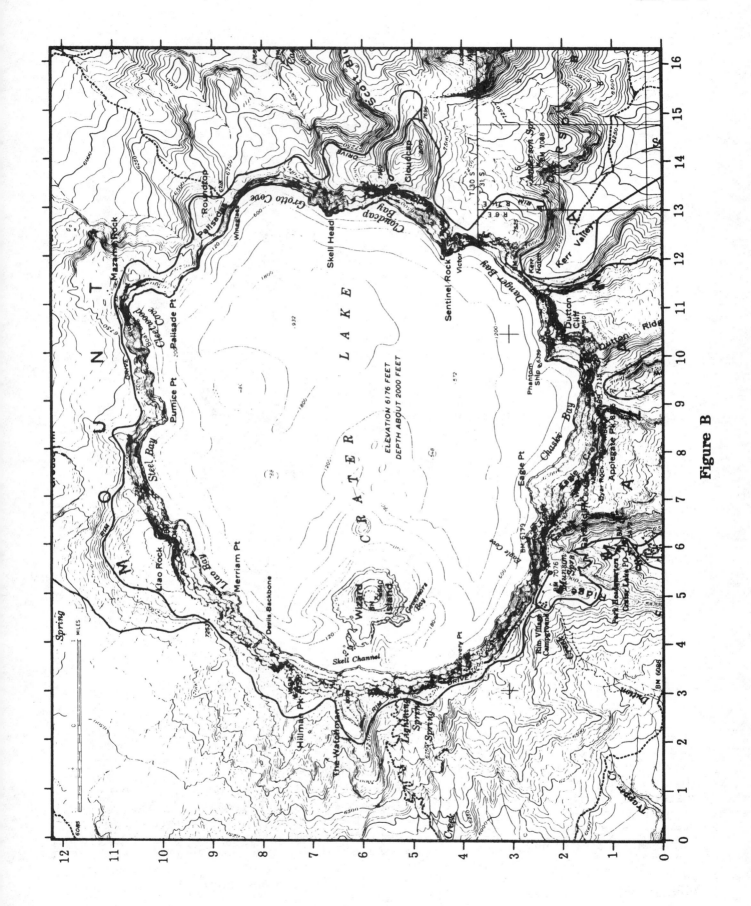

Figure B

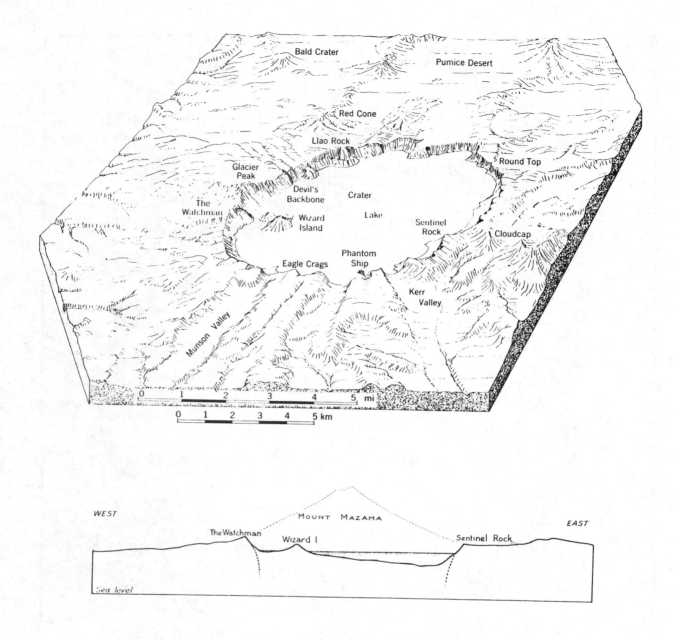

Figure C

A geologist named Howel Williams studied Crater lake in great detail. From his years of research he concluded that the summit of Mount Mazama rose to an elevation of about 12,000 ft. He calculated that the total volume of rock which disappeared to form Crater lake was about 17 cubic miles. He also carefully measured the thickness and area of the pumice and ash that lies in the region surrounding Crater lake. Adding to this the volumes of liquid magma and fragments of rock that were also erupted, he could account for approximately 6.5 cubic miles of the missing material, or substantially less than one half of it. This led Williams to favor the collapse theory, illustrated in our set of cross-sections, Figure A,. Others before him had reached the same conclusion. His position was challenged, but his evidence has remained convincing.

Figure D Crater Lake, Oregon. (U.S. Army Air Force photograph.)

(3) From a study of landforms shown in the map, block diagram, and photograph of Crater Lake, what features can you identify that might support the opinion that a large composite volcano once stood here? Be specific. (Hint: Examine text Figures 18.5, 18.7, and 18.9.)

The high rim of Crater Lake has some deep U-shaped valleys carved into it.

These are shown clearly in the block diagram on the near side of the caldera.

Two are named: Munson Valley and Kerr Valley. A third lies midway between

them. The corresponding gaps in the crater rim are named Sun Notch and Kerr

Notch. These features can be interpreted as rock troughs carved by former

valley glaciers that headed high on the former cone. (These features are called

beheaded glacier troughs.)

(4) Carefully examine the wall of the crater around its entire circumference. Is it smoothly curved in one simple circle or ellipse? What details of its pattern suggest that collapse of large rock masses may have been involved in rim formation? (Hint: Study text Figure 14.21.)

The rim consists of a succession of curved indentations, best described as "scallops," several of which are given the name of "cove" or "bay." (Steel Bay, Cloudcap Bay, Grotto Cove). [Note: These coves and bays have the appearance of curved scarps made by the slipping down of large rock masses form which support was removed by the downsinking of the caldera floor; i.e., they resemble landslide scars.]

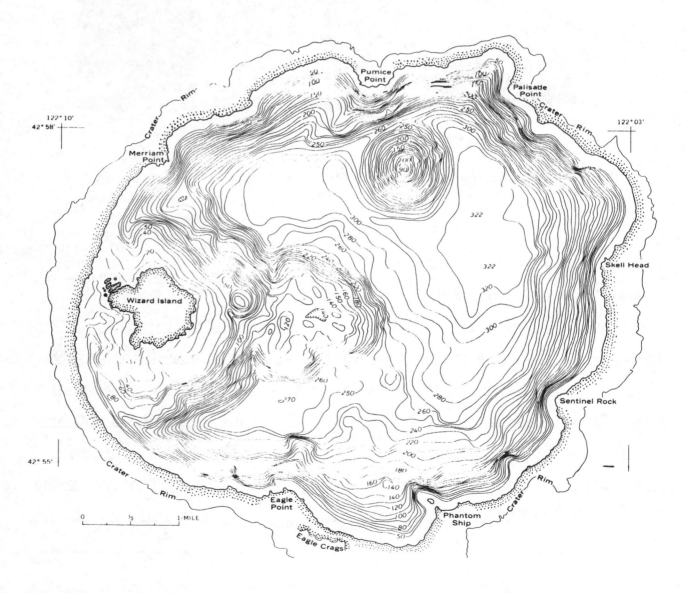

Figure E Contour map of the floor of Crater Lake. Contour interval 10 fathoms. (U.S. Coast & Geodetic Survey and U.S. Geological Survey.)

Name _____ Date _____

_____ _____

Exercise 13-D Shield Volcanoes of Hawaii
[Text p. 290-91, Figure 13.7, 13.8.]*

The Big Island of Hawaii is endowed with the greatest active basaltic shield volcanoes on our globe. Iceland can also claim world-class standing in the competition for active basaltic volcanoes--among them Hekla, with 15 eruptions in historic time. Both Hawaii and Iceland are gigantic piles of basaltic lava built up from ocean floor depths of several thousand feet. But there is a difference: Whereas Iceland lies squarely on the Mid-Atlantic Ridge--the active spreading boundary between two plates--Hawaii and its older family members are far from any active plate boundary. As your textbook explains, they have been generated from a hot spot in the basaltic crust, thought to be fed by a mantle plume of rising magma that stays put while the Pacific plate rides over it, heading westward.

In this exercise, we focus on the summit region of Mauna Loa, largest of the active shield volcanoes. To investigate the rather strange volcanic landforms in this area, we make use of a splendid air photograph taken many years ago by what was then the U.S. Army Air Corps in the pre-World War II period (Figure A). A schematic block diagram (Figure B) and two geologic maps (Figures C and D) provide the information we need to interpret the photo.

(1) Using Figure C, name the active and inactive volcanoes shown on the map. What map evidence suggests that Mauna Kea is extinct, or at least, has long been inactive?

Active: Mauna Loa, Kilauea, Haualalai. Inactive: Mauna Kea, Kohala. No lava

flows of historic date are shown to be present on Mauna Kea.

_____

(2) Give the approximate summit elevations (ft) of Mauna Kea and Mauna Loa.

Mauna Kea: __13,000+__ ft. Mauna Loa: __13,000+__ ft.
(Actual: 13,796   13,677)

_____

*Modern Physical Geography, 3rd Ed., p. 246-48, Figures 14.8, 14.9.

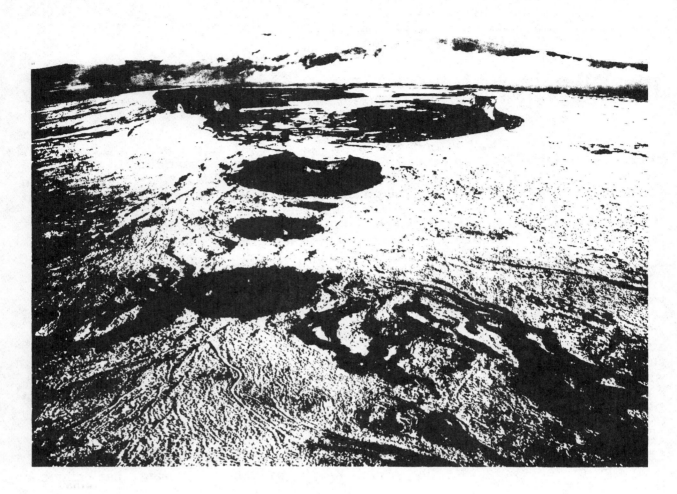

Figure A

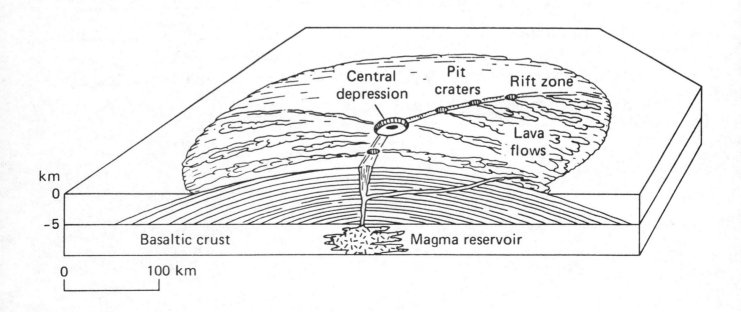

Figure B

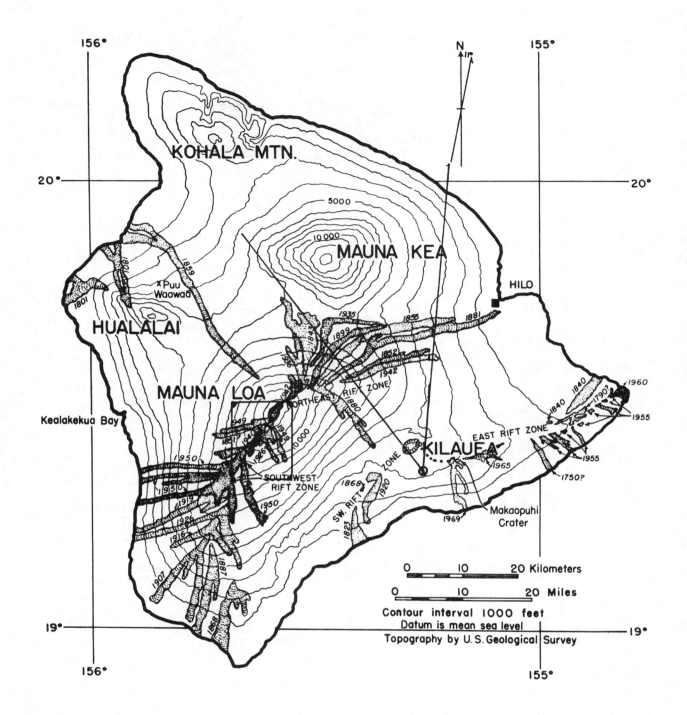

Figure C Contour map of the island of Hawaii. (Based on data of the U.S. Geological Survey. From G.A. Macdonald, A.T. Abbott, and F.L. Peterson, <u>Volcanoes</u> <u>in</u> <u>the</u> <u>Sea</u>, Second Edition. Copyright © 1970, 1983 by the University of Hawaii Press, Honolulu. Used by permission.)

(3) Study the details of text Figure 13.7 and relate them to features on Figure D. Draw two straight radiating lines on the map to show your interpretation of the position of the camera and the field of view of the photograph.

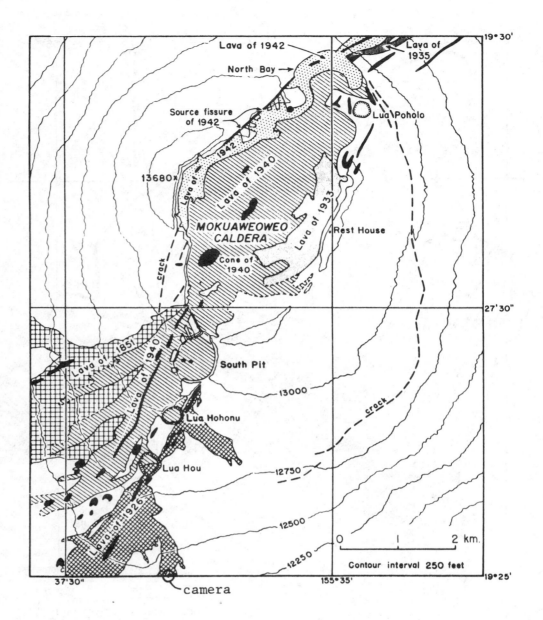

Figure D Geologic map of the summit area of Mauna Loa, showing the lava flows of 1940 and part of the lava of 1942. (Same data source as Figure C. Used by permission.)

(4) Study the distribution of dates (years) given for the lava flows of Mauna Loa. Note where the oldest flows are located and where the youngest occur. What general statement can you make about this distribution? What does it mean in terms of the recent history of volcanic activity?

The oldest flows lie farthest distant from the summit along the rift zone in both

directions. Ages become younger toward the summit. For Mauna Loa, most of the

activity after 1949 lies in a part of the southwest rift zone. Perhaps the souce of

rising magma is becoming more concentrated in a smaller length of rift. There

also sems to be a shift in lava flow sources from the NW (Hualalai) toward the SE

(Kilauea) and the East Rift Zone. (Caution: flows of past three decades are not shown.)

Figure D shows some details of the summit region covered by the photograph. Note that the central depression is called a "caldera." Unlike the explosion caldera of Crater Lake, this depression results from collapse of the volcano summit as magma is withdrawn from below. The caldera floor may also be raised in level by the outpouring of magma.

(5) If the caldera and pit walls (steep cliffs) result from down dropping of solid lava, are these walls actually a kind of fault landform? If so, what kind of fault? (Consult text p. 296.)

Yes, these features are fault scarps. They are expressions of normal faults.

(6) On the southern boundary of the map, Figure C, mark and label a point that seems to lie beneath the aerial camera when the photo was taken.

(7) On the photograph, Figure A, write in the names of the three prominent pit craters and the caldera itself.

(8) What is the meaning of the long black lines on the map, Figure D? Identify and label one such feature clearly shown on the photograph.

These are fissures (cracks) from which magma is extruded to form the flows. One is visible to the left of the line of pit craters.

(9) On the photograph, snow cover appears white. However, flows at the lower right, apparently coming from the area of Lua How, are black. Does this mean that the rock is so hot that it melts the snow?

Not likely, unless the flows are extremely recent. Perhaps the flow surface is extremely rough, with vertical surfaces that do not hold snow. Its blackness may contribute to more rapid heating under direct insolation. Older flows may accumulate lichens and oxide films that reduce the ability of the surface to absorb solar heat.

Name _____ Date _____

_____   _____

Exercise 13-E Cinder Cones
          [Text p. 291-92, Figures 13.9, 13.10.]*

A cinder cone is a mere molehill compared to the great mountain that is the composite cone of Shasta. Actually, a gopher mound is a better simile because it looks like a cinder cone and is constructed in much the same way. A pocket gopher in action is rarely seen, but the shower of soil particles and small stones that emerges from its hole is very visible. As the loose particles fall on the ground, they build up in a rounded pile. In much the same way, gases emerging under great pressure project tephra particles from a deep vent and the coarser cinders and bombs build a loose structure with a rounded profile similar to the one shown in text Figure 13.9.

Most cinder cones develop a central crater with a rounded rim, shown in the photograph, Figure A. In come cases, large numbers of lava bombs emerge in a plastic state to build a sharp-crested rim, which may become even sharper when water erosion on the flanks of the cone removes some of the loose cinders and ash.

Cinder cones often occur in groups. One such group is visible in text Figure 2.23B. Examine this Skylab photo closely in the area of Sunset Crater (a recent cinder cone of near-perfect shape) in the upper-left part of the frame. Dozens of cones are scattered about in the vicinity of Flagstaff and several lie close to the major highways. Some are rapidly disappearing to furnish base material for roads and building foundations.

For our example of a cinder cone we use Menan Butte on the Snake River Plain of southeastern Idaho, a few miles north of Idaho Falls. It is shown on the topographic map, Figure B. Much of the plain is covered by quite recent basaltic lava flows, known locally as "lava beds," and this landform type can be identified in the northwest corner of the map, where numerous small closed depressions (hachured contours) are shown. Washington Irving in 1868 described this lava plain as "an area . . . where nothing meets the eye but a desolate and awful waste; where no grass grows nor water runs, and where nothing is to be seen but lava." Similar cinder cones occur farther to the west, where they are called "Craters of the Moon." Actually, there are two Menan Buttes, side by side and almost perfect twins, but we show only the northern one here.

Modern Physical Geography, 3rd Ed., p. 248, Figures 14.10, 14.11.

Figure A A fresh cinder cone and a recent basaltic lava flow (left). Dixie State Park, near St. George, Utah. (Frank Jensen.)

(1) Measure the approximate (average) diameter of the base of the cone, the diameter of the crater rim, the height of the rim above the base, and the maximum depth of the crater below the rim. (Measure the basal diameter between 0.1-2.0 and 3.8-2.0.)

> Basal diameter: ___3,600___ yds
>
> Crater rim diameter: ___1000___ yds
>
> Height of rim above base: ___800___ ft
>
> Max. depth of crater: ___390___ ft

(2) Compare the above dimensions with those of Wizard Island, Crater Lake (Exercise 13-C). Be quite specific. Take into account that the base of Wizard Island is well below lake surface level.

> Diameter: ___Wizard Island, about 2,500 yds, is about one-third smaller in diameter.___
>
> Crater rim diameter: ___Wizard, about 130 yds, is only about one-eighth as wide as Mennan, i.e., much smaller.___

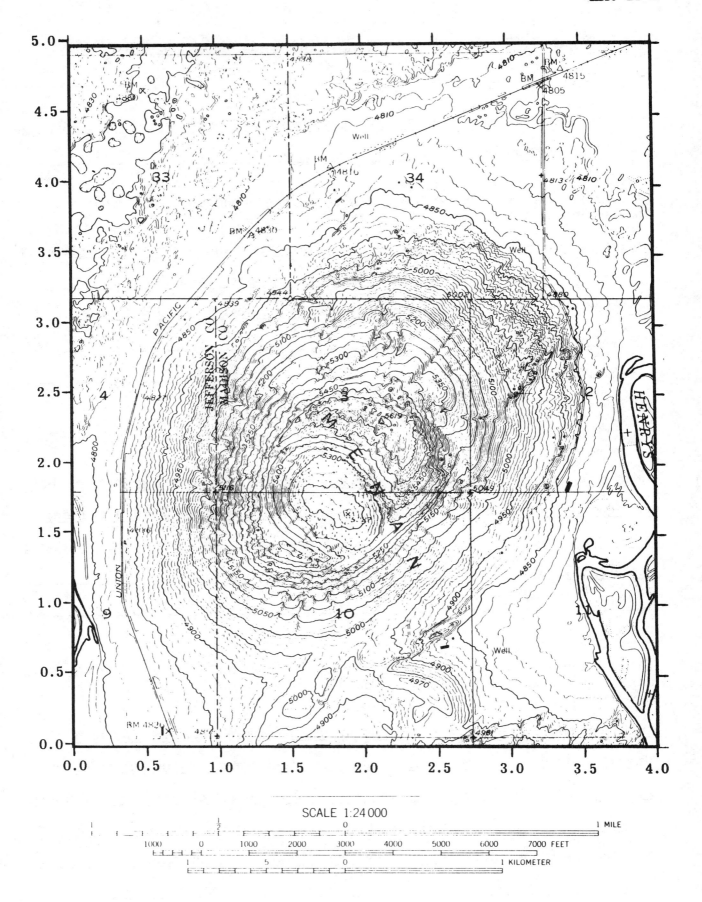

SCALE 1:24 000

Figure B

Height of rim above base: Wizard, 760 ft, is about one-third less.

Max. depth of crater: Wizard, not measurable from map, but probably on order of 100-150 ft, much less than Menan.

(3) Carefully examine the outer and inner slopes of Menan Butte. What features do you find that may relate to the age of the crater?

Numerous narrow valleys are carved into both inner and outer slopes, probably by streams of running water. This suggests the cone is relatively old. On a recent cinder surface, most rainwater quickly percolates into the tephra, preventing runoff from forming.

(4) Can you suggest a reason why the Menan crater is relatively large in comparison with its basal width? Could it have been enlarged by water erosion? By some other process?

Water erosion would act on both outer and inner slopes, lowering the rim, but not necessarily widening the crater. Besides, there is no exit for the eroded debris, so it would accumulate and reduce the crater depth. Perhaps the crater was enlarged by a violent terminal explosion.

(5) The Mennan cone is not perfectly circular. How would you describe its form, both as to its basal outline and the outline and position of the rim? Offer an explanation.

Both base and crater rim are elliptical in outline. Moreover, the long axes of the elipses are in a line. Also interesting is the displacement of the center of the crater to a position southwest of the center of the outer ellipse. These features suggest that the crater was shaped by prevailing winds from the southwest, carrying the finer cinder and ash downwind toward the northeast.

Exercise 13-F   Volcanoes and Plate Tectonics
> [Text p. 271-75, Figures 12.14, 12.16, 12.22;
> p. 299-301, Figure 13.22.]*

The global distribution of volcanoes is important to geographers as part of their need to know the global distribution of all classes of landforms. Volcanoes and volcano chains are a major class of mountains, and mountains are important to humans as a special kind of physical environment. However, we want to know more than just <u>where</u> those volcanic mountains are; we seek to understand <u>why</u> they are there. Plate tectonics gives us most of this information, which, before 1960 simply wasn't available.

Our exercise is based on a world map of volcanoes, Figure A. Dots show the locations of volcanoes known or inferred to have erupted within the past 12,000 years, as determined by radiocarbon dating, geological evidence, and other techniques. Each dot represents a volcano or a cluster of volcanoes.

You are asked to use this map in conjunction with the plate tectonics map, text Figure 12.14, A and B, p. 272-73. First, you should transfer the major plate boundaries to the volcano map using color pens or crayons. Use red for converging (subduction) boundaries and continental sutures; green (or blue) for spreading boundaries. To avoid unnecessary work, transfer only those sections of plate boundaries that coincide with volcano locations. Before you begin, study Figure B, showing the relationships between various forms of volcanic activity and plate tectonics. Note especially that composite volcanoes associated with downplunging plates erupt at some distance from the subduction plate boundary. This means that you will draw the plate boundary parallel with the chain of volcanoes, allowing a space between the two. The plate boundary symbol will be related to the volcano chain in the following manner:

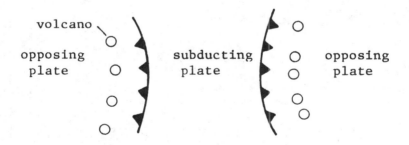

Our questions single out specific localities of interest. Many of the isolated volcanoes shown on the map cannot easily be explained except by applying geological information not available to you.

---

*Modern Physical Geography, 3rd Ed., p. 228-29, Figures 13.16, 13.30; p. 253-54, Figure 14.23.

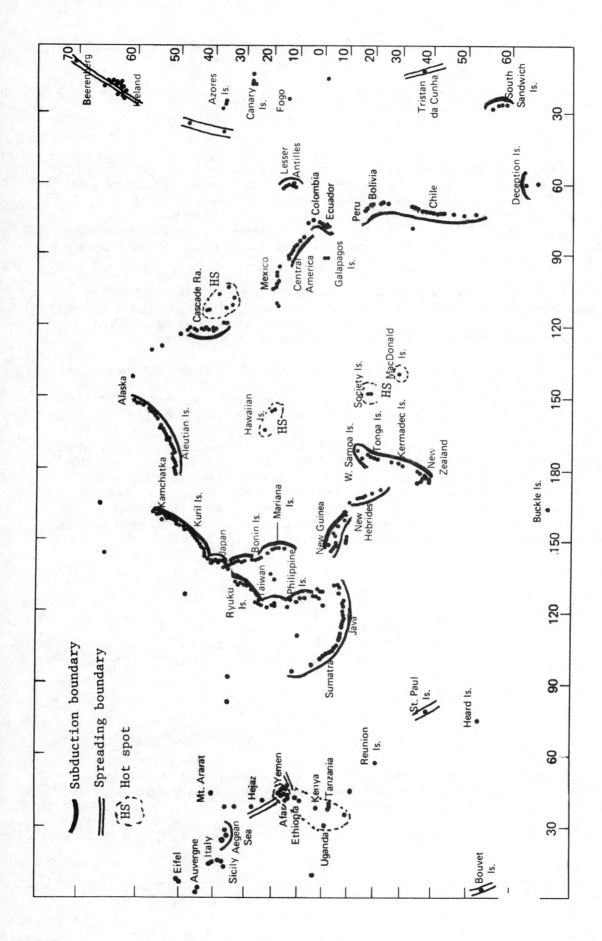

Figure A World map of volcanoes active since the end of the Wisconsinan Epoch. Based on data published in 1979 by NOAA, Boulder, Colorado. (Copyright © 1982 by Arthur N. Strahler.)

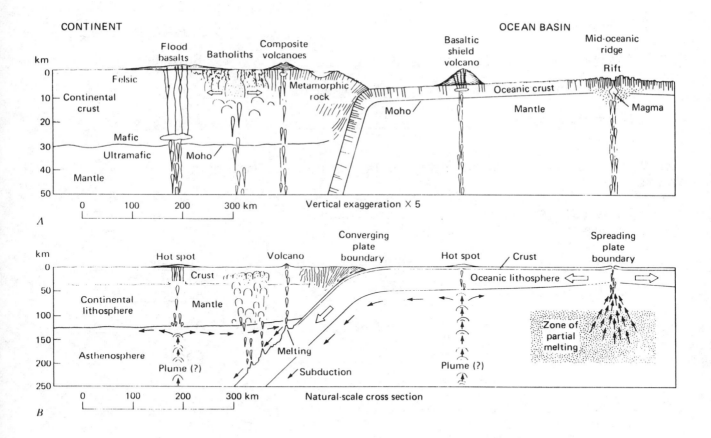

Figure B Schematic cross sections showing the relationship of plate tectonics to various forms of igneous activity. (Copyright © 1982 by Arthur N. Strahler.)

(1) Volcanoes related to plate subduction: With what subduction plate boundary is each of the following volcanic chains associated? Insert the names of the two plates.

Range or arc, as labeled on map:	Subducting plate:	Opposing plate:
Cascade Range:	Juan de Fuca	North American
Aleutian Is.	Pacific	North American
Lesser Antilles	North American	Caribbean
Central America	Cocos	Carib. & N. Amer.
Peru-Bolivia-Chile	Nazca	South American
Mariana	Pacific	Philippine
Kuril	Pacific	Eurasian & N. Am.
Philippine Is.	Philippine	Eurasian
Sumatra-Java	Austral-Indian	Eurasian

Kermadec Is.	Pacific	Austral-Indian
Aegean Sea	African	Eurasian
South Sandwich	South American	Scotia
Deception Island	Scotia	Antarctic

(2) Volcanoes related to spreading plate boundaries: (Same instructions as for Question 1.)

Island(s), as labeled on map:	Plate to W or S:	Plate to E or N:
Iceland	North American	Eurasian
Tristan da Cunha	South American	African
St. Paul Is.	Antarctic	Austral-Indian
Galapagos Is.	Nazca	Cocos

(3) Volcanoes over hot spots within oceanic lithosphere: Locate and name three such islands (or island groups) in the Pacific Ocean. Circle these on the map with a yellow line and label "HS".

Hawaiian Is.   Society Is.   MacDonald Is.

(4) Volcanoes over hot spots within continental lithosphere: Circle in yellow and label "HS" the six unnamed spots shown in the western United States. (Keep clear of the Cascade chain.) These include the Snake River Plain and San Francisco Mountains of Exercise 13-E.

(5) Volcanoes of continental rift zones: Refer to text Figure 13.22 to interpret the volcanoes shown as dots in East Africa, Ethiopia, and the Arabian Peninsula. Circle these dots with a yellow line. (a) Name the plates that are involved in this activity. (b) Name two great volcanoes of the East African Rift system (c) Speculate in a general way as to how these volcanoes are related geologically to the tectonic and igneous processes involved in rifting.

(a) Three plates with a triple junction are involved: African, Somalian, Arabian.

(b) Mt. Kenya, Mt. Kilimanjaro. (c) Continental rifting is accompanied by doming of the plate and underlying mantle. The crust is thus thinned and magma moves upward to reach the surface.

Exercise 13-G   Fault Scarps of the Klamath Lakes Region
        [Text p. 296-98, Figures 3.16, 3.18.]*

The Basin and Range Province of Oregon, Nevada, and other states to the south
and southeast is one of the major geomorphic divisions of the United States. You
can think of it as a "pull-apart" section of the North American plate. Stretched
out in an east-west direction, the brittle crust has fractured into many north-
south normal faults, forming fault blocks. Some have remained high as horsts or
tilted blocks; others have and foundered to form fault valleys and grabens. A
striking example is the Kalamath Lakes region of southwestern Oregon, just east
of the southern end of the Cascade Range.

In this exercise we study the fault landforms of the Kalamath Lakes region by
using a contour topographic map (Figure A) in combination with a finely rendered
block diagram by Erwin Raisz (Figure B).

(1) Using a red pen or crayon, draw on the map all fault lines you can
confidently interpret from steep scarps shown by the crowding together of
contour lines. Write the letters D and U on opposite sides of each fault to
indicate downthrown and upthrown sides, respectively. Find and label an
excellent example of a graben (G). Find and label an example of a horst (H).
Give the locations of these two features.

Upper Klamath Lake (where labeled near 10.0-17.0) is an excellent example of a

graben. (A narrower graben is at 18.0-7.0.) A horst is at 17.0-6.0.

_____

(2) What evidence can you give to conclude that the upthrown block at 20.7-7.0
has been tilted to the east?

The highest elevations occur at the top of the fault scarp. The block surface

slopes away to the east from the top of the scarp.

_____

*Modern Physical Geography, 3rd Ed., p. 251-53, Figure 14.17, 14.19.

EX. 13-G

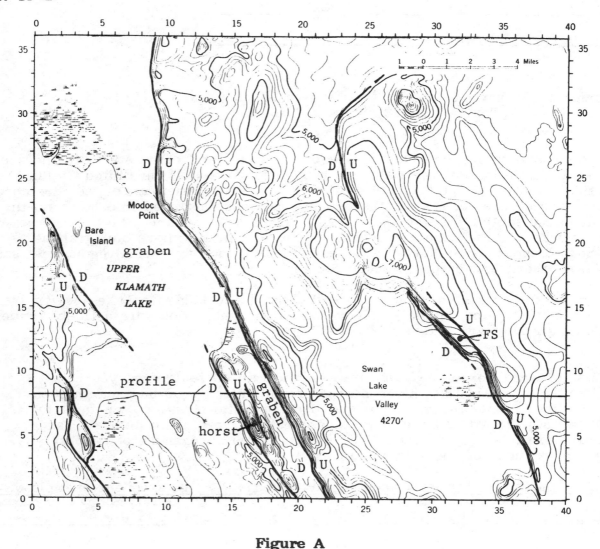

Figure A

(3) Is Swan Lake Valley a good example of a graben? Give evidence for your answer.

Swan Lake Valley is best described as a downtilted fault block, because the western side lacks a steep scarp; instead, the land surface to the west slopes gently eastward toward the valley floor.

---

(4) Estimate the height of the fault scarp at (a) 13.0-20.0 and (b) 35.0-10.0.

(a) __1,200__ ft   (b) __1,400__ ft

(5) Find a good example of a <u>fault</u> <u>splinter</u> and label it "FS" on the map. The concept of a fault splinter is illustrated in the special block diagram, Figure C. Give the map coordinates for this feature.

Fault splinter: __32.0-13.0__

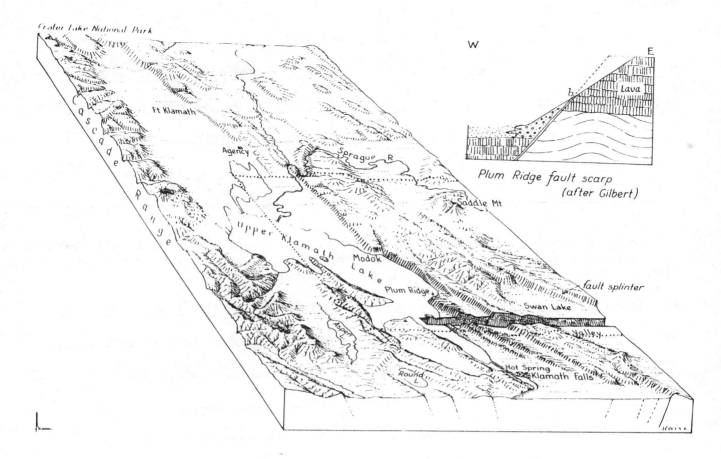

Figure B Perspective diagram of the Klamath Lakes region, Oregon. A dotted line shows the limits of the accompanying topographic map. (Drawn by Erwin Raisz.)

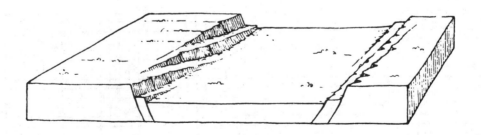

Figure C A fault splinter consists of a narrow sloping fault block between two normal faults, one of which dies out as the other begins. Thus the total displacement is passed gradually from one fault to the other. Two splinters are shown here. (Drawn by Erwin Raisz.)

(6) Offer an explanation for the flatness of the floor of Swan Lake Valley.

The area of the downthrown fault block has been the site of deposition of sediments derived from the surrounding high areas. Possibly a lake once existed here and has been completely filled. Lake floors are typically very flat surfaces.

(7) (Special problem.) Study carefully the contours just east of the fault scarp between 10-27 and 10-34. Describe what you find and relate the features you identify to the progress of up-faulting of the block.

Four shallow stream valleys make notches in the upper part of the scarp. The streams in them flow east-to-west toward the scarp. At the scarp, each valley steepens sharply in gradient. This suggests that valley deepening could not keep pace with the rate of tectonic uplift of the scarp.

(8) (Special project.) Construct a topographic profile across the map from 0.0-8.0 to 40.0-8.0. Use the blank profile, Graph A. Then draw lines down from the profile line to show the fault planes in cross section. Add arrows to show downthrown and upthrown sides.

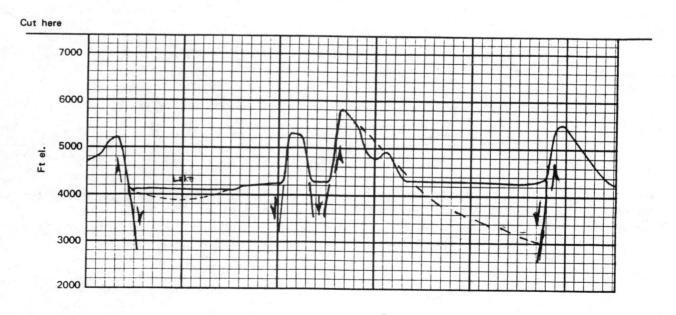

Graph A

Name _____ Date _____

_____   _____

Exercise 13-H   Landforms of Two Great Transform Plate Boundaries
[Text p. 296-98, Figures 13.16, 13.18, 13.20,
13.21, 13.27; p. 304-5, Figure 13.30.]*

A hot item in landform study is called <u>tectonic geomorphology</u>, the study of landforms produced directly by faulting and folding. Examples are the fault scarps pictured in your textbook as Figures 13.17, and 13.20, and the young fold mountains shown in Figure 13.15. Not that such features haven't been investigated before--that's been going on for more than a century--but the new emphasis comes about because people are getting terribly worried about destructive earthquakes. Geologists are disovering new techniques for unraveling the history of great earthquakes by studying the landforms and alluvial deposits formed along major active faults, among them the San Andreas Fault of California. Even tree rings have been called in as witnesses of great quakes. It's now possible to pinpoint the years of occurrence of great quakes as far back as 1,000 years, and to get good quality information for more than 10,000 years before the present. This information yields a sort of timetable on which to base the probability of occurrence of the next big one.

This exercise comes in two parts. Both deal with great transform boundaries between moving plates. The first is the San Andreas Fault and its distinctive small landforms; the second is the western transform boundary of the Arabian plate, coinciding in part with the political boundary between Israel and Jordan, and including the Dead Sea and the Sea of Galilee. Here, we study fault landforms that are on a much larger scale.

Part A   Landforms of the The San Andreas Fault

(1) Review the definition of a <u>strike-slip</u> fault, (text p. 298-99 and Figure 13.16). Then review the definition of a <u>transform</u> fault (p. 270 and Figure 12.12). Do these two names of faults refer to one and the same thing, or is there a distinction in meaning? Explain fully.

<u>Strike-slip fault is a class of faults defined in terms of relative motion of two</u>

<u>crustal blocks (i.e., horizontal motion). Strike-slip faults occur widely and may</u>

<u>bear no relationship to plate boundaries. Transform faults are identified with</u>

<u>tectonic plate boundaries and make up one class of plate boundary. They are also</u>

<u>strike-slip faults by definition of the relative motion.</u>

_____

*Modern Physical Geography, 3rd Ed., p. 252-53, Figures 14.17, 14.20, 14.21; p. 255-256, Figure 14.25; p. 257-58.

Our exercise deals with three kinds of landforms illustrated in text Figures 13.20 and 13.21. First is a fault scarp produced by horizontal motion. Second is the abrupt offsetting of a stream that crosses the fault line. Third is a kind of closed depression found along the fault line.

Figure A Fault scarp on the main San Andreas Fault between Olema and Point Reyes Station, looking southeast. Chunks of sod and soil have fallen from the scarp. (G.K. Gilbert, 1906.)

Fault Scarps

(2) The accompanying photo (Figure A) shows a fresh fault scarp (a small cliff or wall) formed within a few seconds during the great San Francisco earthquake of 1906. It lies along the San Andreas Fault in Marin County, just north of San Francisco Bay. We see one side of a hill crossed by the fault line. Our problem is: How can a vertical scarp be produced by purely horizontal motion of the earth on the two sides of the fault line? Refer to the diagrams in Figure B for help in figuring out an answer.

(a) Upper diagram. What evidence is visible to prove that only horizontal motion has occurred?

First, the crest of the ridge is of the same height on both sides of the fault.

Second, on the flat plain there is no scarp along the fault line.

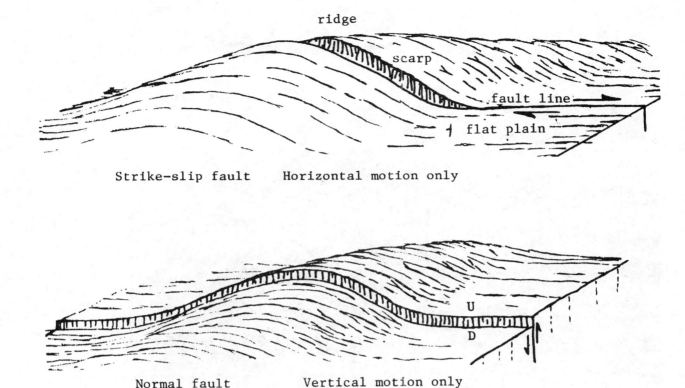

Figure B Sketches of landform expression of two kinds of faults. (A.N. Strahler.)

(**b**) Upper diagram. If you were to cross over the ridge to the far side and look back, what evidence of the fault movement would you see?

There would be a similar scarp on the far side, facing in the direction opposite

to that showing on the near side.

(**c**) Lower diagram. What evidence is visible to prove that the relative motion is purely (or largely) vertical?

The scarp maintains the same height in crossing both the ridge and the flat

plain.

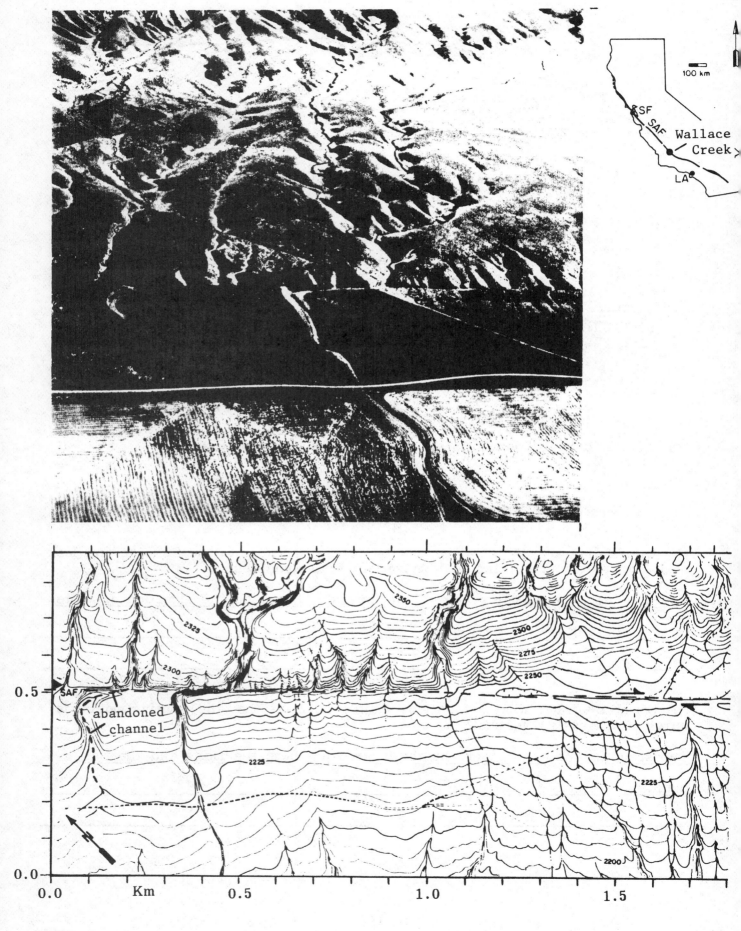

Offset Streams

Examine the oblique air photo, text Figure 13.20. Find a clear example of an offset stream and its valley (near top of photo). [Optional: Using a color pen, highlight the course of the stream and the line of the fault. Then draw arrows on both sides of the fault line to show the directions of motion. Label the Pacific plate and the North American plate.]

Figure C is an oblique air photo of an offset stream similar to the one shown near the top of text Figure 13.20. It is named Wallace Creek, after a leading authority on the San Andreas Fault, geologist Robert E. Wallace. Figure D is a special large-scale topographic contour map including the area shown in the photograph. Actually, the topographic map coincides quite well with the section of fault shown in text Figure 13.20. California geologists made a careful study of the offset and the alluvium (stream-deposited sand and gravel layers) along its course. They were able to date various layers by using the carbon-14 method on organic matter (charcoal) in the beds. This information enabled them to learn when the offsetting took place.

Figure C Oblique air photo of Wallace Creek in the Carrizo Plain of central California. (Courtesy of Dr. Robert E. Wallace, U.S. Geological Survey, 1974.)

Figure D A special large-scale topographic contour map of the San Andreas Fault zone in the vicinity of Wallace Creek. Contours in feet. The grid coordinates are in thousands of meters. (From Kerry Sieh and Robert E. Wallace, Geological Society of America Centennial Field Guide--Cordilleran Section, 1987, Fig. 3. Reproduced by permission.)

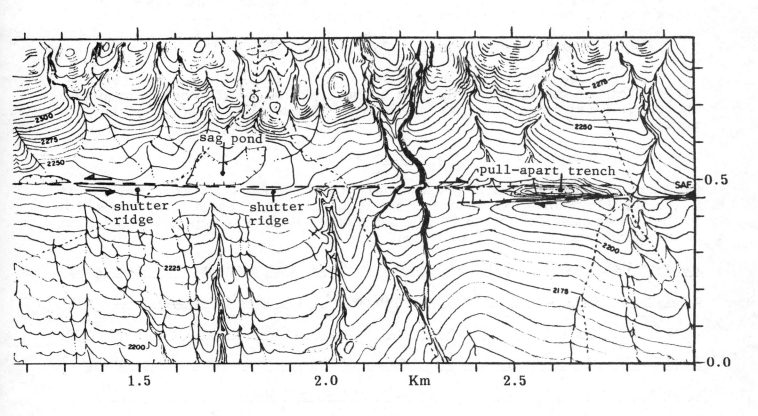

(3) Using a pale color pen or pencil (yellow or pink) draw a straight line along the foot of the steep scarp between ~~0.00-0.81~~ and ~~1.00-0.82~~. This is the approximate position of the San Andreas fault line. Measure the distance of offset of the present channel of Wallace Creek.

Note correction: Grid coordinates should read 0.00-0.51 and 1.00-0.52.

Offset: ___130___ m: ___430___ ft;

(4) Radiocarbon dating shows that approximately 3,800 years were required for the measured offset of the present channel of Wallace Creek (distance given in your answer to Question 4). Calculate the average rate of slip (cm per year) along the fault during this time period.

Average rate of slip: ___3.4___ cm/yr

(5) Did the entire slippage required to produce the present offset occur in a single earthquake? What is your opinion about this? (Hint: The total horizontal slip during the San Francisco 1906 quake was about 6 m (20 ft); that of the Fort Tejon quake of 1857 was about 10 m (33 ft).

Not likely. If the horizontal slip was as great as 10 m (33 ft) per earthquake, at least 13 such quakes were required.

(6) Going back in time, we observe that Wallace Creek formerly occupied another offset channel located to the northwest of the present channel, i.e., toward the left in Figures C and D. Using a color pen or crayon, draw on the map the former stream course from the present channel to the abandoned channel. Label this course "abandoned channel." What is the total amount of the offset of this abandoned channel (including the present channel offset)?

Offset of abandoned channel: ___380___ m; ___1280___ ft

(7) Radiocarbon dating shows that the abandonment of the old channel occurred about 13,250 years ago. What was the average rate of slip in cm/yr for that period?

Average rate of slip: ___2.8 to 2.9___ cm/yr

(8) The location of Wallace Creek is shown in the small map accompanying Figure D. Locate this point on the San Andreas fault map, text Figure 13.30. Does it lie in the locked sector or in the Parkfield-Hollister sector of creep movement? What is the significance of your observation?

Because Wallace Creek is in the locked Fort Tejon sector, it has not experienced a major slip since the Fort Tejon earthquake of 1857, which was about 135 years ago. Based on an estimate of an average slip rate of 3.4 cm/yr (from answer to question 5) another quake of 10 m displacement would be most likely to occur after a lapse of 300 years (i.e. about in the year 2,160).

Closed Depressions

Closed depressions along the San Andreas Fault are known as "sag ponds." In the arid climate prevailing in the central and southern parts of the fault, these small, shallow depressions hold water only after heavy winter rains. Even the word "sag," implying a downsinking of the crust, is not apt in most cases. Tectonic depressions along the fault are of two basic kinds. The first consists of a depressions formed behind <u>shutter</u> <u>ridges</u>.

Figure E is a perspective diagram to show the concept of formation of shutter ridges by strike-slip faulting. The general slope of the land surface is toward you, so that streams in the valleys shown in the upper diagram are flowing toward you. Next, we imagine that a strike-slip fault develops along the dotted line, with the part of the crustal mass on your side moving to the left. The lower diagram shows the landscape after lateral displacement has occurred. Each ridge in the nearer side has moved left to a position that forms a blockade or dam across a valley beyond it on the other side. It's as if a shutter were drawn across the valley. Behind the shutter ridge lies a closed depression. If water accumulates in the depression is becomes one kind of "sag pond," but actually there has been no sagging (sinking) of the crust beneath the depression.

(9) To analyze this process of shutter ridge development, use the hypothetical map, Figure F. Cut the entire page in to two parts along the dashed line that bisects the map. Slide the upper half to the right relative to the lower half by the distance equal to the length of the arrow. Tape the two halves together in this position. (Apply tape to reverse side of sheet.) Find the place where a stream valley is now blocked by higher ground of a ridge. Outline this topographic depression and label as "sag pond."

Solution on p. 360.

(10) (Advanced problem for special credit). Using a color pencil or pen, attempt to draw new contour lines along the fault zone so that no contour line ends within the map area. Allow for weathering and erosion to soften the steep slopes along the fault line.

Solution on p. 360.

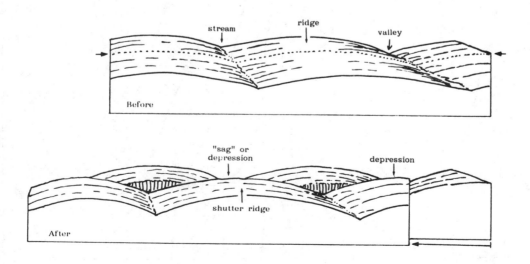

Figure E Perspective diagram of formation of shutter ridges. (A.N. Strahler.)

(11) Next, look for shutter ridges on the map, Figure D. They are near the center of the map. Look for a "sag pond," indicated a closed hachured contour line. Label two shutter ridges and the sag pond. In the spaces below, give the grid coordinates of these features. Locate and label these features on your text photograph, Figure 13.20. Draw in the continuation of your line marking the position of the San Andreas Fault, and continue it further to the right.

Shutter ridge extending from <u>1.30-0.48</u> to <u>1.65-0.47</u>

Shutter ridge extending from <u>1.75-0.47</u> to <u>1.90-0.47</u>

Center of closed depression <u>1.72-0.53</u>

A second kind of closed depression is the <u>pull-apart</u> <u>trench</u> illustrated in Figure G. Its evolution is shown in the schematic maps <u>a</u> through <u>d</u>. The strike-slip fault develops a small kink (a) that becomes an abrupt offset. As the offset widens it also lengthens, causing a steep-sided trench to open up (b, c, and d). The trench floor is partly filled by blocks that have slid in from the ends, and by sediment washed into the trench from outside (d'). Use the paper cut-out model to demonstrate the process of trench opening.

(12) Study carefully the lower part of text photograph Figure 13.20. The long dark shadow shows a deep pull-apart trench. Notice that the fault line is offset below and beyond the trench. Label this feature "pull-apart trench." Find this same trench on the topographic map, Figure D. Give the grid coordinates of the center of the deepest part of the depression. Draw in the trace of the offset ends of the fault and continue the line to the edge of the map. What is the separating distance between the offset branches?

Coordinates of deepest point: <u>2.60-0.47</u>

Distance between offset fault lines: <u>50</u> m

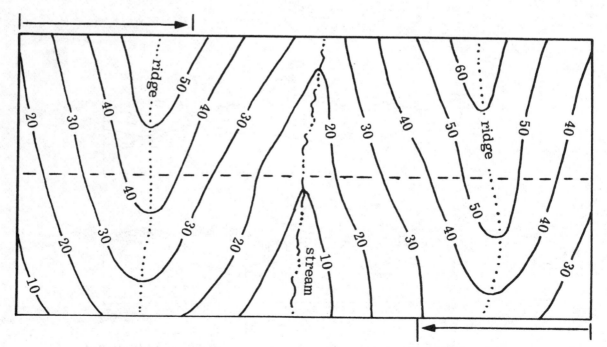

Figure F Map model of formation of shutter ridges.

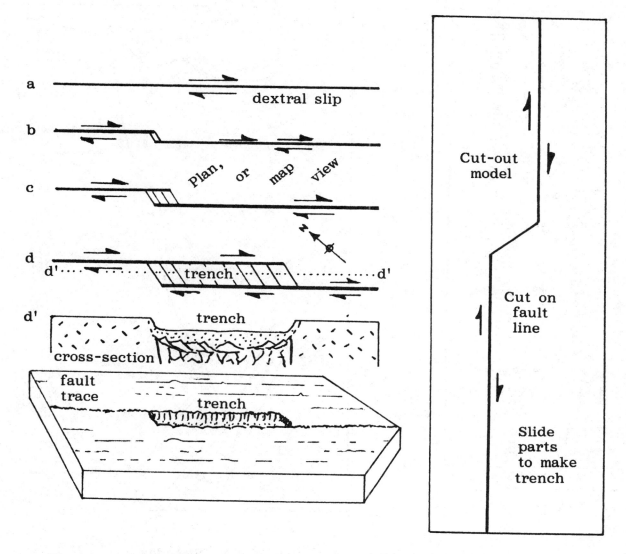

Figure G Stages in the evolution of a pull-apart trench.

Part B   The Dead Sea Rift Valley--A Transform Plate Boundary

The Dead Sea, shown in the sketch map, Figure H, lies on the floor of what has long been called a graben, or simple down-faulted block, such as that illustrated in text Figure 13.18. With the advent of plate tectonics, however, the explanation has been greatly revised. The surface of the Dead Sea lies at an elevation of about 400 m below sea level. The topographic trench it occupies continues far to the south to join the Gulf of Aqaba and far to the north to include the Jordan River valley and the Sea of Galilee. See Figure I.

The ancient city of Jericho, destroyed several times during warfare, lay at the western edge of the trench, north of the Dead Sea. It is thought that the "evil" cities of Sodom and Gomorrah were located near the Dead Sea. They and three other cities lay in the Vale of Siddim, described as a fertile plain--probably the floor of the trench. The biblical account of the destruction of these cities, perhaps around 2000 B.C., uses the words "brimstone and fire," which may be a reference to a local volcanic eruption. The floor of the valley may have subsided to become what is now the Dead Sea.

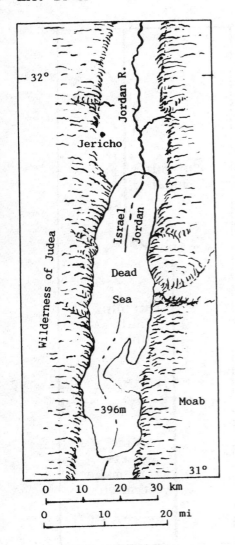

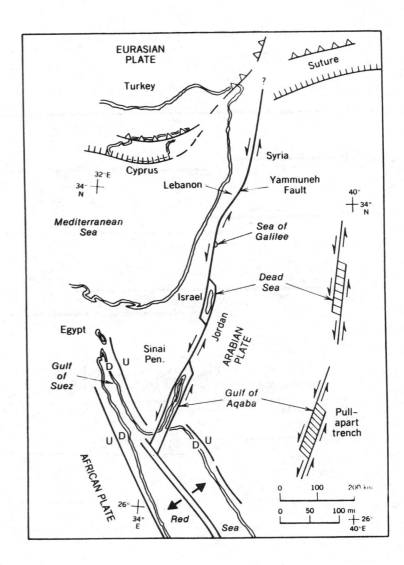

Figure H Sketch map of the Dead Sea trench. (A.N. Strahler.)

Figure I Tectonic sketch map of the western boundary of the Arabian plate.

Figure I is a rough map showing the setting of the entire western plate boundary of the Arabian plate. North of the Sea of Galilee, the transform fault has been named the Yammuneh Fault. It ends in the complex converging plate boundary with the Eurasian plate. The Arabian plate has moved more than 100 km northward, relative to the portion of the African plate to the west, sometimes referred to as the Palestine block.

We will propose the hypothesis that the Jordan-Dead Sea-Aqaba trench is a series of pull-apart trenches. First, however, we need to clear up a technical point about strike-slip faults in general. Relative motion on a strike-slip fault obviously occurs in one of two directions. Standing on one side of the fault line and looking across to the other side, the opposing block can move either to the right or to the left. On this basis, the fault can be described as dextral (moves to right) or sinistral (moves to left). These two are also called "right-lateral and "left lateral," respectively.

(13) In the space below, sketch with a line and two arrows (a) a dextral strike-slip fault and (b) a sinistral strike-slip fault. Examine both the San Andreas fault and the Jordan-Dead Sea fault. Under each sketch write the name of the fault with the corresponding motion.

a. Dextral
(right-lateral)

b. Sinistral
(left-lateral

Name: San Andreas   Jordan-Dead Sea

(14) Using only the evidence available on the map, Figure I, estimate the maximum northward displacement that has occurred on the transform boundary. Explain how you arrived at the estimate.

The northward motion is limited to the amount of spreading that the Red Sea has undergone since its rifting began. Measuring from the Red Sea spreading boundary NNE to the normal fault on the northeastern side of the Red Sea, we get a distance of about 150 km. (Geologic evidence from aeromagnetic data yields a displacement of 110 km.)

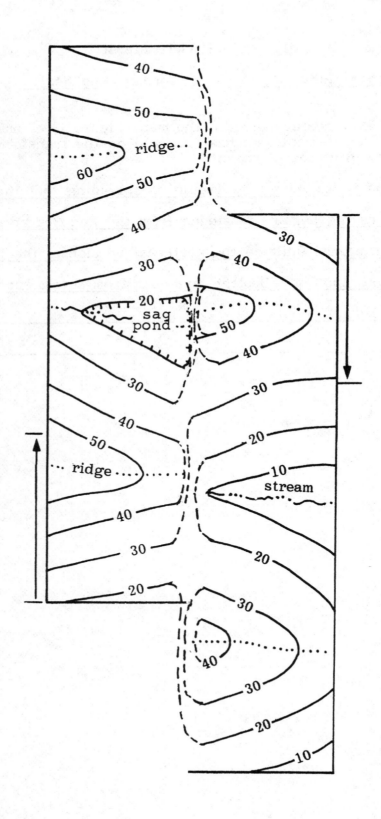

Name _____   Date _____

_____   _____

Exercise 14-A Exfoliation Domes of the Yosemite Valley
[Text p. 310-11, Figure 14.10.]*

Deep mining for ores is hazardous work, made particularly dangerous by what miners call "popping rock." Without warning, from the ceiling and walls of a tunnel, or "drift," a massive slab of rock bursts free, falling on the unwary victims. A similar thing occurs in rock quarries, where the quarry floor may suddenly jump upward, as a thick slab of granite rifts loose. This phenomenon, which we call unloading, is caused by a spontaneous expansion of solid bedrock, previously somewhat contracted in volume under great pressure of overlying rock.

We include unloading under the heading of rock weathering, even though it is not caused by atmospheric agents, solar heat, or ice. In nature, unloading accompanies the overall process of landscape denudation, causing sheeting structure in massive rocks, such as granite, and often producing exfoliation domes. Nowhere are such domes more wonderfully displayed as landforms than in the Yosemite Valley of Calfornia, site of one of our oldest and most popular national parks.

We will investigate the Yosemite domes with the help of a photograph (Figure A) and a topographic contour map (Figure B). Two excellent specimens of exfoliation domes are shown: North Dome and Basket Dome. Label these two domes on the photograph.

Take some time to compare the map with the photograph, noting how the photo features are expressed in the contours.

(1) What contour interval is used on the map?   100   ft

(2) To get an idea how big these domes are, calculate the difference in elevation between the summit of Basket Dome and the floor of Tenaya Canyon.

Basket Dome summit   7602   ft

Elevation of lowest contour in Tanaya Canyon   4100   ft

Difference: 3502   ft

(3) Lay out a trail from the floor of Tenaya Canyon to the summit of Basket Dome, using switchbacks (zig-gags) where needed, so that a rise of 100 ft uses no less than 300 ft of trail (horizontal measure). Start at point "P" on the stream. Draw your trail in pencil or pen directly on the map.

---

*Modern Physical Geography, 3rd Ed., p. 261, Plate G.2

Figure A North Dome and Basket Dome, viewed from the south. (Douglas Johnson)

(4) Find the steepest slope shown on the map and mark it with a short line drawn across the contours. Determine the least horizontal distance of this line in a vertical distance of 400 ft (four contour intervals). Using the blank graph, construct a triangle with these dimensions for legs and draw the hypotenuse. Using a protractor measure the angle that the hypotenuse makes with the horizontal. Enter your answer on the graph.

(5) Explain the succession of curious angular zigzag (sawtooth) bends in the contours at 0.5-0.4 and 0.2-0.2. Connect the nested zigzags in each group with a smooth curving line. Find these features on the photo and mark them with a color pencil or pen.

Each group of nested sharp bends represents a cliff surface that lies on the side of the dome. Each is the edge of a rock shell, part of which has fallen away, exposing the surface of the next shell beneath.

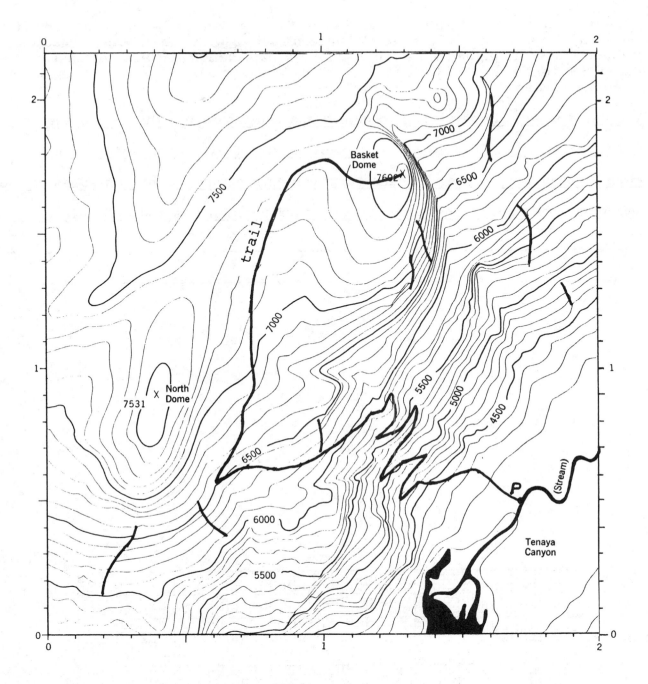

Figure B Portion of the Yosemite National Park topographic map, scale 1:24,000, U.S. Geological Survey.

(6) Find at least five other such groups of zigzag contour bends. Draw a color line along each. In the spaces below, give the map coordinates of each group.

<u>1.0-0.7</u>   <u>1.3-1.4</u>   <u>1.35-1.5</u>   <u>1.6-1.9</u>   <u>1.85-1.5</u>   <u>1.9-1.3</u>

(7) Note that the topography of the wall of Tenaya Canyon below about 6500 ft elevation is steep, rough, and blocky in contrast to with the smooth, broadly rounded slopes at higher elevations. Offer an explanation for this contrast. (Hint: See text Figures 18.5 and 18.8. Tenaya Canyon was formerly occupied by an alpine glacier.)

A valley glacier (alpine glacier) occupied the canyon during the Pleistocene Epoch, or Ice Age. It stripped away many layers of exfoliation shells from the lower walls of the canyon. The irregular contours depict the broken edges of the remaining rock shells. The upper limit of glacier action was about 6500 feet.

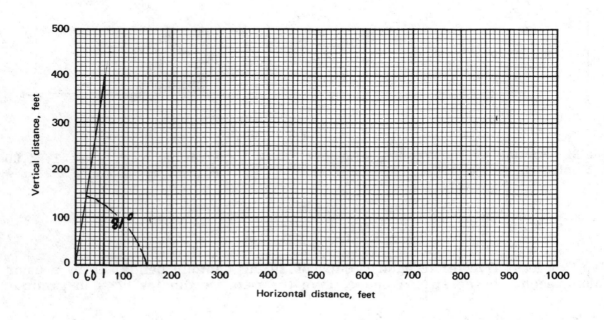

Name _____   Date _____

_____   _____

Exercise 14-B   Landforms of the Kentucky Karst Region
[Text, p. 312-15, Figures 14.15, 14.17, 14.18.]*

The karst landscapes of Kentucky and southern Indiana have been developed in a moist continental climate in areas where thick limestones of Mississippian geologic age lie at or close to the surface. Mammoth Cave, established as a national park in 1936, lies in southcentral Kentucky, a few miles northeast of Bowling Green. It is one of the largest cavern systems in the world, with about 150 miles (240 km) of known passageways. Five separate levels of passageways have been formed, the lowest of which carries an underground river, Echo River, that drains into the the Green River. Kentucky pioneers discovered the cave in 1799, and during the War of 1812 saltpeter deposits formed from bat dung were mined to make gunpowder.

Caverns don't show on topographic maps, but their presence at depth is often indicated by a karst landscape with numerous sinkholes. We begin by examining a small area of a contour map in a nearby area, using another of Erwin Raisz's fine block diagrams for guidance. We can then turn to a section of a quadrangle that includes a large part of Mammoth Cave National Park.

Sinkholes on the Pennyroyal

The sinkhole plain that lies south of Mammoth Cave has long been known as the "Pennyroyal." (Pennyroyal is an aromatic American mint with blue or violet flowers.) The small portion shown in Figures A and B is typical of this plain. You will also find it across the southern one-third of the map in Figure C, where the upland surface elevation is about 600 ft.

_____

*Modern Physical Geography, 3rd Ed., p. 262-63 , Figure 15.8.

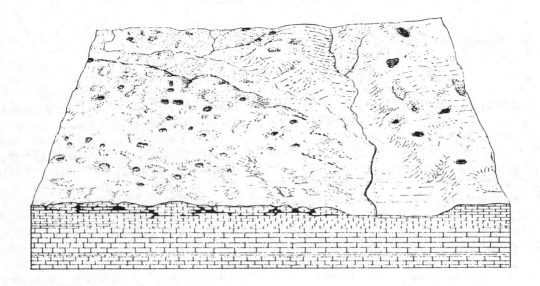

Figure A Block diagram of the area shown in Figure B (Drawn by Erwin Raisz.)

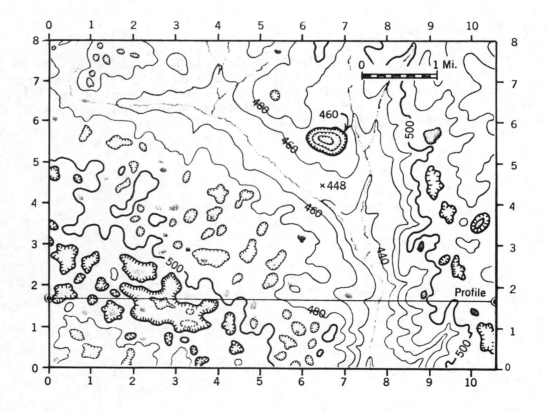

Figure B Contour topographic map of a portion of the Sinkhole Plain of southcentral Kentucky. Ponds are shown in black. (Princeton, Ky., Quadrangle, 1:62,500, U.S. Geological Survey.)

(1) What contour interval is used on the map, Figure B? (Refer to the numbered heavy contour and to the spot height at 6.5-4.4.)

Contour interval: __20__ ft

Depth of a sinkhole can be estimated from its depression countours. By "depth" we mean the difference between the lowest outlet point on the rim of the depression and the deepest point on the bottom of the depression. To illustrate this concept, the profile diagram below may be of help.

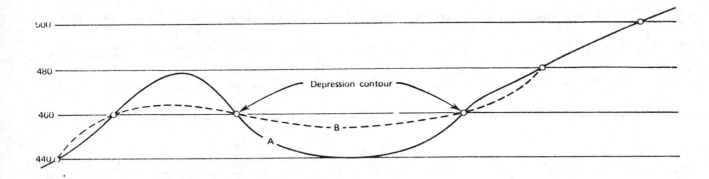

Given one set of contours, the true profile may approach Profile A (maximum depth) or Profile B (minimum depth). Thus we can only say for sure that the depresssion depth (d) is more than zero but less than 40 ft (two contour intervals). So we write: 0 < d < 4

(2) Estimate the depth of each of the sinkholes for which the location is given below by grid coordinates.

Grid coord.	Depth range
2.7-1.5	20 < d < 60
2.2-2.3	0 < d < 40
9.1-4.5	0 < d < 40

(3) On the broad divide located at 6.5-5.5, draw contours to show a sinkhole with depth more than 40 ft but less than 80 ft. Label the elevation of the outermost contour.

(4) (Extra credit) Using the blank graph provided (Graph A), construct an east-west profile from 0.0-1.6 to 10.6-1.6.

Abandoned Valleys of the Mammoth Cave Area

Turn now to a topographic map of the Mammoth Cave area, Figure C. It is on the scale of 1:62,500. The contour interval is 20 ft. (The regional map, Figure D, shows the location of the topographic map.) Make a careful study of the topography in the northern two-thirds of this map.

(5) Focus your attention on Woolsie Hollow in the center of the map. Describe this hollow and what lies in it.

The hollow is a deep valley, about 3 miles long in an east-west direction. It has branch valleys coming into the main valley axis from both sides. However, the hollow has no stream flowing in its floor, nor does it have an outlet. A chain of small sinkholes lies in the valley bottom.

(6) Describe the general pattern or trend of the deep depressions in this part of the map. Is there a typical "grain" that indicates a trend or direction? What does this pattern mean?

The short, steep side valleys on opposite sides of each depression come in at angles resembling the feathers of an arrow (<<<<<<<). They seem to be pointing generally toward the west and northwest, in the direction of the Green River. This suggests that there formerly existed here a system of connected open stream valleys.

(7) Why are there no flowing streams in these deep valleys? What history can you reconstruct for the valleys.?

After the system of valleys was carved by surface streams, the water of the streams began to be diverted down into holes ("swallow holes") to flow into vertical shafts and in the caverns below.

Figure C Portion of the Mammoth Cave, Ky., Quadrangle, 1:62,500. Ponds are shown in black. (U.S. Geological Survey.)

Contour interval 50 feet.

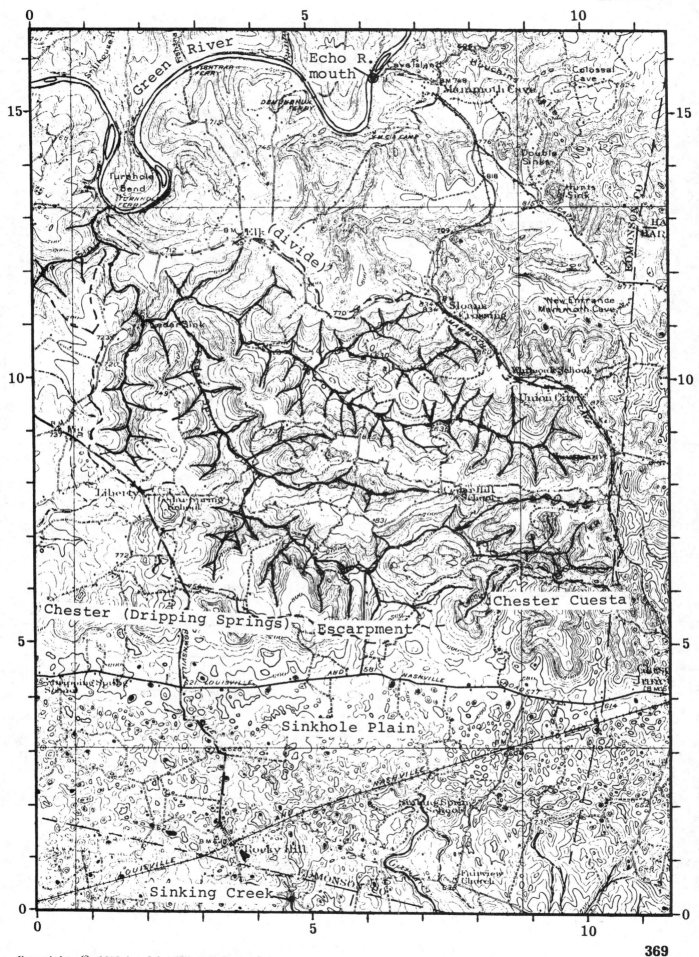

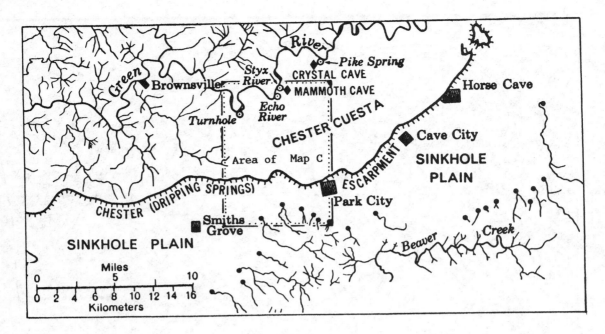

Figure D Outline map of the southcentral Kentucky karst region. (From W.B. White, R.A. Watson, E.R. Pohl, and R. Brucker, 1970, The Central Kentucky Karst, <u>Geographical</u> <u>Review</u>, vol. 60, p. 88-115. Copyright © by the American Geographical Society. Used by permission.)

(8) Can you work into your scenario the deep, winding gorge of the Green River? (Refer to text Figure 14.15.)

The cavern system developed and deepened as the Green River cut more deeply downward, lowering the ground water table. This left a cavern system above it to serve as a conduit system for the water entering the sinkholes in the valley floors.

(9) Why are the "empty" valleys now blocked by divides?

The valleys have continued to deepen by action of limestone solution and some water erosion in the short side valleys. This material has been carried down into the cavern system.

(10) Why are there very few sinkholes on the crests of the major ridges at elevations above 800 ft? (Refer to the geologic cross section in the accompanying block diagram, Figure E.)

The higher parts of the ridges are underlain by a sandstone formation, shown in stippled pattern on the cross section. This sandstone forms a caprock that is not susceptible to solution removal.

(11) Place a sheet of tracing paper over the topographic, Figure C. Tape it down securely along the top edge only. Trace the rectangular map boundary. Reconstruct as best you can the paths of the former surface streams flowing northwest to the Green River. Connect all the streams into a single branching network. Then, using a dashed line, draw in the drainage divide (watershed) between this stream system and those around it.

(12) Using information on the location map, Figure D, and the geologic cross section, Figure E, find and label on the topographic map, Figure C, the following features:

Green River

Chester (Dripping Springs) Escarpment

Chester Cuesta (see text p. 355, Figure 16.9)

Sinkhole Plain

Sinking Creek

Echo River (mouth)

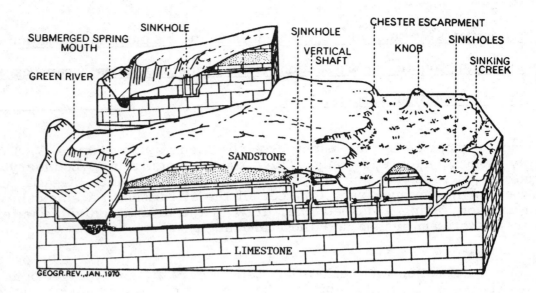

Figure E Block diagram showing the relation of topography to geology in the area covered by Figure C. (Same source as Figure D. Used by permission.)

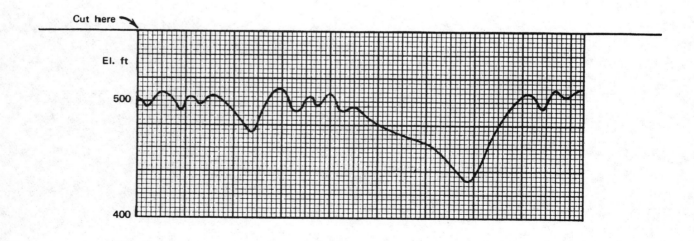

Graph A

Name _____   Date _____

_____   _____

Exercise 14-C The Great Turtle Mountain Landslide
[Text p. 317-18, Figure 14.26.]*

The great Madison Slide, described in your textbook, was rivaled in size and destructive force by a Canadian landslide that occurred 53 years earlier at another place in the Rocky Mountains--in the southwestern corner of Alberta. Named the Turtle Mountain slide, after the mountain peak that broke free in 1906, it destroyed a large part of the town of Frank, which lay in the valley at the base of the mountain. The accompanying drawing (Figure A), made from a photo taken shortly after the slide occurred, shows the mass of broken rock (left) and the great scar the slide left behind. The block diagram (Figure B) shows the scene from a different perspective. A detailed map of the landslide area (Figure C) was made during an investigation by geologists of the cause of the slide.

The great rock mass began its descent at a point over 3,000 ft (1000 m) above the valley floor. It quickly broke up into a rubble of huge boulders that roared across the channel of the Crow's Nest River. Engulfing part of the town, it killed and entombed about 70 persons. So great was the momentum of the mass that it traveled upgrade 400 ft (120 m) to reach a point about 1 mi (1.6 km) beyond the river bed. (Figure 14.26 of your textbook shows a similar Canadian landslide.) In terms of volume of rock in motion, that of the Turtle Mountain slide was about the same as for the Madison Slide, or perhaps a bit greater. Another section of the crest of Turtle Mountain (North Peak on the diagram) remains poised as a threat to the surviving part of the town. Similar geologic conditions favor another slide, which may come at any time.

Figure A Sketch from a photograph taken shortly after the Turtle Mountain landslide. The view is toward the south from a high point to the right (north) of Frank, as shown in the block diagram.

---

*Modern Physical Geography, 3rd Ed., p. 268, Figures 15-18, 15-19.

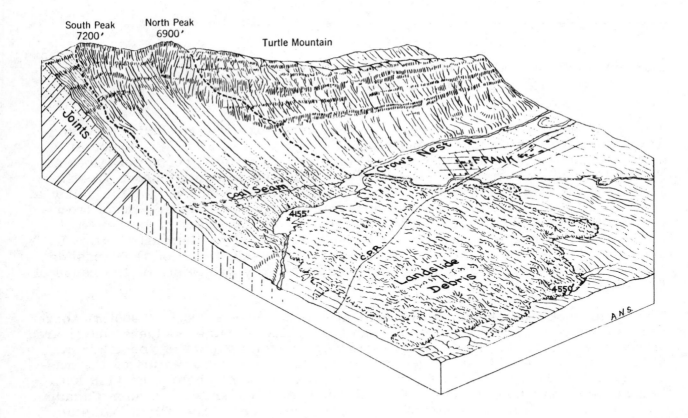

Figure B Block diagram of the area of the map, Figure C. (Drawn by A.N. Strahler.)

(1) Notice that different contour intervals are used on different parts of the map. Describe and explain this usage.

On the mountain, the interval is 100 ft; on the slide area east of the river it is

20 ft. The smaller interval is necessary to show details of the debris

topography.

(2) What is noteworthy about the appearance of the contours on the slide debris? Explain.

Contours on the debris are extremely crenulated (crinkled) on a small scale,

depicting the small mounds and large boulders that characterize the surface.

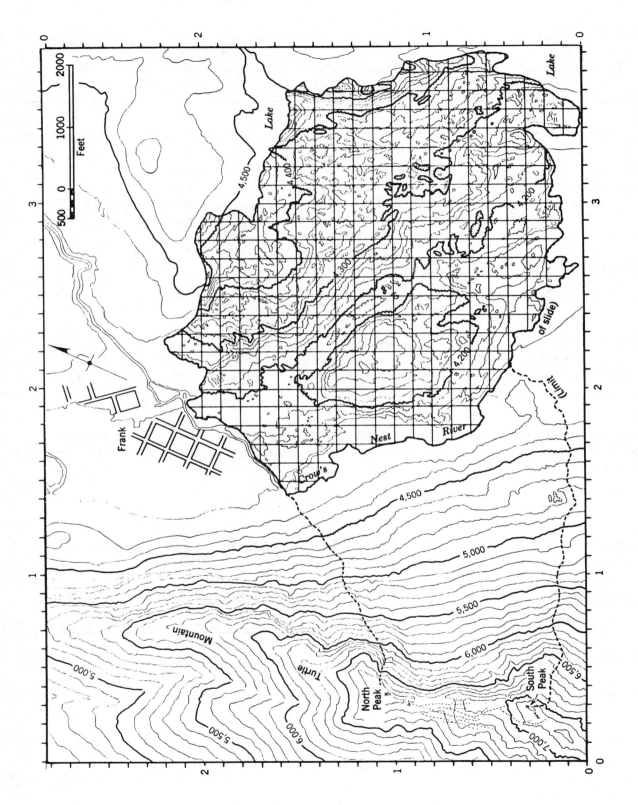

Figure C Contour topographic map of the Turtle Mountain landslide. (Canada Department of Mines, Geological Survey Branch, Memoir No. 27, 1911.)

(3) Numerous closed depressions, shown by hachured contours, are present in the area of slide debris. How did these features originate?

The depressions may lie over sections of former valleys that were blocked by the debris. Some may simply be random effects from the irregular motions and thicknesses of the moving debris.

_____

(4) Attempt an approximate measurement of the average thickness of the landslide debris east of the Crow's Nest River. Take the volume of this debris to be 30 million cubic yards. Draw in the 500-yd grid lines and count the squares within the slide limit. Include fractions of part-squares around the edges. For a more accurate answer, use 100-yd grid squares. Show your data below.

     Area of debris: 3,100,000   sq yds

     Average thickness of debris: 10   yds;   9   m

Geological features. Refer to the geological cross section on the block diagram, Figure B. Turtle Mountain consists of dipping (inclined) beds of limestone. Beneath the limestone beds are near-vertical beds of shale and thin sandstone layers. A seam of coal is shown.

(5) In what direction, and about how steep is the dip of the massive limestone beds? What is the strike of these beds? Give the approximate compass bearing. (Refer to text p. 352 and Figure 16.3.)

     Dip: 45   degrees.   Strike: about N 25 W

(6) Examine the line of contact between the limestone beds and the shale beds beneath. Describe this contact. What geologic event is indicated here?

The beds are not parallel, but make an angle at the line of contact. The arrow shows that the contact is an overthrust fault, in which the limestone mass has been thrust upward over the shale mass.

_____

(7) On the cross section, dotted parallel lines show a set of joints cutting across the limestone beds at about right angles. Could these joints have played a part in the occurrence of the great landslide? Suggest an explanation.

The joints are roughly parallel with the mountain face and are planes of weakness in the rock. The mass could have had its initial sliding surface along one of these joint planes.

(8) Investigators of the cause of the slide suggested that the mining of coal from the coal seam (indicated on the diagram) may have been a factor in setting off the slide. Speculate about how this process worked.

Removal of coal would have tended to remove support from the limestone mass above it. As a result, a small downhill movement may have first occurred along one of the dipping joint planes, predisposing the mass to start its rapid descent.

Name _____   Date _____

_____   _____

Exercise 15-A   Two Famous Waterfalls and How They Differ
[Text, p. 332-36, Figures 15.12, 15.13, 16.16, 15.17.]*

Waterfalls are young landforms with short life spans. If Nature abhors a vacuum, Nature also abhors a waterfall. The enormous concentration of kinetic energy at the brink of a fall guarantees that rapid lowering will occur, unless exceptional geological conditions are present. In this exercise, we examine two well-known scenic falls of North America. Waterfalls of the Yellowstone Canyon (there are two of them--Upper and Lower) lie in a gorgeous natural setting of colorful rock and a richly forested high plateau (text Figure 15.13). Niagara Falls, also in two parts--the Canadian, or Horseshoe, Falls and the American Falls--dominate the viewer by sheer power and enormous water volume. Both sets of falls were formed in late Pleistocene time, and both experienced involvement with glacial ice and crustal tilting. Our emphasis here is on these falls and their spectacular gorges as landforms produced by stream erosion.

Part A   Falls of the Grand Canyon of the Yellowstone River

The oblique air photo (Figure A), aimed upstream, shows the Lower Falls in the distance; The Upper Falls is out of sight around a sharp bend. To either side we see the undulating surface of a plateau ranging in the elevation range between 8,000 and 8,500 feet. The gorge and plateau are also shown on the topographic contour map, Figure B. The plateau came into existence by the outpouring of a great flood of rhyolite lava. As the plateau was tilted slightly toward the northeast, the Yellowstone River took its new course northeastward over the undulating lava surface. Canyon cutting began in the late Pleistocene at a point off the map to the northeast and rapidly progressed upstream.

(1) By comparing photograph and map, locate and label on the photo the following features: Lower Falls, Artist Point, Grandview Point, Inspiration Point, Point Sublime. Find and label the Yellowstone River above the Upper Falls.

(2) What contour interval is used on the map? About how deep is the river gorge at a point between Inspiration Point and Point Sublime? (Subtract lowest contour in gorge from contour on canyon rim.)

Contour interval: ___20___ ft

Rim elevation: __7,900__ minus __6,900__ equals __1,000__ ft

---

*Modern Physical Geography, 3rd Ed., p. 282-85, Figures 16.14, 16.16, 16.17.

Figure A The Grand Canyon of the Yellowstone River, Yellowstone National Park, Wyoming. (U.S. Army Air Force photo.)

(3) Is the gradient of the Yellowstone River steeper or less steep below the Lower Falls than above the Upper Falls in the SW corner of the map? What evidence can you cite?

<u>Much steeper below Lower Falls. Contour crossing occurs at distances of less</u>

<u>than 200 m below falls, but no contour crossings are shown in a distance of 1,000</u>

<u>m in the SW corner.</u>

Figure B Contour topographic map of the Grand Canyon of the Yellowstone River. Scale 1:24,000, The squares are 1 Km on a side. (U.S. Geological Survey.)

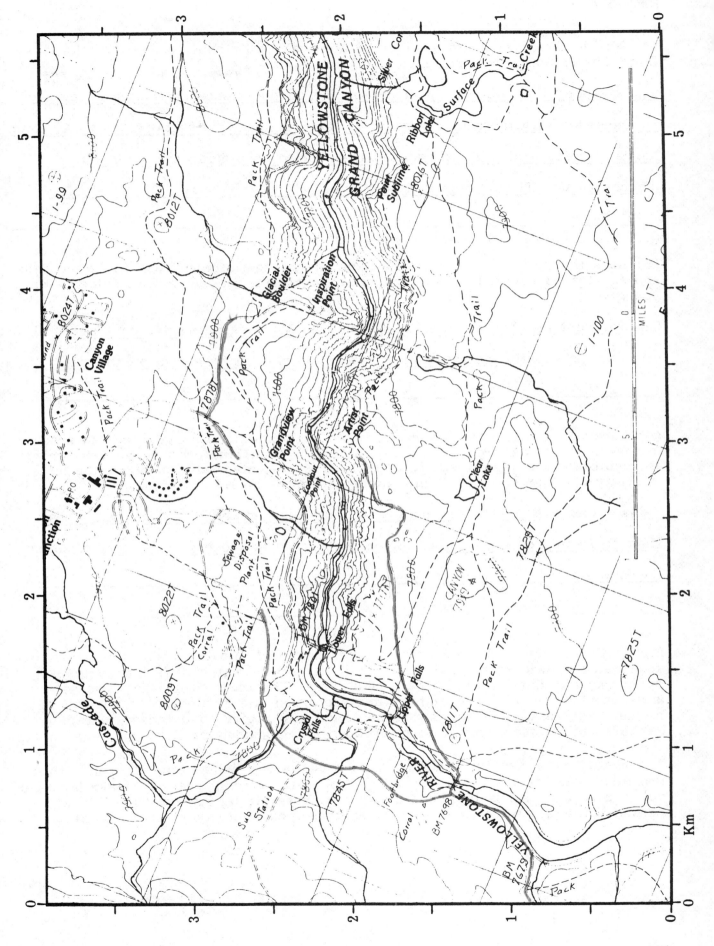

(4) Why is the river channel so broad above the falls, but so narrow below the falls? (Hint: Refer back to text p. 226 and Figure 10.10.)

The steeper the gradient, the narrower the channel, because the mean velocity, V, increases with increase in gradient, causing the cross-sectional area, A, (and hence also the width) to be reduced in order to pass along the same water discharge, Q. The formula is $Q = A \cdot V$.

(5) Does the gradient of the Yellowstone River, downstream from the Lower Falls and the rapids below it, steepen, lessen, or remain constant? What evidence can you give?

The gradient decreases downstream. Distance between successive contour crossings is greatly increased in the final 700 m of river shown on the map.

(6) Carefully examine the lower slopes of the Yellowstone gorge, both as shown in the air photo (A) and in text Figure 15.13. What landforms do you recognize as resulting from mass wasting? What do these features tell you about the intensity of river erosion? (Hint: Refer to Figure 14.3.)

Talus slopes are clearly shown as extending down to the river channel. These show that undercutting of the canyon walls is an active process, oversteepening the side slopes.

Figure C is a set of three schematic cross-sectional drawings. The uppermost (a) shows an idealized situation in which a high vertical waterfall is formed quickly by normal faulting across the path of a river. The rock within the cross section is assumed to be perfectly uniform in composition and hardness. As the fall retreats upstream, it is quickly transformed into rapids, and these are rapidly cut down and replaced by a smoothly descending graded profile.

The middle cross section in Figure C represents the Yellowstone Canyon with its two falls. Each fall is upheld by a layer of massive, dense lava. Weaker layers of volcanic rock materials lie above and below these strong layers. The falls will persist, although retreating, so long as these resistant layers are present.

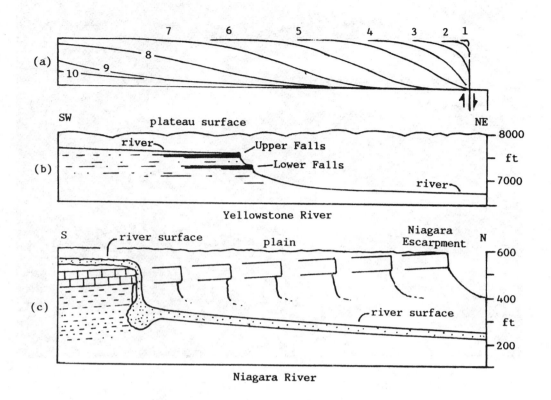

Figure C Schematic cross sections of the Yellowstone and Niagrara river gorges and waterfalls.

Part B   Niagara Falls

The two falls of the Niagara River are seen in an oblique air photograph (Figure D). Compare the features of the photograph with those shown in the text diagram, Figure 15.17. The topographic contour map, Figure E, shows the entire stretch of the Niagara River from above the falls to where it emerges from the Niagara Escarpment. Compare the map area with the block diagram, text Figure 15.16. With a color pen, sketch in the map area on the diagram.

The lower cross section of Figure C shows the geologic features related to the Niagara River gorge and its falls. The major feature is a massive layer of dolomitic limestone of a Silurian formation called the Lockport Dolomite. Beneath this resistant cap is a weak shale formation--the Rochester Shale--and below that a succession of interbedded sandstones and shales. On either side of the Niagara gorge the Lockport dolomite forms a cuesta that drops off abruptly along the Niagara Escarpment. The dip of the strata is gently southward from the escarpment, so that as the falls retreat, their elevation becomes lower.

Figure F reproduces an original drawing by the noted geologist, G.K. Gilbert, it shows his concept of the plunge pool of the Horseshoe Falls and the water currents responsible for eroding the base of the falls. (Note: Text Figure 15.17, drawn by Erwin Raisz, is in error for showing a deep plunge pool at the foot of the American Falls. Perhaps he did this because he couldn't show the plunge pool as lying at the foot of the Horseshoe Falls, where it belongs. Shouldn't we allow Raisz this bit of artistic license?)

Figure D Oblique air photograph of Niagara Falls. (Geological Survey of Canada, Atlas No. 101.)

(7) The Horseshoe (Canadian) Falls, as its name implies, has a deeply concave outline and is located at the head of the gorge. Moreover, a deep plunge pool lies below this curved rock wall. The lip of the American Falls is more nearly straight across (disregarding a great chunk of the caprock that fell out in 1954) and lies along the side wall of the gorge. Besides that, instead of a plunge pool, a great accumulation of limestone blocks lies at the base of the American Falls. Offer an explanation for these differences between the two falls.

Clearly the Horseshoe Falls is eroding and retreating more rapidly. This leads to the conclusion that the volume of flow (discharge) is much the greater over the Horseshoe Falls. (By international agreement, weirs placed upstream divert part of the river flow to the American Falls to supply adequate water power to the U.S. generating plants.)

Figure E Contour topographic map of the Niagara Falls and the gorge of the Niagara River. Scale 1:62,500. (U.S. Geological Survey, 1986.)

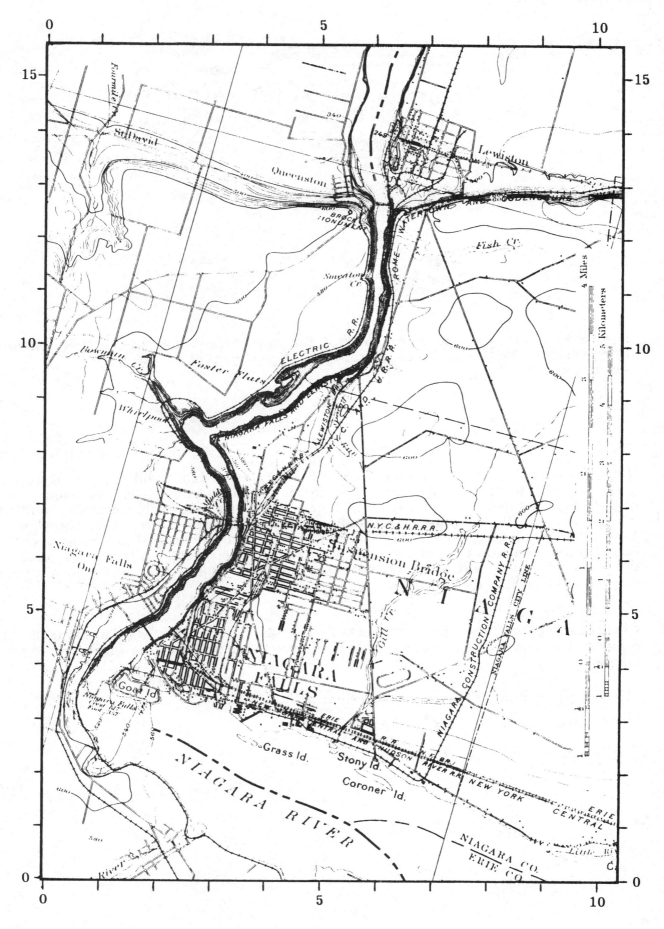

Figure F A cross-sectional drawing of the Horseshoe Falls. (From G.K. Gilbert, 1896, Niagara Falls and Their History, in <u>Physiography</u> <u>of</u> <u>the</u> <u>United</u> <u>States</u>, American Book Co.)

Rate of retreat of the Horseshoe Falls. In 1827, a British naval captain made a detailed drawing of the cataract at the lip of the Horseshoe Falls. He used a <u>camera</u> <u>lucida</u> which provides a fixed frame to get an extremely accurate rendition of the details. His drawing is reproduced here as Figure G. In 1842, the New York State Geologist, James Hall, made instrumental surveys of the same scene, showing that substantial retreat had occurred. Other precise surveys followed. A recession of 220 ft had occurred between 1842 and 1896, for an average of about 4.6 feet per year.

(8) Assuming that the above calculated rate has held since the falls began at the Niagara Escarpment, calculate how long it has taken to form the entire river gorge.

Length of gorge: __7__ mi, or __37,000__ ft.

Total time for gorge to form: __8,040__ years

Figure G Captain Basil Hall's 1827 camera lucida drawing of the Horseshoe Falls, seen from Forsyth's Hotel. (Same source as Figure F.)

(9) (Optional) Radiocarbon dates of beach deposits at Lewiston establish the date of start of the ancestral Niagara Falls at the escarpment as -12,000 years. (See text p. 420 for explanation of dating method.) Using that figure, calculate the average annual rate of retreat for the entire gorge. Comment on the large discrepancy between this rate and the one based on recession of the lip of the falls.

Average rate of retreat: _____ 3.1 _____ ft/yr

The observed rate of 4.6 ft/yr for 1843-1890 constitutes a very small sample and cannot be safely extrapolated to such a long period. Factors that might have caused much slower rates are reductions in river discharge and changes of thickness and resistance of the caprock from place to place in the limestone formation. [Note: As G.K. Gilbert recognized, and is now generally accepted, there were long periods in post-Valders and post-Algonquin times when much of the drainage of the Michigan and Huron basins was diverted through the Trent Valley, causing a dramatic drop in the Niagara discharge.]

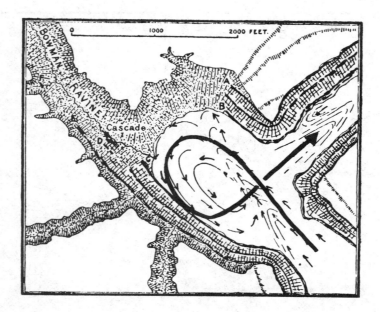

Figure H Gilbert's sketch map of The Whirlpool. (Same source as Figure F.)

The Whirlpool. Examine the strange right-angle bend in the Niagara gorge, where the gorge has been opened out into a cul-de-sac. The American geologist, G.K. Gilbert, in his report of 1896, gave the following explanation. We reproduce his sketch-map here as Figure H. He noted that the rock walls of the gorge disappear at the points A-B and C-D. Between those points, the northwest wall of the Whirlpool cavity is made up of loose sand, gravel, and boulders. In its retreat, the waterfall had encountered an earlier bedrock valley (of a pre-glacial stream) filled with alluvium. The large cavity was quickly created, but in time a mass of huge boulders--too large to be moved by the river current--blocked further expansion. Gilbert observed that the river current coming into the present bowl crosses to the opposite side, strikes the northern bank, and is turned left to become a reverse flow along the southwest bank. The water then sinks to the floor of the gorge and passes under the incoming surface current to exit in a northeast direction.

(10) Using a pen or pencil, draw arrows on the map of the Whirlpool to show current directions of the surface current system described above. Use small, short arrows. Then add a single continuous bold line to show how the current loops around and under itself to escape.

Name _____   Date _____

_____   _____

Exercise 15-B Alluvial Terraces
[Text p. 337-39, Figures 15.20, 15.21.]*

Alluvial terraces made life a great deal easier for white settlers in New England and the Middle West, who established new towns and carved out new networks of roads and railroads connecting them. Advances and retreats of the great North American ice sheets in the Pleistocene Epoch had made things both worse and better for the newcomers. The bad news was a cover of bouldery glacial till that defied the plow on the uplands. The good news was the filling of valleys with alluvium, carved by post-glacial streams into strips of flat land running for tens of miles at a stretch along the valley sides of the major streams. These natural pathways came already graded and ready for roads and railroads. They also provided prime agricultural land and well drained sites for towns. As a bonus, there were many points at which rapids and falls in the stream bed could be put to use to turn the waterwheels that powered grinding and textile mills.

River terraces also occur in the semiarid steppes of western North America, and those shown in Figure A are a prime example. They are found in valleys and basins never covered by Pleistocene glaciers, but the same climatic controls were probably responsible for them. With only desert shrubs to cover them, and no forests to hide them, these western terraces are a delight to the landscape photographer.

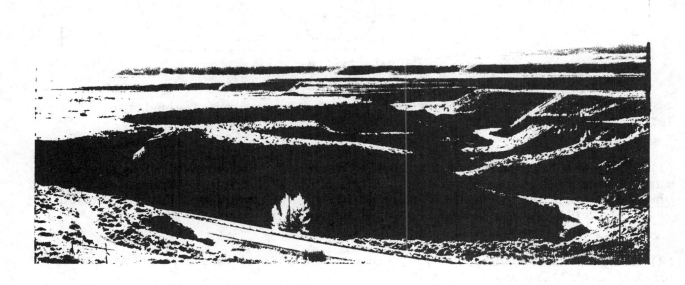

Figure A Terraces of the Shoshone River, west of Cody. Wyoming. (Frank J. Wright.)

*Modern Physical Geography, 3rd Ed., p. 285-86, Figure 16.19.

Alluvial terraces, pictured in text Figures 15.20 and 15.21, belong to one major class, or variety, of alluvial terraces. Another class, not presented in your textbook, is also important, and we think you should know about it. The first class is a sequence of <u>unpaired</u> terraces; the other, of <u>paired</u> terraces. These are compared in Figure B.

Part A   Unpaired Alluvial Terraces

The concept of unpaired (or unmatched) terraces is laid out in the upper diagram of Figure B. We call the flat land of the terrace itself the <u>tread</u>. The steep, undercut slope that bounds the tread is called the <u>scarp</u>. As in a staircase, each terrace tread is bounded on one side by a rising scarp and on the other by a descending scarp.

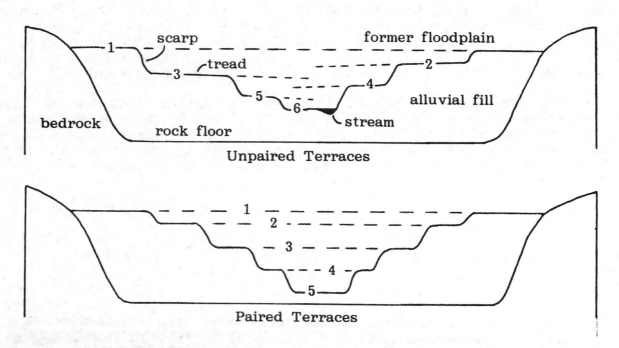

Figure B Unpaired alluvial terraces (above) compared with paired alluvial terraces (below).

(1) Study the text photo of New Zealand terraces, Figure 15.21. Identify as many treads and scarps as you can. Tape a piece of tracing paper over the text photo and mark the corners. Outline and label the treads and scarps. Transfer your completed tracing to the rectangle provided on page 399.

(2) Compare the two diagrams of Figure B. Do they have the same number of terrace levels? Explain.

<u>There is one more terrace level in the unpaired sequence. The assumption is that non-matching (unpaired) levels were formed at different times. Thus the time sequence of terrace-forming events alternates from one side to the other.</u>

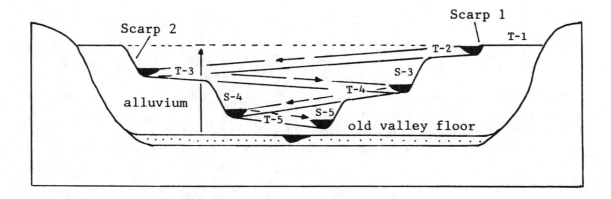

Figure C Schematic cross sectional diagram showing stages in the formation of unpaired alluvial terraces. (A.N. Strahler.)

(3) Figure C is a more detailed diagram of the development of unpaired terraces. Review and explain the plan of development shown here, starting with the former open valley with its bedrock floor.

Initial condition: <u>An open, flat-floored valley has been formed by the stream in a</u>

<u>graded condition.</u>

_____

Aggradational event: <u>Aggradation sets in, filling the valley with alluvium until</u>

<u>the valley floor is a floodplain at the uppermost level across the valley.</u>

_____

First downcutting event, including scarp and terrace formation: <u>As the river</u>

<u>channel cuts slowly down, the channel shifts across the valley to the opposite</u>

<u>side, leaving behind scarp S-1 and terrace T-1. Scarp S-2 is higher than S-1.</u>

<u>Terrace T-3 is lower in level than T-2.</u>

_____

(4) Text Figure 15.20, diagrams B and C, are reproduced here as Figure D. Identify and label the terraces and scarps in the manner shown in Figure C. (i.e., "T-1, T-2," etc. "S-1, S-2", etc.). Where space is tight, place the label outside the diagram and run an arrow to the feature.

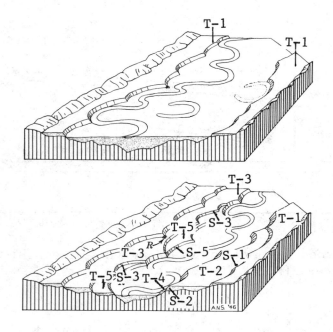

Figure D Formation of unpaired alluvial terraces (A.N. Strahler.)

(5) Special think-tank question. In Figures B and C, each of the unpaired terrace scarps is shown to be located closer to the mid-line of the valley than the previous one on the same side. The obvious implication is that each time the stream shifted across the valley it moved a shorter distance laterally than in the previous move. Can you offer two hypotheses, each of which will explain why this pattern of events occurred?

(a) As the stream was cutting its way down through the alluvium, its volume (discharge) was diminishing and its power to cut sideways was decreasing. (b) The bedrock floor beneath the alluvium was not flat and deep as shown in the diagrams, but was closer to the surface, as shown in text Figure 15.20. With each sidewise sweep, the stream encountered hard bedrock, which prevented it from further widening its floodplain.

Figure D is a synthetic topographic contour map, not representing any real area, but serving as a diagram on which terrace features are idealized. It illustrates certain typical features of New England river terraces. The contour interval is 20 ft; the scale is 1:17,000.

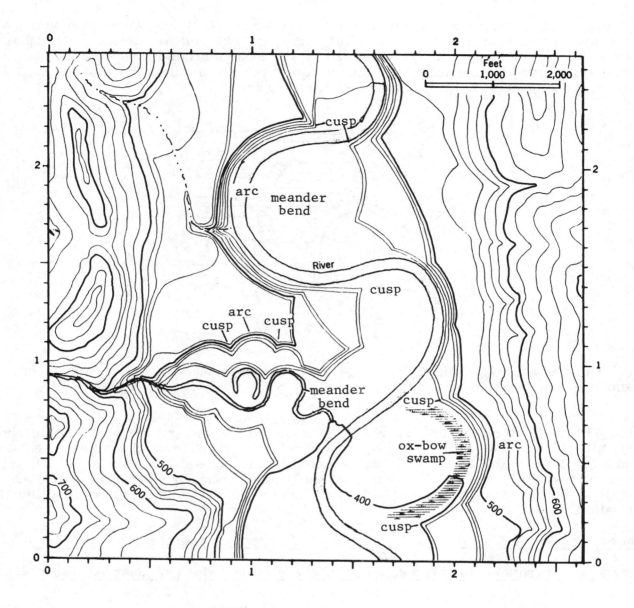

Figure E Contour topographic map of imaginary alluvial terraces. (A.N. Strahler.)

(6) Determine the height of the terrace scarp at each of the following points. Give limiting values. For example, "20 < h < 60" means "height greater than 20 ft but less than 60 ft." (Conceivably, the scarp might extend upward almost a full 20 ft and downward almost another full 20-ft. Although not likely, the possibility needs to be allowed for.)

Scarp at 1.1-2.2:   __80__ < h < __120__

Scarp at 1.6-1.7:   __20__ < h < __60__

Scarp at 1.4-1.0:   __60__ < h < __100__

Where terrace scarps are being carved by a stream that has well developed meanders, the scarps tend to take the form of <u>arcs</u>, concave toward the valley axis. These arcs are fitted to the size of the meander bend (more or less). Adjoining arcs are separated by sharp <u>cusps</u> pointed toward the valley axis. The accompanying map-diagram illustrates these terms:

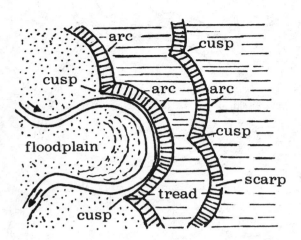

(7) On the map, find and label three good examples of an arc, its two cusps, and a closely related meander bend.

(8) Study the terraces and scarps on either side of the tributary stream in the vicinity of 1.0-0.8. Note the close match between the size of the stream meanders and the size of the arcuate terrace scarps north and south of the stream. Then note the size of the meander bends of the large stream in comparison with the arcuate terrace scarps it has produced. Make a general statement to describe this relationship.

<u>Meander size, measured by the radius or diameter of the curve, is related in</u>

<u>some direct manner to the discharge of the stream. Width of channel also</u>

<u>increases with discharge, so that meander size will also increase with increasing</u>

<u>channel width as shown on the map.</u>

(9) What is the origin of the semicircular swamp extending from 1.7-0.2 to 1.8-0.8? Name this feature and label it on the map. Find and label another similar floodplain feature.

<u>This is an oxbow swamp within a cutoff-meander of the main stream. An ox-bow</u>

<u>lake located at 1.0-0.9 was formed by a meander cutoff of the small tributary</u>

<u>stream.</u>

(10) Why does the small stream have such a steep gradient at 0.5-0.9, but such a low gradient at 0.9-0.8.

The stream is flowing over resistant bedrock at 0.5-0.9, whereas it is on a

floodplain at 0.9-0.8. [Note: The gradient is somewhat less at 0.1-0.9, upstream

from the rapids.

(11) Why does the contour line at 1.95-1.9 bulge westward in this place instead of bending east in a sharp V as the higher contours do just east of it? What landform is present?

This contour reveals a small alluvial fan built upon the terrace by the small

tributary.

(12) If deep pits were to be dug in the vicinity of 1.0-1.2, what composition and texture of material would be revealed, and why?

Probably sand and gravel at least 100 ft thick would be found here. It was

deposited by a braided meltwater stream during the process of deglaciation of the

region.

Part B   Paired Terraces

Figure F is a composite cross section showing several stages of alternate valley excavation (degradation) and filling (aggradation), giving rise to paired (matched) terraces. The sequence starts with a broad, open valley in which a bedrock floor is veneered with alluvium spread by the graded stream that occupied the valley (Level O). Vertical arrows indicate alternating aggradation (up) and degradation (down) in a numbered sequence.

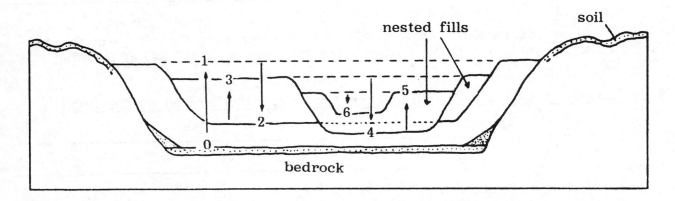

Figure F Stages in the development of paired alluvial terraces and nested fills. (A. N. Strahler.)

The alternation of aggradation and degradation might be attributed to various causes, including cycles of climate change from arid to humid, or crustal tilting. A favored explanation is that the cycles are locked into alternating glaciations and interglaciations (explained in text Chapter 18, and shown in Table 18.1). Briefly, the idea is that during a glaciation, sea level fell many tens of meters, causing streams to deepen their valleys and excavate alluvium, whereas, duing interglaciations, sea level rose to previous levels and the streams responded by aggradation and valley filling. The result, as shown in Figure F is a complex cross section including <u>nested</u> <u>fills</u>.

To analyze the steps shown in Figure F, you are asked to take the changes step by step, using the cross sections provided on following pages. In the blank spaces below each section, enter the following information:

> Flood-plain level (same as number in circle)

> Activity occurring between this stage and previous stage (i.e., aggradation or degradation)

Label the level on the cross section and draw a vertical arrow from the previous level. To enhance the cross sections, you may wish to color each age of fill with a different color of your choice. Apply the same colors to Figure F.

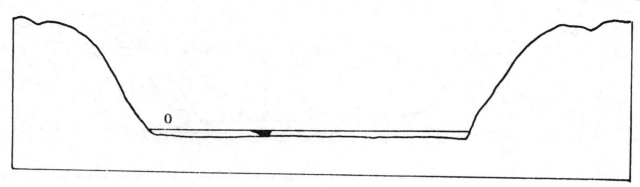

Stage a. Level: __0__   Activity: (none)

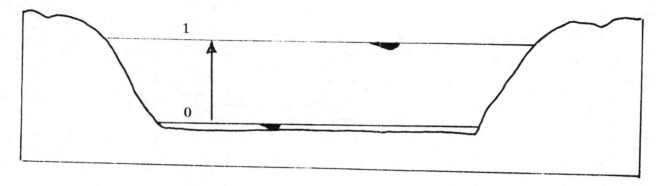

Stage b. Level: __1__   Activity: _aggradation_

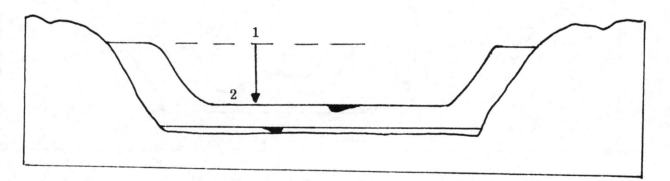

Stage c. Level __2__   Activity: _degradation_

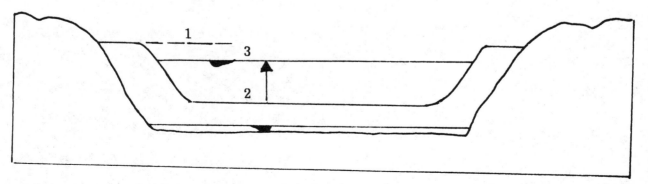

Stage d. Level __3__   Activity: _aggradation_

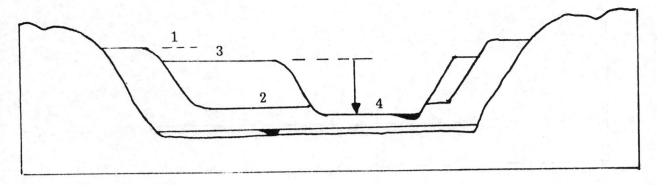

Stage e. Level ___4___   Activity: __degradation__

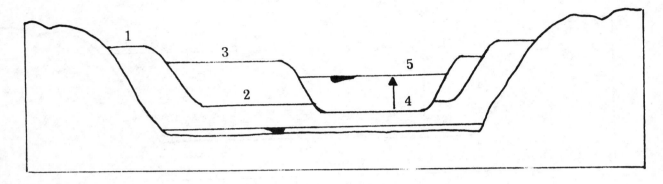

Stage f. Level ___5___   Activity: __aggradation__

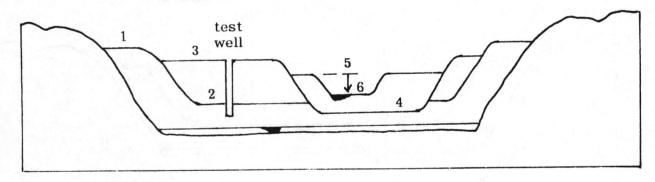

Stage g. Level ___6___   Activity: __degradation__

(13) (Optional) The base of each nested fill is completely concealed from surface observation. Suppose, however, that as a research geologist you drill an exploratory well at the location shown in the cross section for Stage g. Samples of the alluvium are taken for every foot of depth. What sequence of materials might you expect to encounter as the drill passes through the base of the upper nested fill and enters the top of the underlying alluvium?

Well-washed and sorted sand and gravel would make up the upper fill. The top of the older underlying alluvium would probably consist of fine muds and clays, possibly with organic matter, representing the overbank floodplain deposits of the meandering stream that flowed on Level 2. Below this fine-textured layer would be more sand and gravel produced during the earlier period of aggradation.

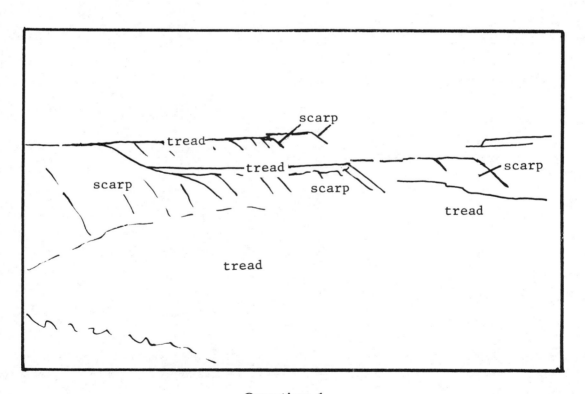

Question 1

Name _____   Date _____

_____   _____

Exercise 15-C Floodplains and Their Meanders
[Text p. 334, Figures 15.12, 15.14; p. 340-43,
Figures 15.22. 15.23. 15.24.]*

Alluvial meanders, those snakelike bends of a river channel that require it to
travel so long a distance to make its way to the sea, build the character of the
entire floodplain. As meander loops grow, are cut off, abandoned, and replaced
by new ones, the river shifts back and forth over the floodplain, constantly
reworking the alluvium and shaping it into new landforms. As much as engineers
have tried to eliminate those meander bends and replace them with straight
reaches, they just won't stay straight. Nature establishes a steady river regime
that, for all its continual change and seeming inefficiency, holds the total length
of its meandering course more or less constant. Wouldn't it be better for humans
to learn to live with that natural regimen?

Part A A Floodplain in the Making

We examine first a floodplain in a rather early stage of its development, much
like that shown in Block C of text Figure 15.12. and in stage D of Figure 15.14.
Our contour topographic map, Figure A, is a portion of the Kanawha River valley
in West Virginia. The floodplain is easily identified as a belt of land free of
contour lines. It contrasts strongly with the surrounding hills, which rise to
heights as much as 500 feet above the floodplain.

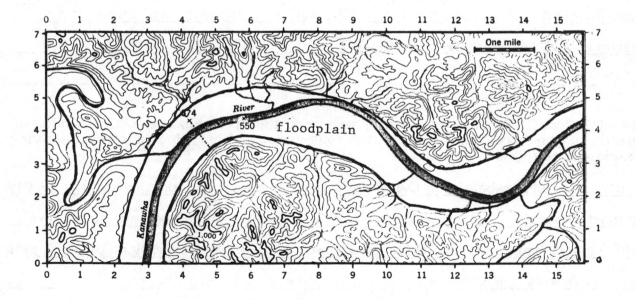

Figure A Map of a portion of the St. Albans, WV, Quadrangle. Scale 1:62,500.
(U.S. Geological Survey.)

*Modern Physical Geography, 3rd Ed., p. 287-89, Figures 16.21-16.24.

EX. 15-C

(1) Using a color pencil or pen, draw lines to show the limits of the floodplain. Label the floodplain. Measure the width of the river channel and of the floodplain. Make your measurements near the elevation number "574."

    River width: _250_   yds

    Floodplain width: _1,400_   yds

(2) In what direction is the Kanahwa River flowing? State the evidence for your answer. (North is to the right.)

Flow is toward the right (north). Evidence lies in the acute angle that a tributary makes when entering the river. The acute angle points upstream. There are exceptions on this map, but the large tributary at the far left makes a convincing case.

(3) From map data alone, can the gradient of the river be measured? Explain what problem is involved here.

Only a single elevation figure is given on the river itself (550 ft). No contour crosses the river within the limits of the map. We can't make a determination of the gradient. (A maximum value can be set by using the map length of the channel and the contour interval.)

(4) (Optional) Compare the size of bends of the river channel with the size of bends of the floodplain, i.e., of the valley itself. What kind of stream activity is suggested by this relationship?

Within the larger bends of the valley, the river swings from one margin to the opposite margin. The river bends are thus shorter in wavelength than those of the valley. This suggests that the river is widening its floodplain by impinging on the foot of the valley wall at the outsides of the river bends.

402

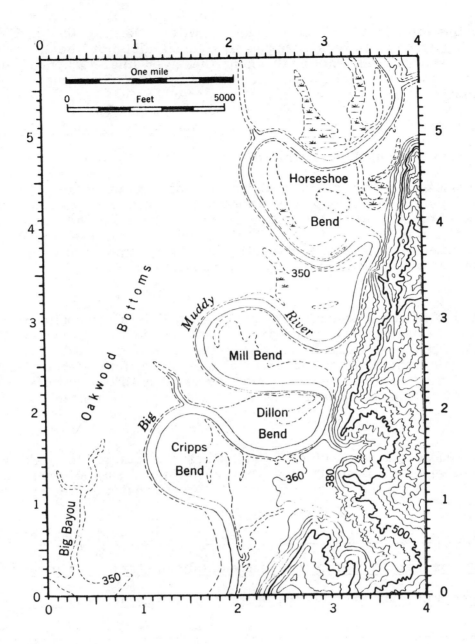

Figure B Map of a portion of the Gorham, IL, Quadrangle. Scale 1:24,000. (U.S. Geological Survey.)

Part B   Meanders on a Floodplain

Fully developed meanders on a broad floodplain are illustrated in the accompanying map, Figure B, of the Big Muddy River in Illinois. The channel lies close to the valley bluffs on the east, while to the west there extends a broad, flat floodplain, Oakwood Bottoms. The river channel is shown by the shaded band. An additional contour (350 ft) brings out details that would otherwise be lost. (Note also that three 20-ft contours have been omitted between 380 and 500 ft.) The gradient of Big Muddy River cannot be calculated from this map alone, but is on the order of 2 ft per mile, or less.

(5) Measure the total length of river channel within the map limits. This can be done by setting the points of a compass to a short distance, such as 1000 ft, and swinging the compass back and forth along the river line. Compare this figure with a straight line from start to finish. Give the ratio of channel length to airline distance.

Channel length: __42,000__ ft.   Airline distance: __18,000__ ft.

Ratio, channel length to airline distance: __2.33__ to 1

Note: The Mississippi River between Cairo, IL, and Baton Rouge, LA, held to a more or less constant river length of about 850 mi from 1765 to 1930, despite having 19 major cutoffs and many minor channel shifts. For an airline distance of 470 mi, the ratio was about 1.8 to 1. The higher ratio for the Big Muddy River is partly explained by the fact that section of river shown was specially selected for its display of a sequence of nicely formed meander loops.

Channel changes in the Mississippi River. Figure C is a composite map of a portion of the Mississippi River showing its channel in four surveys. The first, by Lieutenant Ross, was made in 1765; the last shown was made 135 years later. During this period the river was free to change its course without human intervention. Then, in the 1930s, a great series of artificial cutoffs was begun by the U.S. Army Corps of Engineers to shorten and straighten the river course.

(6) Study the Moss Island meander bend, which was cut off in 1821. How wide is the river channel here? Describe how the narrowing of the meander neck took place. In what part of the neck did most of the channel shifting take place, and in what direction?

The channel is is about 0.8 mi wide. Narrowing of the neck took place largely by the downvalley movement of the bend immediately upstream, whereas the channel below the bend was shifted only westward, but not downstream.

(7) Study the growth of the two new meander bends indicated by the letters a, b, and c on the map. What event seems to have set off the growth of bend a?

The cutoff of the Moss Island bend would have made a very tight turn in the river, which would have rapidly enlarged into a more open meander curve shown in the 1881-1893 survey.

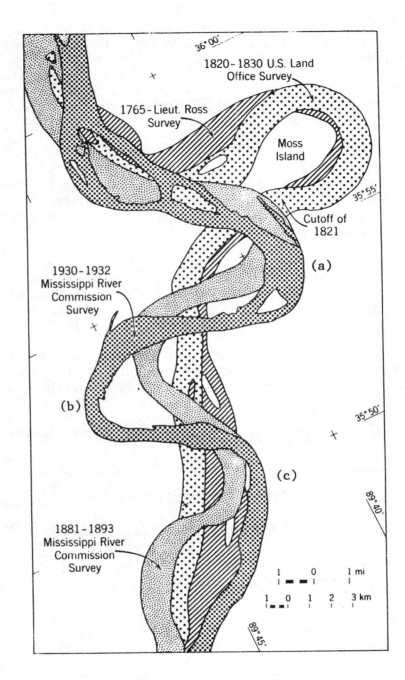

Figure C Four surveys of the Mississippi River superimposed. (After U.S. Army Corps of Engineers.)

(8) Downvalley motion of bends, accompanying their outward (sideward) growth is typical of alluvial river migrations; it is called <u>downvalley sweep</u>. (Consult Figure E for diagram of what happens.) What evidence of sweep can you find in the growth of bends <u>a</u>, and <u>b</u>? About how much distance of sweep is indicated and in what period of time?

In each west-east (east-west) channel reach that connects one meander bend to the next, the channel has moved sideways down valley. Where bend a connects with bend b, the amount of sweep is about one mile, occurring in about 4 decades.

Oxbow lakes and marshes. Figure D is a vertical air photograph taken at an altitude of about 20,000 ft (6,100 m). It shows the meandering channel of the Hay River in Alberta, replete with cutoff bends and the resulting oxbow lakes, which in turn have been partly obliterated by new bends and their remnants filled by marsh (bog) vegetation. Compare what you see with the color photopgraph, text Figure 15.23.

(9) Fasten a sheet of tracing paper over the photograph and trace the map border. Using a color pen or pencil, trace the course of the present river channel across the photo. Using the same color, trace the outlines of what appear to be oxbow lakes (open water surfaces). In another color, trace all segments of older, filled channels, presumed to be bogs.

(Solution on p. 412.)

(10) Locate the most recent cutoff; label it as C-1. Locate a point at which the next cutoff is due to take place; label it as C-2. Find a likely candidate for the next cutoff and label it C-3.

(Solution on p. 412.)

(11) Study the surface features lying inside the narrowed bend in the lower right portion of the photo. What is the significance of the set of fine concentric curved lines lying within the bend? Trace some of these lines to establish their pattern. (Compare with the small inset photo of text Figure 15.24.)

The lines show bar-and-swale topography, consisting of low ridges (bars) and intervening narrow troughs (swales). These are like growth lines on the shell of a clam, and each may show the annual increment of growth accompanying the spring flood. (These features are called point-bar-deposits, shown in Figure E.)

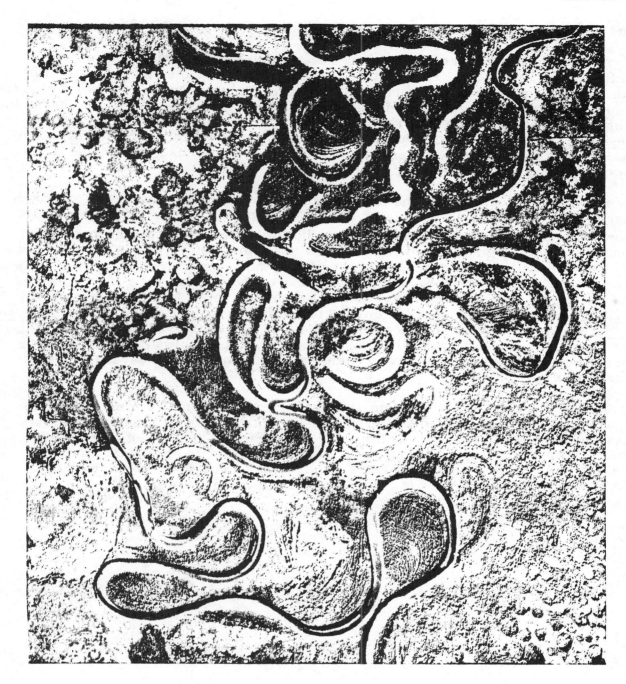

Figure D Vertical air photograph of the Hay River, Alberta. (Lat. 58°55'N; long. 118°10'W.) (National Air Photo Library, Surveys and Mapping Branch, Canada Department of Energy, Mines and Resources. Photo No. A1-5183-38.)

Part C   Submerged Landforms of the River Bed

A great meandering river such as the Mississippi has a special kind of configuration of its bed, related to its meander bends. Safe navigation of Old Man River by stern-wheel steamboats depended on quite precise knowledge of the channel--where the bars lay, and where the safe deep pools were located. Samuel Langhorne Clemens described these problems in his classic book, Life on the Mississippi (1883). That he chose "Mark Twain" (the call of two-fathoms) as his nom-de-plume testifies to his adaptation to that treacherous river in his apprentice days as an aspiring steamboat pilot.

To assist river boat pilots, the Mississippi River Commission prepared detailed charts of the channel on a scale of 1:20,000. Figure F is part of such a chart. It shows topographic contours on the river bottom; they are given as negative numbers (minus signs) to tell depth below mean low water (M.L.W.), which is a level established as the average height of low-water gauge readings over a 40-year period. Our map shows the M.L.W. line on both sides by a heavy dashed line; in terms of depth contours, it has a value of zero. Contour lines above M.L.W. are on an entirely different system, giving values in feet above mean sea level (M.S.L). In this reference frame, M.L.W. has an elevation of 35 ft. This system sounds strange, but it is very practical, because what counts in river navigation is depth below a locally standardized river surface level (M.L.W.) recognizable by pilots.

To help you interpret and explain the bottom contours on the chart, refer to Figure E. Arrows show where the threads of fast surface current are located. As the two cross-profiles show, a deep pool lies near the outside of each bend, where there is a steep undercut bank. From one bend to the next, the fast current shifts across to the opposite bank. This is the dangerous crossing, where the river-bed profile is shallow and has numerous bars.

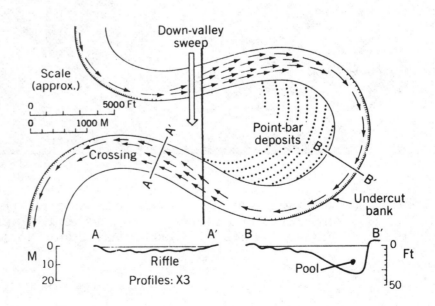

Figure E Details of a meander bend. Arrows show surface currents. (A.N. Strahler.)

Figure F Contour map of the channel in a portion of Rodney Bend of the Mississippi River. Scale 1:20,000. Mean low water (M.L.W.) is based on the mean of 40-year low-water gauge readings, 1891-1930. (Mississippi River Commission.)

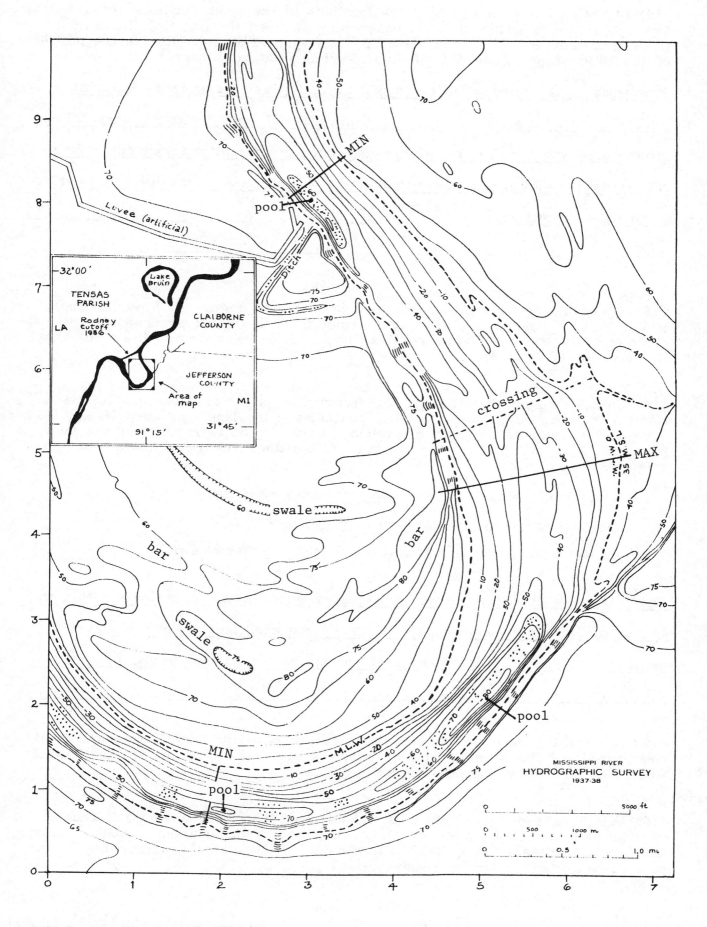

MISSISSIPPI RIVER
HYDROGRAPHIC SURVEY
1937·38

(12) Study carefully in Figure E the locations of the pools and crossings. Do the two successive crossings lie on a center-line between the limits of the bends? Draw that center-line. (It lies near the bold vertical arrow.) How is the position of the bank-hugging current affected? Explain what you observe.

The crossing is "delayed" to a position beyond (downstream from) the center-line. The bank-hugging current also continues well past the center-line. These effects are explained by the down-valley slope of the entire floodplain, which causes down-valley sweep. A small fraction of the force of gravity acts in the down-valley direction.

(13) Study the bottom contours on the contour map, Figure F. Find the deep pools. Using a color pencil or crayon, color in dark red all areas within the -60 contour. Label each of these areas as "Pool". Then color orange the areas that lie between -60 and -40 ft.

(14) We have stated that the M.L.W. dashed line is the zero contour for the depth contours, but is also the 35-ft contour for the land contours. We will now convert the value of the deepest bottom contour to the same units as the land contours. Use the spaces below for your calculations. What interesting fact does this calculation reveal?

Elevation of M.L.W. line: ___35___ ft above mean sea level (M.S.L.)

Deepest bottom contour: ___-80___ ft below M.L.W.

Bottom contour relative to M.S.L.: ___-45___ ft below M.S.L.

The bottoms of the pools lie well below mean sealevel. All of the area within the -40 bottom contour is below mean sealevel. (This condition of below-sealevel elevation exists as far upstream as 470 miles above the river mouth.)

(15) Study the bottom contours in the middle part of the map. Locate and label the "crossing." To do this, find by trial and error that line across the channel where the shallowest current threads are located.

(16) Using the two M.L.W. lines as the limits of the channel, where is the channel narrowest? Where widest? Draw lines across the channel at these points and label "MAX" and "MIN" (two MIN places). (Disregard the narrow bay in the M.L.W. line where a small stream enters.)

(17) What relationship can you establish between channel width and bottom depth? Explain in terms of the channel cross-section and stream discharge. (Use the information on text p. 226 and Figure 10.10.)

Clearly, the narrowest channel widths correspond with the deep pools at the outsides of the bend, while the greatest channel width is in the region of the crossing, where depths are least. The deeper and narrower channel allows the same discharge to occur as for a shallower and broader channel, according to the equation Q = AV.

(18) Carefully study the land contours bordering the M.L.W. line. Look for evidence of natural levees. (Refer to text p. 340-41 and Figures 15.22 and 15.24.) Describe the evidence and give specific locations (grid coordinates) of at least three good examples.

Evidence lies in closed contours shaped as long, narrow elliptical areas parallel and close to the river banks. From these, land elevation descends in a direction away from the river. Examples: (a) 75-ft contour at 5.5-1.7. (b) 80-ft contour at 4.5-4.5. (c) 75-ft contour at 2.4-8.1.

(19) Examine the region of the map centered at about 3.0-3.0, within the river bend. Compare what you find with the features shown in text Figures 5.23 and 5.24. What kind of topography is represented here? Find and label a bar and a swale. How can the swale be identified on the contour map?

The curved contours describe point bar deposits, consisting of bars and swales. A good example of a swale is located at 2.0-4.5, as shown by a depression contour.

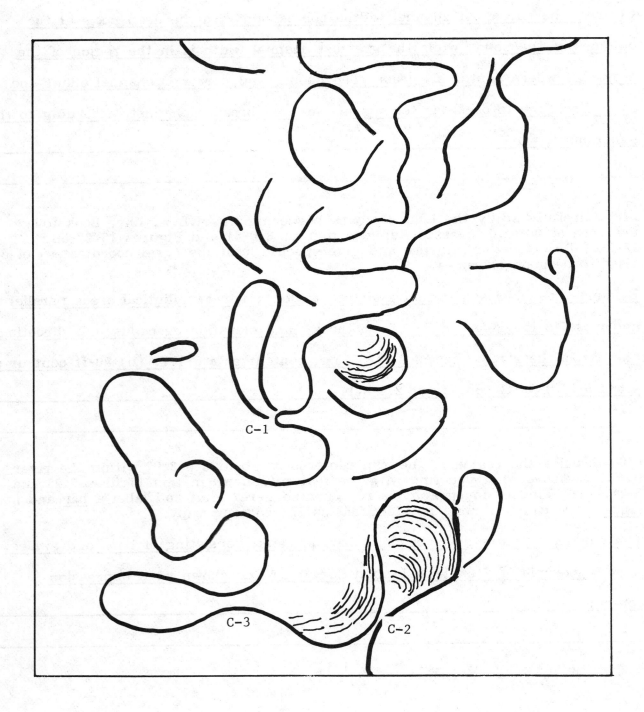

Name _____   Date _____

_____   _____

Exercise 15-D   Entrenched Meanders and Their Goosenecks
        [Text p. 209, Figure 9.28.]*

When is a meander not a meander? A river can appear to have floodplain meanders when you see only a map of its sinuous, looping channel, i.e., as shown on a planimetric map, such as that in text Figure 1.21C. But perhaps when you examine a topographic contour map of the same river, you discover that the channel lies at the bottom of a steep-walled gorge, carved in bedrock. One particularly good example is the San Juan River in Utah, pictured in Figure A. The tortuous bends of its gorge were named "goosenecks" by the local people. It would be quite correct to say that the river gorge itself has meanders--it's a meandering gorge.

Another striking example is the winding gorge of the Mosel River in western Germany, pictured in text Figure 9.28. What you see is a great sweeping curve toward the left, as the river passes out of sight in the distance. The steep outside wall of the curve, seen at the right, resulted from the undercutting action of the river on the outside of its huge bend. Mostly, it supports vineyards, yielding the fine wines for which the region is famous. The meandering gorge runs for some 35 mi (60 km) as a crow flies, with a dozen big loops. One of the most picturesque of these is shown in Figure B, reproducing a portion of an old German topographic map that uses hachures to depict the slopes of the land surface. Holiday travelers taking in the scenery by river boat were allowed to disembark at the foot of the narrow meander neck (just north of Punderich), and to climb to the narrow divide, where they found a delightful tavern. Fully refreshed, they descended the north slope to the river bank, to reboard their waiting craft. That was a pleasant way to perform a "meander cutoff." (Notice that the railroad executes the same maneuver by means of a tunnel through the neck.)

The sinuous bends of a river gorge are called <u>entrenched</u> <u>meanders</u>. The implication is that they originated as floodplain meanders, as pictured in text Figure 15.22. The larger overview of what happens is shown in text Figure 15.18, in which the floodplain meanders of Block D respond to landmass rejuvenation by trenching their valleys into the underlying bedrock, as shown in Block E. If conditions are just right, the meandering pattern is retained, as shown in Figure C. Of course, the meanders continue to grow as they incise the rock, and eventually a cutoff occurs. In rare cases, a cutoff leaves a <u>natural</u> <u>bridge</u>. One well known example is Rainbow Bridge in southeastern Utah.

---

*Modern Physical Geography, 3rd Ed., p. 290-91, Figures 16.28, 16.29.

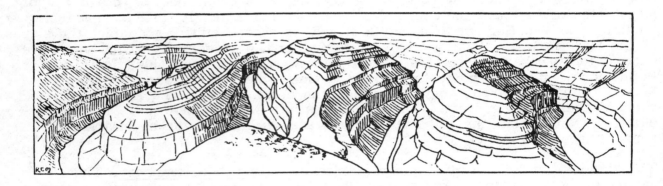

Figure A Oblique air photo of the Gooosenecks of the San Juan River. The meander belt is a little over one mile wide and the canyon is about 1,300 ft deep. The strata are of Carboniferous age. (Spence Air Photos.) Below: A field sketch of the same subject by Raymond C. Moore. (U.S. Geological Survey.)

(1) Find and label the following features in Figure C. Place the label outside the drawing and run an arrow to the feature: entrenched meander, recent cutoff, natural bridge.

Figure B Portion of Sheet 504, Cochem, Germany, 1886. Original scale 1:100,000.

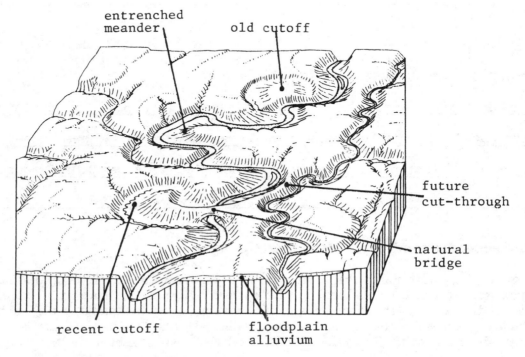

Figure C Block diagram of entrenched meanders and related landform features. (Drawn by Erwin Raisz.)

(2) Near the rear of the diagram (Figure C) find evidence of a meander cutoff that took place in the geologic past. Describe the feature. What evidence is there that this feature was a former course of the river?

A circular valley is shown, with radius comparable to that of the other valley meander bends. A central hill is clearly shown as the former upland within the bend. The present river channel is shown to be well below the level of the circular valley, indicating a long lapse of time since the cutoff occurred.

(3) (Special credit question.) Gaze into your crystal ball and find a spot on the diagram, Figure C, where a geomorphic event of great importance is foreseeable. What is the event and what will be its consequences? Mark and label the place on the diagram.

At one spot the winding valley of the smaller stream at the right comes very close to that of the main stream. As the meander bend of the latter enlarges, it will cut through the barrier between the two streams. Because the smaller stream runs at a higher elevtion, it will be diverted into the larger one. The downstream reach of the smaller stream will be abandoned. (This is a form of stream capture known as intercision.)

(4) In the cross section at the front of the block diagram, Figure C, the artist has shown a thin layer of loose sediment capping the upland surface. What interpretation do you offer for this layer?

This is a layer of floodplain deposits (sand, silt, clay) formed by the meandering river at its original level prior to the entrenchment. Where present, it is field evidence of that previous cycle in which the river was fully graded.

Entrenched Meanders of Mahoning Creek

Figure D is a topographic contour map of an area in the Appalachian Plateau Province of western central Pennsylvania. Above the map is a carefully drawn block diagram of the same area. Take some time to compare and match features on the diagram with those on the map. The stream on the left is Mahoning Creek; that on the right, Redbank Creek. Label these streams on the map.

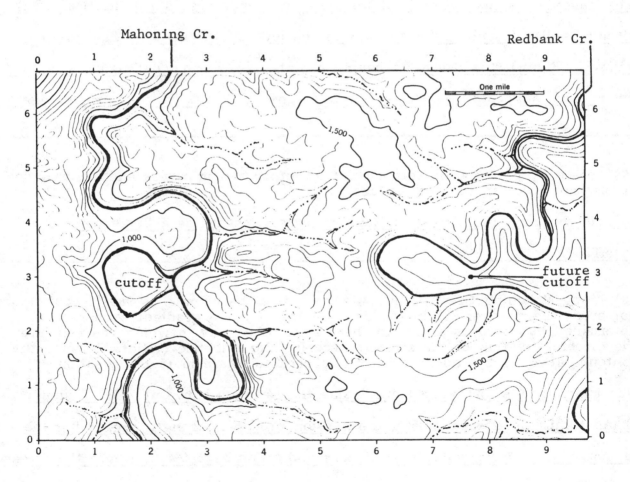

Figure D Entrenched meanders of the Appalachian Plateau Province. Above: Block diagram by Erwin Raisz. Below: Portion of Rural Valley, PA, Quadrangle, scale 1:62,500, U.S. Geological Survey.

(5) Do the two streams appear to be quite similar in size (discharge), or do they differ greatly in size? Give your evidence.

Size of meander bends (measured as radius of curvature) appears quite similar in both sets of entrenched meanders, although possibly those of Redbank Creek are somewhat smaller on the average. In any case, they are streams of comparable discharge.

(6) Find and label a meander cutoff. Draw the former course of the stream in the meander loop. Compare this example with that in Figure C, Question 2. What is the important historical difference in these two events? Give evidence.

The Mahoning Creek cutoff is of more recent occurrence, since the floor of the circular valley is only a few feet above the level of the stream. The circular valley in Figure C is drawn as having a high, steep scarp between it and the river. This suggests a much greater difference in level, hence a greater age.

(7) Locate by grid coordinates a point on either stream where another meander cutoff may occur in the near future.

The narrow neck at 7.7-3.0 on Redbank creek appears as a likely point of cutoff.

(8) Along Mahoning Creek, the contour next to the channel is 980 ft, but it does not cross the channel within the limits of the map, so gradient cannot be determined by contours alone. Is there some other way in which you can infer the direction of flow of the stream? Describe the evidence. (Hint: Examine the contours at 2.2-1.2, and at 2.5-4.0. Go back to Exercise 15-B, Question 8.)

The long, gentle northward slope inside the meander curve at 2.2-1.2 shows channel shift northward during the incision process, strongly suggesting the phenomenon of down-valley sweep. Same indication at 2.4-4.2, where the stream is undercutting its north bank in an east-west section of channel. This, too, is a clear sign of down-valley sweep. Clearly, the stream flows northward.

Exercise 15-E   Alluvial Fans of Southern California
      [Text p. 345-46, Figures 15.28, 15.29, 15.30.]*

Alluvial fans are especially important landscape features for Californians who dwell in a special piedmont belt that runs west to east at the base of the great San Gabriel mountain block. Towns and cities sited on this series of alluvial fans begin on the west with La Crescenta, La Canada, Altadena, and Pasadena, then on through Glendora and Cucamonga (skipping a few), and ending on the east with San Bernardino. In the 1930s the western section, close to Hollywood, was prime suburban real estate, and many fine homes were built on the higher ground. Orange groves and vineyards occupied much of the eastern belt, then considered remote from Los Angeles. Migrants in a great stream, including many from the Dust Bowl, gazed in wonder at this idyllic landscape as they drove west across the alluvial fans on old Highway 66. However, things were not all that good on the fans. During winter rains, terrible floods of water and rock debris could spill across a fan, wiping out homes and roads and, if you were more fortunate, leaving a mixture of mud and boulders reaching halfway up the side of your house. Today the unending stream of newcomers to California uses Interstate 10, parallel to old 66, but they are puzzled by the strange odor of L.A. smog, and in its murk they rarely get a glimpse of the mighty San Gabriel range lying so close by.

Saved from such urbanization are the gigantic alluvial fans of Death Valley. They have been protected through the establishment of Death Valley National Monument and by nature through its hot, dry climate. In this exercise, we look first at a fan in Death Valley, then turn our attention to the alluvial fans of the Cucamonga district.

Part A   The Death Valley Fans

The oblique air photo, Figure 15.29 in your text, shows the general relationship of the Death Valley fans to the deeply dissected mountain range which feeds them with water and debris It shows well the braided channels of the main fan-building streams and the myriad of smaller channels that are carving up the older fan surfaces. For closer study, we refer to Figure A, another oblique air photo, but taken nearer to the mountain base. To assist you in finding specific features on the photo, we use a letter/number grid. The Panamint Range in the background rises to summit elevations of 10,000 to 11,000 ft (3,000 to 3,300 m). Surface elevations in the foreground are close to zero (sea level). Hanaupah Canyon is the main watershed on which the photo is centered.

*Modern Physical Geography, 3rd Ed., p. 297-301, Figures 17.8-17.11.

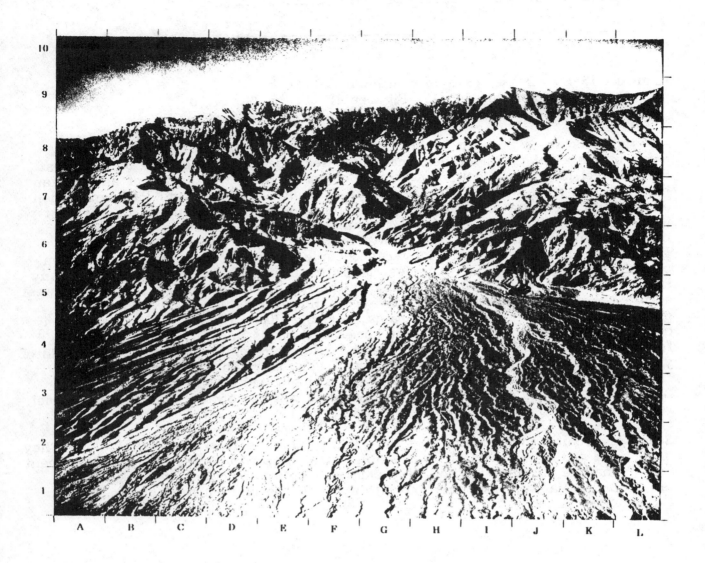

Figure A The Hanaupah Canyon alluvial fan in Death Valley, Inyo County, California. The view is west toward Telescope Peak in the Panamint Range. (Spence Air Photos.)

(1) Follow the course of the main channel from its mouth at G-6 to E-3. Describe this channel and its boundaries.

The channel is broad, flat-floored, and bounded by steep banks. The braided pattern of the bars in the channel floor is clearly shown. The active part of the channel is a narrower zone within the broad trench and appears to swing alternately from one bank to the other.

(2) What conspicuous change does this main channel undergo downstream from about E-3? Describe what you see.

The channel broadens greatly, forming a fan-like expanse of braided channels. This appears to be a new or modern section of the fan, where water spreading and alluvial deposition is occurring.

(3) Another conspicuous channel can be identified at J-3. Where does this channel originate? Why doesn't it join with main channel?

This secondary channel originates at the mouth of a smaller canyon (near I-6) that lies to the right of the main canyon mouth. The water flow must follow the radius of the fan, so it diverges from the main channel.

(4) Between the two channels referred to above, there is a broad area of fan surface (vicinity of H-3) scarred by many narrow "wiggly" ravines. Where do these minor channels arise and what geomorphic activity are they performing?

These channels originate on the fan surface, fed by direct runoff from storm rainfall on the fan. They are carving the fan into deep narrow valleys, and are receiving tributary channels as they flow down the fan.

(5) Where else on the fan are similar minor channels present?

To the left of the main channel, in the vicinity of D-4, are similar deeply-cut valleys of minor channels.

(6) What feature can you identify at F-6? Describe the activity taking place here.

A small alluvial fan has been built into the main channel near the mouth of the canyon, but has been cut back by the main stream, forming a high undercut bank or cliff.

(7) Summarize as best you can the geomorphic history of this fan as related to the features you have identified. What sequence of events can you postulate?

We see here the remains of an older alluvial fan built on a higher level and with a steeper gradient than that of the present main channel. Since this earlier fan surface was constructed, the main stream has entrenched itself into the head of the fan. Cut off from the supply of water and debris of the main stream, the old fan surface has become a surface of erosional removal. The main stream is now bulding a new fan with its head farther out on the fan and at a lower level.

Part B   Alluvial Fans of the Cucamonga District

We investigate the alluvial fans of the Cucamonga district by means of two contour topographic maps. The first is an older map on a smaller scale (Figure B); the second is a more recent map on a larger scale (Figure C).

Map B shows the fan of San Antonio Creek, a comparatively large canyon with a large watershed area. The concentric contours on the fan are well developed. To the right (east) is the fan of Cucamonga Creek, a much smaller canyon and watershed, for which the fan form is only weakly developed.

(8) Channels of San Antonio and Cucamonga creeks are shown by dotted lines on the map. Compare the upper reaches of these fan channels with their lower reaches, paying special attention to the form of the contour lines as they cross the channels. Describe and interpret what you observe.

The upper reaches consist of a single broad channel of box-like form, with steep, parallel sidewalls. This is evident in the broad notch made by each contour as it crosses the channel. The lower reaches consist of numerous branches in a radiating pattern. These are narrow and V-shaped in cross section. Evidently, the streams have trenched the upper part of each fan and are evolving newer fans at a lower level and farther out from the mountain base.

(9) Compare your interpretation with that given for the Hanaupah fan of Death Valley (Figure A).

They are similar in having cut trenches into the upper parts of the fans and developing newer, lower fans farther out at a lower level.

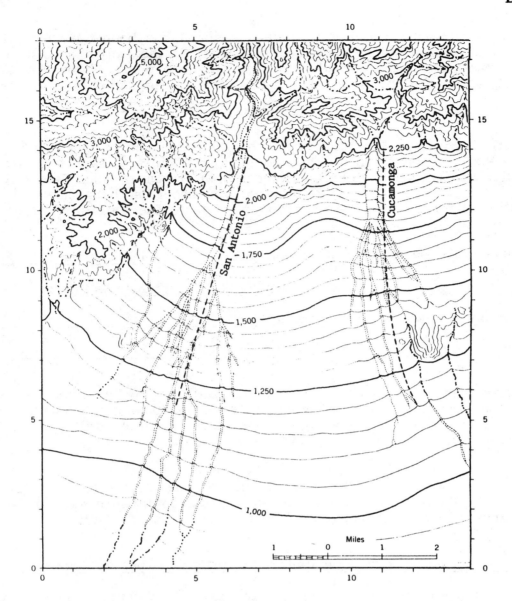

Figure B Topographic map of the San Antonio and Cucamonga alluvial fans at the foot of the San Gabriel Mountains, Los Angeles County. (Cucamonga Quadrangle, 1:62,500. U.S. Geological Survey.)

(10) Study the spacing of contour lines on the San Antonio and Cucamonga fans, starting at the canyon mouth and traveling south to the limit of the map. What changes occur in this distance?

In a general way the spacing increases from north to south, showing that on the average, the gradient of the streams is decreasing as distance from the canyon mouth increases.

(11) For each stream, a north-south line (dashed) has been drawn on the map. Using a compass or dividers set to 0.5 mi, start at the canyon mouth and work southward, estimating the elevation at each half-mile point. Enter the data in Table A. Use the contour given in the first line of the table as your starting point.

(12) Plot the elevation/distance data of the table columns on the blank graph, Figure D. Plot the points first, then connect the points of each set with straight-line segments, producing a longitudinal stream profile. Select two different colors for these lines. How would you describe the geometrical form of these two profiles? How do the differences relate to the watershed areas, also given in the table?

The profiles are upwardly concave (concave-up) in form. The degree of up-concavity is greater for Cucamonga, which has the smaller watershed area, than for San Antonio, with the larger watershed.

Special assignment, for advanced study. To continue our investigation and reinforce the tentative conclusion derived from the above two examples, we look for more fan profile data. Your table gives data for three additional fans in this area: Deer Canyon, Day Canyon, and Lytle Canyon.

(13) Study the map of the Lytle Canyon fan, Figure C. First, label the fan contours in bold numerals. (Note that the contour interval is 80 ft.) Do the contours show a decreasing surface gradient from canyon mouth outward toward the fan base?

Yes, the contour spacing widens uniformly, provided that we disregard the contours at the fan apex and start with the 1840-ft contour. This change is even more clearly shown than for the Cucamonga and San Antonio fans.

Plot the profiles of Deer Canyon, Day Canyon, and Lytle Canyon on the graph, Figure D. We now investigate the relationship between profile form and three other measures: (a) elevation drop in the first mile, (b) watershed area in square miles, and (c) average diameter of the largest fragments in the stream bed at the apex. The data for (b) and (c) were determined by a geologist, Rollin Eckis, in a pioneering quantitative study of alluvial fans of the Cucamonga district. His figures are given in Table A for each of the five fans.

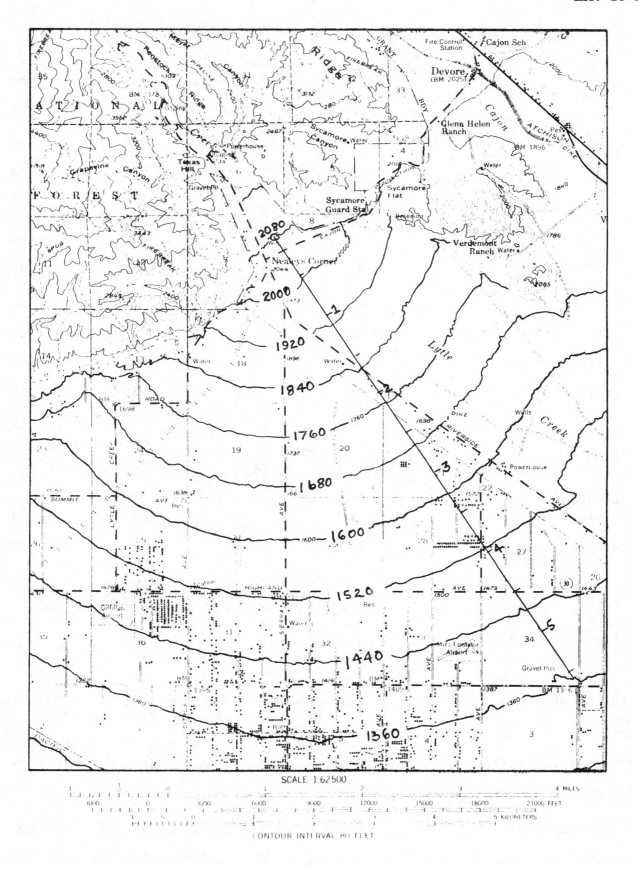

Figure C Topographic map of the Lytle Canyon fan, San Bernardino County. (San Bernardino Quadrangle, 1:62,500. U.S. Geological Survey.)

Table A   Alluvial Fans of the Cucamonga District

Distance, miles	San Antonio Canyon	Cucamonga Canyon	Deer Canyon	Day Canyon	Lytle Canyon
0.0	2250	2300	2750	2350	2080
0.5	2080	2075	2450	2120	2000
1.0	1960	1925	2140	1900	1935
1.5	1840	1775	1910	1700	1860
2.0	1720	1670	1680	1540	1780
2.5	1620	1550	1525	1430	1705
3.0	1525	1450	1420	1370	1640
3.5	1430	1360	1340	1310	1575
4.0	1325	1280	1280	1250	1515
4.5	1240	1220	1225	1205	1465
5.0	1160	1150	1175	1170	1415
Elevation drop, 1st mi.	290	375	610	550	145
Watershed area, mi^2	26.2	10.6	3.4	4.9	47.9
Boulder diam., in., at apex*	69	90	157	112	87

*Median length of 10 largest boulders in distance of 50 yds. along channel.

(Data sources: Rollin Eckis and U.S. Geological Survey.)

(14) On the table, enter the elevation drop within the first mile for all five fans. Compare your figures with the figures for watershed area. What relationship do you find? How do you explain this relationship?

The smallest elevation drop corresponds with the largest basin area; the largest elevation drop with the smallest basin area. The same relationship holds in sequence for all five basins. Explanation: Stream gradients as a rule decrease downstream as a watershed area (and hence also the discharge) increases. The greater the stream discharge, the more efficient is the stream, hence requiring a lesser gradient in the graded condition.

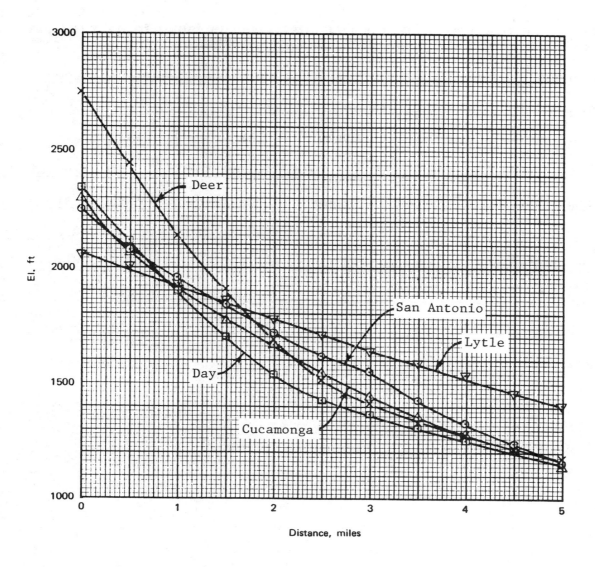

Figure D Graph of relationship of elevation to distance for five alluvial fans of the Cucamonga district.

(15) Compare the profile curvatures of the five fans. Which is the most strongly curved; which the least? Then relate this information to the steepness of the gradient in the first mile. Is there a correlation?

Most strongly up-concave is Deer Canyon, which has the steepest gradient in the first mile. Least strongly curved is Lytle Canyon with the smallest elevation drop. This relationship agrees with that discovered in Question 13. A small watershed area is associated with small stream discharge, requiring a steep gradient for the transport of the bed load. Stream profiles tend to flatten with increasing distance downstream.

(16) Study the relationship of fragment size (given in table) to both fan gradient and watershed area. What principle is demonstrated here?

Largest fragment size at the apex belongs to the smallest watershed (Deer Canyon), which as the lowest gradient. Smallest fragment size correlates with largest watershed (Lytle Canyon). The other three examples fall into order in the same way. The smaller the stream and its discharge, the less is its capability to move fragments of a given size. Consequently, the smallest stream has the largest fragments lying immovable in its bed. (These large boulders were put in place by mudflows or have descended from adjacent channel banks or steep ravines.)

Name _____   Date _____

_____   _____

Exercise 16-A   Mesas, Buttes, and Cliffs
           [Text p. 352-55, Figures 16.4, 16.5, 16.6.]*

An arid climate leaves much of the land surface barren of large plants. Consequently, we find that vertical cliffs, flat-topped mesas and plateaus, and isolated buttes stand out boldly, giving us some of the finest scenery in the American west. Torrential rains rapidly remove the soft, weak shales from under the edges of hard caprock layers, creating badlands that resemble mountains in miniature.

Part A   Buttes and Mesas of Monument Valley

Monument Valley, found in the Navajo Country of Arizona and Utah, ranks close to first as a landscape of buttes and mesas. Figure A shows the two "most photographed" buttes in the valley, if not in all the world. Called the Mitten Buttes--for obvious reasons--they are even recognizable as a matching left-hand and right-hand pair. That on the far right, Merrick Butte, is more conventional; it is one large rock mass. All three buttes are remnants of a single thick layer of massive sandstone that once extended over the entire area. Hollywood quickly discovered Monument Valley and capitalized on its beauty.

(1) Study the vertical sides of the three buttes in the photograph. What do you see that gives information on the process of cliff retreat and butte formation?

The massive sandstone formation shows vertical cracks that are joints dividing

the layer into a system of natural vertical rock columns. As the sandstone is

undermined, these columns fall away one at a time, leaving behind other vertical

joint surfaces. (Note the angular boulders strewn on the slopes below; fragments

of fallen columns.)

_____

Figure B is a contour topographic map of mesas and buttes in the central part of Monument Valley. Cliffs are indicated by the bands of closely crowded contour lines. Keep in mind that, if vertical cliff faces are correctly represented, contours on the cliff face will be "stacked" one upon the other to make a single line.

(2) Locate as exactly as possible the point on the map where the photograph, Figure A, was taken (along the road to Lookout Point). Draw rays from the point to indicate the area encompassed in the photo.

*Modern Physical Geography, 3rd Ed., p. 305-6, Figures 18.3, 18.4, Plate I-3.

Figure A Buttes of Monument Valley, northern Arizona. Mitten Buttes at left, Merrick Butte at right. (Infrared photograph by A.N. Strahler.)

(3) About how high do these buttes rise above the level of the surrounding lowland?

Summit elevation: __6,200__ ft. Contour near base: __5,200__ ft.

Difference: __1,000__ ft.

(4) The height of the vertical sides (cliffs) of these buttes is difficult to determine from the map contours. Judging from the photo the proportion of the height (Question 3) that can be assigned to the cliff, how thick is the massive sandstone formation?

About half, for a thickness of about 500 ft for the massive formation. [Note: Locally, this massive layer was originally named the De Chelly sandstone; the pedestals below it belong to the Moenkopi Formation, consisting of shales with interbedded sandstone layers. Both are grouped more generally under the Chinle Formation in Arizona. Age is Upper Triassic.]

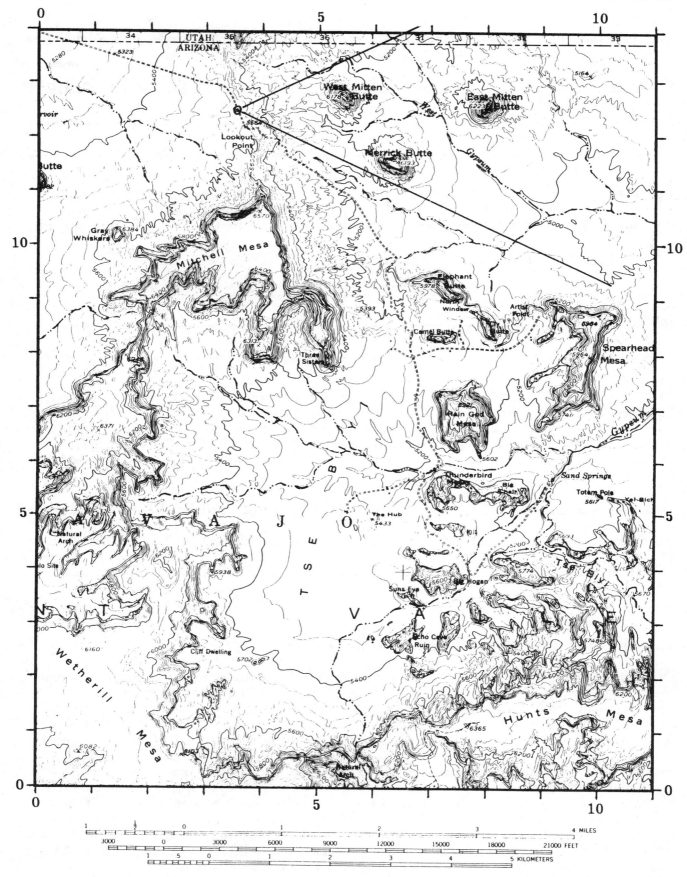

Figure B Portion of the Agathla Peak, AZ-UT, Quadrangle. Scale 1:62,500. Contour interval 40 ft. (U.S. Geological Survey.)

(5) Judging from its appearance in the photo, what kind or kinds of sedimentary rock make up the pedestals of the buttes?

The pedestals appear to be made up of weak shale layers, with thinner hard beds (sandstones?) present at intervals.

(6) Judging from the appearance of the slopes in the foreground of the photo, what kind of rock forms the valley floor surrounding the buttes? Do the map contours support this conclusion? What name is given to this kind of topography?

The valley floor appears to be underlain by a soft rock, probably shale.

Contours in this area are highly crenulated, suggesting the presence of

badlands.

(7) Three mesas conspicuous on the map for their broad, flat tops, are named Mitchell, Wetherill, and Hunts. Read and record the elevation of a contour near the middle of each of these mesas. Assuming that these elevations signify the upper surface of the sandstone formation, what can you conclude about the attitude of that rock layer? Is it horizontal or does it slope (dip)? If the latter, in what direction?

Mitchell: 6,520. Wetherill: 6,120. Hunts: 6,320. These values suggest that the

stratum dips very gently toward the southwest. [Dip would be on the order of

200 ft in 5 mi, or about 40 ft/mi.]

Part B   Redwall--The Great Wall of Grand Canyon

Cliff/slope/bench topography is nowhere better illustrated or on a grander scale than in the walls of the Grand Canyon of the Colorado River. You would encounter at least five major sequences of this kind in the 5,000-ft climb out of the Inner Gorge and up to the canyon rim. None of these cliffs is easy to scale on foot, but one in particular presents a formidable obstacle: the great Redwall.

Figure C is a woodcut drawing of the Redwall by the landscape artist and geologist William H. Holmes, who accompanied geologist Clarence E. Dutton on a pioneering field study of the Grand Canyon during the summer of 1880. Holmes knew his geology well, and it shows in his art. As you can see, the Redwall is not only a sheer rock wall several hundred feet high, but in many places it takes on the expression of alcove-like recesses with overhanging ceilings. The Redwall can be crossed at only a few places, nearly all of which are ancient fault zones where the limestone was crushed. At these points a narrow chute filled with rock rubble can serve as the site of a trail, such as the mule trails tourists follow in descending from the canyon rim to the Inner Gorge.

Figure C The great cliff of Redwall limestone, Grand Canyon, Arizona. Drawn by William H. Holmes. (From C.E. Dutton, 1882, <u>Tertiary History of the Grand Canyon Region</u>, U.S. Government Printing Office, Washington, D.C., Plate 41.)

For our study of the Redwall and other cliffs of Grand Canyon we use a vertical air photo and a topographic contour map of the same area (Figure D). (Note that both photo and map are inverted, so that north is toward the bottom. This is done to give a correct visual image of the terrain on the photo; i.e. shadows lie at lower right of the cliff.) These are supplemented by a schematic geologic profile of the Canyon wall (Figure E) and a generalized topographic plan (Figure F).

Cliffs appear as deep shadows in the air photo. The Redwall is recognized by its typical scalloped outline, each lunate arc being an alcove in the cliff. On the map, the major cliffs are shown as a solid bands of contours. The widest of these is the Redwall, and its curves reveal several alcoves.

(8) On the map, identify and label the Redwall cliff. Place the label outside the map, with a line pointing to a place where the cliff leaves the map area. Approximately what is the elevation of the bench immediately above the Redwall?

Bench elevation: <u>5,000</u>   to   <u>5,100</u>   ft

[Note: Toward the right (west) the bench descends to lower elevation.]

(9) Study the color photograph of Grand Canyon, text Figure 16.5. Locate the Redwall cliff in several places. Attach a piece of tracing paper over the photo and color in red all the exposures of the Redwall you can find.

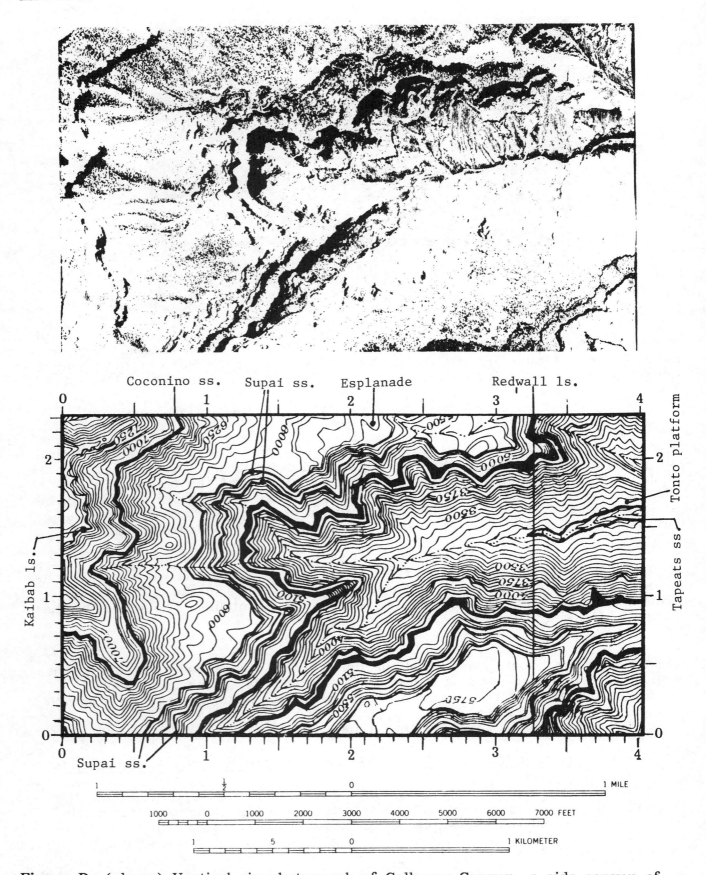

Figure D (<u>above</u>) Vertical air photograph of Galloway Canyon, a side canyon of the Grand Canyon. North is toward the bottom. Lat. 36°20'N, long. 112°25'W. (U.S. Forest Service.) (<u>below</u>) Topographic contour map of the same area. Scale 1:24,000. Contour interval 50 ft. (U.S. Geological Survey.)

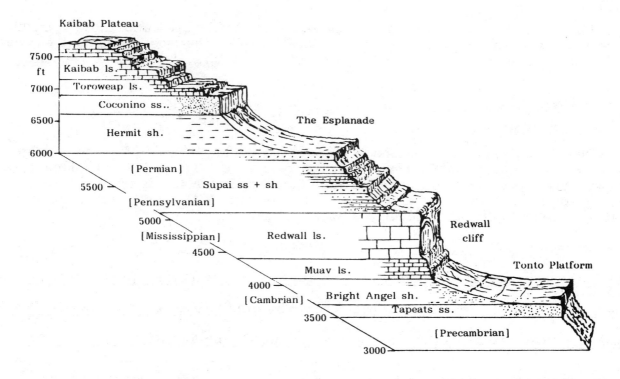

Figure E Diagram of the strata and topography shown in the photograph and map of Figure D. (A.N. Strahler.)

(10) Refer to the cross-sectional diagram, Figure E. What formation lies immediately below the Redwall limestone? How is it expressed in the topography? What is its thickness?

The Muav limestone lies directly below the Redwall limestone. The Muav makes a distinct cliff, separate from the Redwall in many places. For example, at 2.0-0.60 and 2.00-1.55. The Muav is about 300 ft thick.

(11) A second great cliff is found higher in the canyon sequence. It is made up of the Coconino sandstone, a massive accumulation of desert sand of Permian age, about 300 ft thick. Study text Figure 11.10, showing the cliff at close range. On the map, Figure D, find and label the Coconino cliff. Give its location in terms of grid coordinates at two points. Give the approximate elevation of the bench above the cliff.

Coordinates: ___0.60-0.55___ and ___0.40-0.1.8___

Elevation of bench: ___6,900___ ft

(12) The Supai formation contains massive sandstone layers, and these form cliffs in certain parts of the map, Figure D. Locate two of the most prominent of these cliffs, using grid coordinates. Using arrows as pointers, label these two cliffs as "Supai ss."

Coordinates of higher cliff: __0.9-1.2__ ; of lower cliff: __1.1-1.4__ .

(13) The word "esplanade" means a terrace paralleling a drop-off, and it might take the form of a broad balcony or porch of a hotel. There is an Esplanade (a proper name here) in this part of Grand Canyon. Refer to Figure E for its stratigraphic location. Find and label it on the map. With what stratigraphic units is it associated? Why is it developed where it is?

Coordinates: __1.2-0.8__ and __1.5-2.0__ and __2.8-0.4__ .

The Esplanade is a broad bench on the upper surface of the uppermost sandstone

layer of the Supai formation. It has formed by rapid removal of the weak Hermit

shale that overlies it, along with retreat of the Coconino cliff.

(14) Another prominent cliff is that formed by the Tapeats sandstone of Cambrian age. Using Figure E as a guide to its location, find the Tapeats cliff on the map. You will need to look very closely at contours in the lowest elevations on the map, Figure D. Give coordinates. What is the name of the esplanade formed on the upper surface of the Tapeats sandstone? Explain how this esplanade or bench was formed. Find the Tapeats cliff and its esplanade in the color photo, text Figure 16.5; color it brown and label it on the tracing overlay.

Coordinates: __3.4-1.4__ to __4.0-1.6__

The bench or esplanade is named the Tonto Platform. It was formed by rapid

erosional removal of the overlying Bright Angel shale and retreat of the Redwall-

Muav cliff. [Note: The Tonto Platform is not shown in the area of the air photo

and map, Figure D.]

(15) Look for an area on the air photo and map (Figure D) where the Kaibab limestone is represented by minor cliffs. Give coordinates. What name is given to the plateau formed by the exposed upper surface of the Kaibab limestone? Label it on the map. Find the Kaibab cliffs in the color photo, text Figure 16.5, both in the distance and in the foreground. Label and color (orange) this formation on the tracing overlay.

Cliff coordinates: __0.2-1.6__

This plateau surface is named the Kaibab Plateau. Its elevation here is about

7,600 ft. It is the rimrock of the Grand Canyon on both north and south sides.

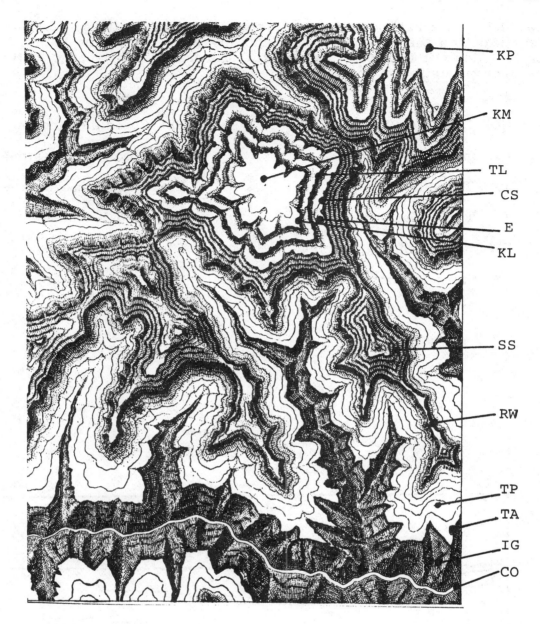

KP
KM
TL
CS
E
KL
SS
RW
TP
TA
IG
CO

Figure F Generalized map of topographic elements of the Grand Canyon, within an area extending from the Colorado River at the south to Shiva Temple and the North Rim at the north. (Plate 42 of same source as Figure C.)

Figure F is a greatly generalized map of a portion of the Grand Canyon a few miles to the south and east of the area shown in Figure D. It appears as a photoengraving in Major Dutton's report of 1882 with the following legend: "Plastic map showing the horizontal projections of canyon topography, and especially illustrating the inward rounded form of all the recesses, great and small, with projecting cusps between." You should now have at your disposal all the information needed to identify and name the cliffs and benches shown on this map.

(16) Below is a list of features and their code designations. For each feature, place a dot on the map (Figure F) showing where it may be found, run a straight line out to a point beyond the map limits, and label it in the margin with the appropriate letter code.

CO	Colorado River	CS	Coconino sandstone cliff
IG	Inner Gorge	TL	Toroweap limestone cliff
TA	Tapeats sandstone cliff	KL	Kaibab limestone cliff
TP	Tonto Platform	KP	Kaibab Plateau
RW	Redwall limestone cliff	E	Esplanade
SS	Supai sandstone cliffs (3)	KM	Kaibab limestone mesa

For questions 9, 14, and 15.

Name _____   Date _____

_____   _____

Exercise 16-B   Stream Channel Systems As Trees
                [Text p. 227-28, Figures 10.12 and 10.13;
                 p. 353, Figure 16.7 and 16.31.]*

A stream channel network, such as that shown in text Figure 10.13, is a branching system of lines best described in pure geometry as a <u>tree</u>. In the world of living things, tree-like systems are essential carrying out certain physiological processes, such as moving air (oxygen and carbon dioxide), and vital liquids, such as nutrients (sap) in plants and blood in animals. Each of these trees has a single trunk that repeatedly branches into smaller and smaller pathways ending in numerous terminals or tips.

Think of a common deciduous tree, such a the maple or oak. In the above-ground part of the tree, water and nutrients flow from the base of the trunk upward to the twigs, where they feed into the leaves. This is a <u>diverging</u> <u>system</u>. Below ground is the root system--a tree in reverse--in which water and nutrients enter the hairlike root tips and pass through a <u>converging</u> <u>system</u> of progressively larger and larger passageways, extending to the base of the trunk, and merging with the diverging system. The stream channel in Figure 10.13 belongs to the same converging class as the root system of the maple.

Perhaps another comparison can prove helpful. Our urban culture requires that we install branching flow systems of two kinds. One is a diverging system that brings utilities from a common source to each of a large number of end points. Examples are electric, gas, and cable-TV supply systems. Quite the opposite is the sanitary sewer system that transports its input of wastes from the tips of the branches through progressively larger pipes to reach the one central treatment plant. The analogy of the sewer system with that of a stream system is remarkably close in another way: both need a downhill gradient to sustain the flow. Not susprisingly, environmentalists refer to heavily polluted rivers, such as the Tiber, Thames, Rhine, and Seine, as "urban sewer systems."

This exercise and the next is based on four concepts: (a) the concept of natural variation, (b) the concept of space-filling, (c) the concept of order, and (d) the concept of pattern. Look for these as we go along.

Our investigation of stream systems begins with a look at the channel tree itself, shown on a planimetric ("flat") map--as if there were no gradients within the system. We are interested in the pure geometry of the tree. Text Figure 10.13 will do nicely for this purpose; we reproduce it here to a larger size as Figure A.

Concentrate first on the fingertip tributaries, which mark the uppermost limit of the channel system. The length of a fingertip tributary is the distance between the upper or free end of the channel and the point where it joins another; i.e. to its first downstream junction. (We will use "F" to designate fingertip tributary.)

---

*Modern Physical Geography, 3rd Ed., p. 191, Figure 11.12, 11.13;
  p. 306, Figure 18.5.

439

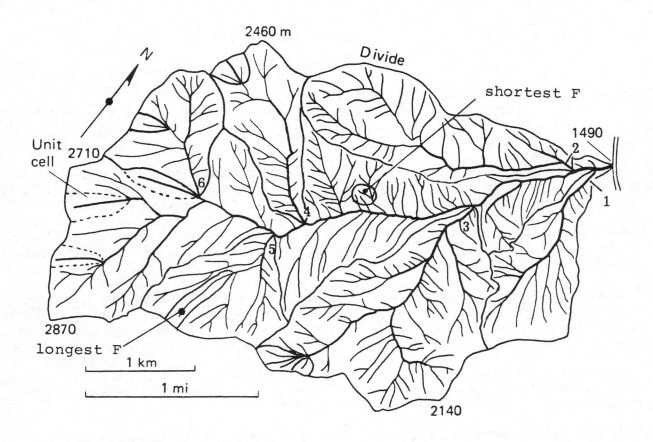

Figure A Channel system of a small drainage basin. (Data of U.S. Geological Survey and Mark A. Melton.)

(1) Are the F-lengths about the same over the entire drainage basin, or do they vary greatly? What maximum range of variation is shown? Give a simple ratio, such as "1/8," meaning "the shortest F is one-eight the length of the longest." On the map label the shortest and longest Fs. (Use arrows.) Offer one or two possible explanations (hypotheses) for this natural variation in F-lengths.

The shortest F on the map is about 1/20 as long as the longest. (Very hard to

measure accurately). Hypothesis: Each F begins as a new sprout from some point

on a channel, after which it grows steadily in length. At some critical length, a

new sprout forms along the length of the older one, causing it to be shortened.

Thus at any one time, many lengths will be represented. Hypothesis: As some Fs

grow longer, they push back others that oppose them but are weaker, so some

are getting shorter as others are getting longer.

Be sure you understand the definition of <u>unit cell</u>, described on text p. 228 and shown on the map, Figure A. All surplus rainfall landing on the surface of the unit cell runs off down the valley-side slopes to enter the channel. This means that a drainage divide coincides with the outline (perimeter) of each unit cell.

(2) On the map, draw a drainage divide around each of several Fs, some large, some medium and some small. Is area of the unit cell related in some direct way to length of F? Explain this relationship.

<u>Generally, a short F is limited to a small cell, because of the proximity of other</u>

<u>channels and divides. For the long Fs, there is much more space available to</u>

<u>draw the limits of the cell. (Some exceptions are present.) This makes sense,</u>

<u>because a longer channel needs a larger supply of runoff to sustain it.</u>

(3) Examine closely many examples of the joining of two Fs to form a single channel, and of a single F joining with a larger channel. Describe the characteristic or most common angle of the joining in one of the following terms: acute angle, right angle, obtuse angle. Explain your finding in terms of efficiency of operation of the drainage basin.

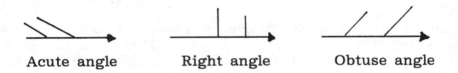

Acute angle   Right angle   Obtuse angle

<u>The acute angle is by far the most common. Right angle is present, but rare. No</u>

<u>obtuse cases observed. The drainage basin is most efficient when the stream</u>

<u>discharge reaches the basin mouth in the most direct way, i.e. by the shortest</u>

<u>pathways possible and still preserve the system of unit cells. Acute angle</u>

<u>junctions accomplish this activity than right angles, which would lengthen the</u>

<u>travel paths. Obtuse angles would greatly reduce the efficiency.</u>

(4) Just by looking at this drainage system of channels, you can see that it consists of several subsystems, each of which is a complete drainage basin in itself. Using a color pencil, draw the complete drainage divide for each major stream branch designated by a   number.

Thus far, we have touched on the concept of natural variation (in lengths of channels and areas of unit cells), and hinted at the concept of order (subsystems within systems). We will now want to develop the concept of space-filling by drainage systems. At the same time, we add the third dimension of relief, absent on a planimetric map. Downhill slope, or gradient, is needed for a gravity-flow system to operate. Not only must the land surface slope downward to the channels, but also the channels must have a continuous down-gradient toward the mouth of the drainage basin.

Figure B is a portion of a contour topographic map, adding the vertical dimension. Superimposed on the topgraphic contours are two sets of lines. The solid lines indicate permanent stream channels. Most are intermittent streams, flowing only after heavy local rains or when snowmelt releases a large supply of overland flow. At times of bankfull flow, the channel is scoured and kept free of soil and rock debris that may be carried into the channel by mass wasting.

The dashed lines indicate the drainage divides of complete drainage basins of various sizes. Two main trunk streams cross the map: Northwater Creek in the northern third, and another stream that cuts through the southeast corner. We focus our attention on the ground between these two trunk channels. Note that all the available space is completely filled with drainage basins of various sizes. One group flows generally northward toward one trunk stream; the other generally southward toward the other trunk stream. You might say that, in the past, these two groups contested for the available space. The contest ended in a draw, or stalemate, and both groups "live in harmony with each other," so to speak, like two nations at peace along their common boundary. Geomorphologists would call this a <u>mature</u> <u>landscape</u>. Erosion of the land surface is continually occurring, of course, but the drainage systems are holding to a stable regime.

(**5**) Focus attention on three drainage basins: Raspberry Creek, Yellowjacket Creek, and Third Water Gulch. Describe their general outline, disregarding the minor curves and bumps, as resembling a familiar fruit, for example, banana, pear, or apple. In the space below, sketch what you think is typical or characteristic basin outline.

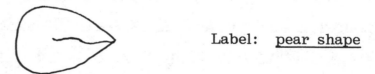

Label: <u>pear shape</u>

Figure B   Portion of the Anvil Points, CO, Quadrangle, 1:24,000. (U.S. Geological Survey.)

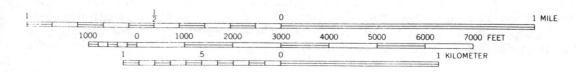

CONTOUR INTERVAL 40 FEET

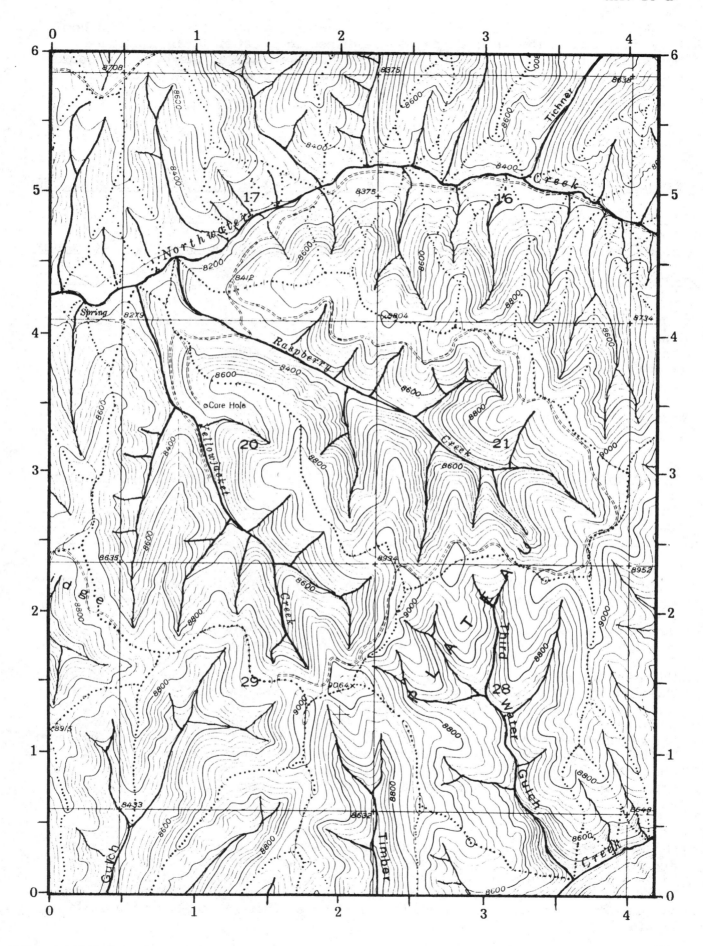

(6) Examine the outlines of various other smaller basins on the map. Note how they depart from the ideal shape you described in question 5. In the space below, sketch the typical outline of basins such as those at the following points on the map: 0.2-3.5, 2.5-5.5, 3.4-4.5. Explain the shape you have drawn.

These basins taper toward the upper end, which is sharply pointed. This form is

forced upon the small basin by much larger basins on either side. This form

occurs where the basin contains a single channel (F) and joins much larger trunk

stream.

(7) Find examples of small, triangular patches of land surface that lie between two divides and the channel of a larger or trunk stream. Give the coordinates of three examples. Why is no channel present in these triangles? In the space below, make a sketch of a typical triangular patch.

Examples: 0.3-4.3, 3.0-5.3, 3.7-0.3. The area of ground that produces overland

flow (runoff) is too small to sustain a permanent channel.

A hierarchy of stream orders. We are now ready to set up an ordering system--a hierarchy, that is--starting with the simplest element and combining elements into successively larger and more complex groups. The process is something like designing a national armed force. We start with the individual foot soldier, combine them into platoons, companies, regiments, divisions, and armies.

Figure C shows how a tree of stream channels within a single basin is ordered. The fingertip tributary is Order 1 (first order). Any two first-order channels upon joining produce a channel of Order 2 (second order); two of Order 2 join to produce one of Order 3, etc. Notice, however, that when a first-order channel joins a channel of second order or higher order, there is no change in the order of the channel it joins. This ordering process ends up with only one trunk channel of the highest order. Ideally, it is a doubling progression, because on the average, when two channels of the same order join, their discharge will be doubled.

(8) Apply the ordering system to the three large complete basins on the map, Figure B. Attach a tracing overlay, and trace the rectangular map boundary. Trace off the channel network for Raspberry Creek, Yellowjacket Creek, and Third Water Gulch. Trace first the first-order channels (solid line); then the second-order (dashed line); then the third-order (beaded line), as shown in Figure C.

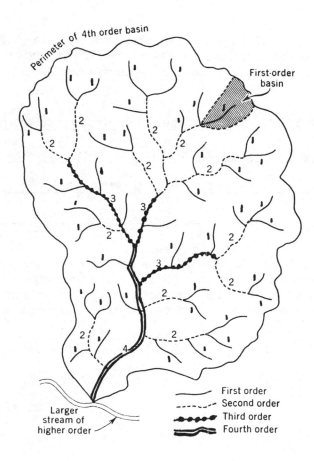

Figure C Method of assigning orders to channel segments in a drainage basin. (A.N. Strahler.)

(9) Count the number of channel segments of each order in each basin, enter in the spaces below, and total the numbers for each order.

	Raspberry Creek	Yellowjacket Creek	Third Water Gulch	Sums of orders:
First order	14	15	10	39
Second order	3	4	4	11
Third Order	1	1	1	3

(10) Obviously, the number of channel segments increases as order number decreases. Obviously, the rate of increase in numbers is greater than a simple doubling. What rough value expresses the rate of increase shown in the table above? (Give to nearest simple fraction.)

Rate of increase in number with decrease in order: $3\frac{1}{2}$
[Note: The increase rate, or ratio, is typically greater then 3 but less than 5 for horizontal strata.]

(11) Based on close visual examination, but without taking measurements, how is the average length of channel segments of the first order related to that for the second order, and for the third order? What general statement can you make about this relationship?

Length seems to average least for the first order, somewhat greater for the

second order and definitely greatest for the third order. Therefore, segment

length increases with increasing order, downstream.

In this exercise, we have developed the concepts of natural variation in size and shape of the elements of drainage systems, the concept of space filling under competition of the parts, and the concept of order or hierarchy within a large drainage system. There remains the concept of pattern, and it is the subject of our next exercise (16-C).

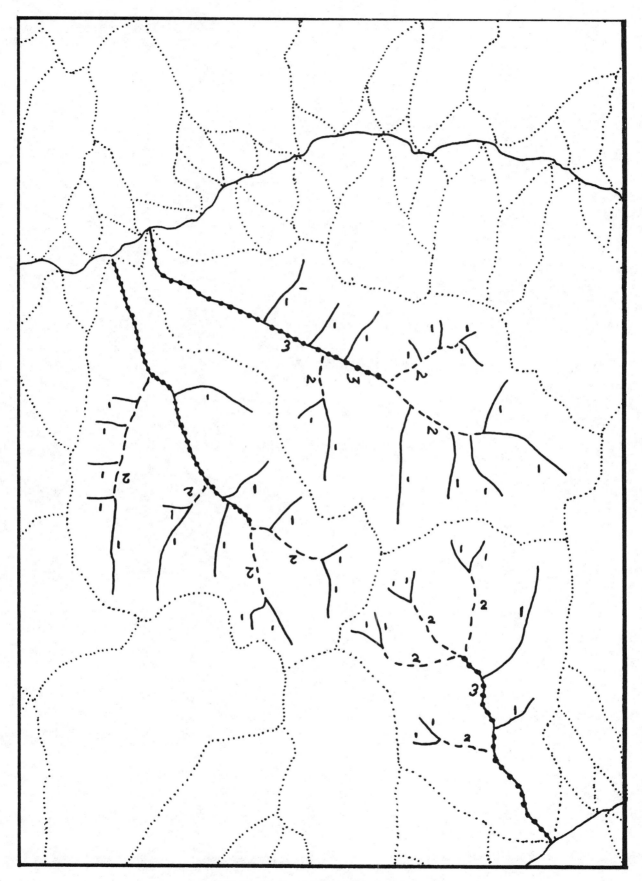

(Attach tracing sheet here.)

Name _____   Date _____

_____   _____

Exercise 16-C   Drainage Patterns as Fractal Systems
[Text p. 353, Figure 16.7.]*

Fractal geometry, a subject of great interest in mathematics, has also caught the interest of geographers who study tree-like systems of various kinds. The basic idea of a <u>fractal</u> is quite simple and easily illustrated: a <u>fractal</u> <u>pattern</u> is one that in every part of itself contains perfect repetitions of its complete form. The leading proponent of fractals, Szolem Mandelbrot, is said to enjoy describing his basic idea in the words of Jonathan Swift: " So, Nat'ralists observe, a Flea/Hath smaller Fleas that on him prey,/ And these have smaller fleas to bite 'em,/ And so proceed ad infinitum." A key term here is <u>self-similarity</u>. Parts of the system, whatever their size, are similar to all smaller and larger parts.

One example of self-similarity in organisms is the shell of the modern chambered nautilus. Figure A shows two examples of molluscan shells from the geologic past. The one on the left is an ancient ancestor of the nautilus. Each turn of the coil repeats exactly the pattern of the larger and smaller coils that are next in order above and below it. A perfect or ideal fractal system would require that the turns of the coil continue forever, so that no matter how powerful a microscope we used to see them, there would always be more similar parts too small to see.

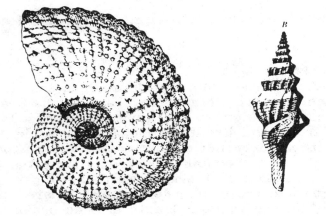

Figure A   Fossil molluscan shells, illustrating self-similarity in a logarithmic spiral. (From K. von Zittel, 1900.)

For drainage systems on the land, however, we need a very different model--a tree, of course--which grows by branching in such a way that each tip of each twig divides into two exactly similar twigs. Add to this, the requirement that each generation of twigs is always shorter, by the same ratio, than the parent set of twigs. Using acute branching angles, what you get is shown in Figure B. Enlarge each of the circled areas in turn and you find another perfect tree. For real drainage systems, the self similarity can't go to infinity, because the outermost twigs--the first-order channels, from Exercise 16-B--are of finite length.

---

*Modern Physical Geography, 3rd Ed., p. 306, Figure 18.5.

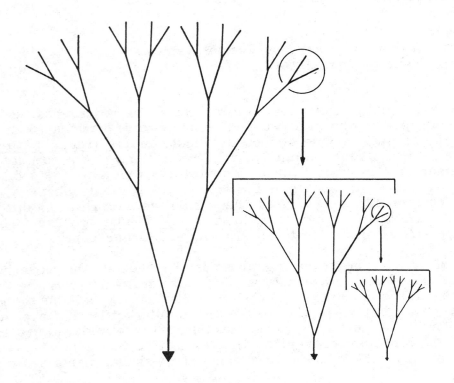

Figure B The concept of self-similarity within a tree-like fractal pattern. (A.N. Strahler.)

Another shortcoming of the tree shown in Figure B is that it fails to fill the available space completely and uniformly. So, back to the drawing board, and we come up with a revised model, shown in Figure C. We specify a branching angle of 180°, so our first trunk and its two branches form a big **T**. At each end of the cross there grow two similar but smaller trees, and so forth. What happens is that the T's turn right or left, doubling back on themselves until they fill all the space uniformly. The extension of T arms toward the upper right is intended to show stream orders down to the first order, where we are required to stop. We designed this tree so that the segment length of each order increases in the ratio of the square-root of 2 ($\sqrt{2}$), which is numerically equal to a ratio of 1/1.414...

As we look at this tree, a remarkable form comes to view--the spiral of the nautilus and its ancestors! You have only to connect the ends of any single right or left succession of T crosses by a smooth curve and there emerges the same spiral. (In the language of mathematics, it is a logarithmic spiral.) Space filling for the drainage system is basically the same as for the nautilus.

Of course, our model is unreal, but it tells us that each of four flow directions is equally probable: north, east, south, and west. Our scientific hypothesis will be that in a complete drainage pattern where all space is occupied, every compass direction of finger-tip tributaries (first-order channels) is equally possible, and therefore equally probable. Can we test this hypothesis? Let's try!

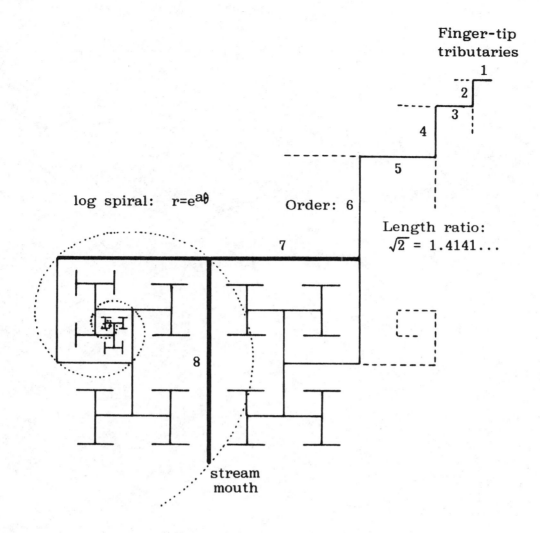

Finger-tip tributaries

1
2
3
4
5

Order: 6

Length ratio:
$\sqrt{2}$ = 1.4141...

log spiral: $r = e^{a\theta}$

7

8

stream mouth

Figure C A fractal tree pattern that fills the available space. (A.N. Strahler.)

Part A   Analyzing a Dendritic Drainage Pattern

The drainage pattern we will analyze has long been known to geomorphologists as the <u>dendritic</u> <u>pattern</u>. (See text p. 353 and Figure 16.7; p. 365 and Figure 16.31.) A good example is shown in Figure D. It was prepared from a modern topographic contour map, and the first-order stream segments were established by a precise formula. The area consists of thin horizontal layers of shale and sandstone; the climate is a moist midlatitude climate, where forest cover is the natural vegetation.

Study closely the drainage pattern on the map. Notice how stream order is depicted with a different line symbol for each order. Notice that the one channel of largest order shown, 5, flows northeastward toward the upper-right corner of the map.

We set up the following test for our hypothesis, stated above. It is simple, to be sure, but may prove meaningful. You record the number of first-order channel segments that fall along each of the four cardinal directions: N, S, E, and W. Use a drawing triangle, preferably of clear plastic. It needs to be somewhat larger than the map area. (If none is available, cut out a true square from a

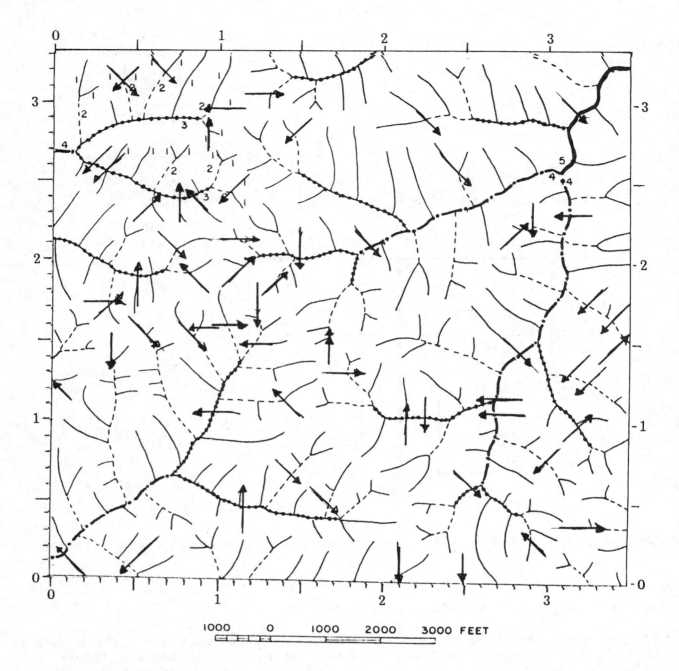

Figure D Ordered channel system of a portion of the Monterey, KY, Quadrangle, U.S. Geological Survey. (After Joan K. Lubowe, 1964, <u>Amer</u>. <u>Jour</u>. of <u>Science</u>, vol. 262, p. 329, Figure 3.)

piece of light cardboard, such as a file folder.) The procedure is to start at one edge of the map and slowly slide the triangle across the map area, always keeping the bottom of the square exactly on the bottom line of the map. Look for first-order channels that parallel the vertical edge. As each appears, stop and draw over it a short arrow pointing in the downstream direction. For each arrow, enter a tally mark in the table below, for either N or S; that is, the direction in which the arrow points. When the N/S set is done, sweep from top to bottom in the same manner, finding, marking, and tallying the streams that flow either E or W.

The following rules need to be strictly obeyed:

(a) Only first-order channels are selected.

(b) Only straight segments, or those with a straight lower portion are selected. Measure the compass direction of the straight portion immediately above the point of junction. Skip curved channels.

(c) The channel must not deviate more then 5° from the true direction.

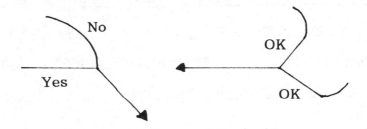

Data Table

Set 1			Set 2			
	Tally	No.		Tally	No.	Sums:
N:	_____	6	NE:	_____	8	14
S:	_____	7	SW:	_____	13	20
E:	_____	6	NW:	_____	6	12
W:	_____	7	SE:	_____	10	17
Sums:		26			37	63

(1) Describe the data you have collected in the first set. In your judgment, do the data tend to support the hypothesis that all directions of flow are equally likely? Or, on the other hand, is one direction clearly predominant? Or two of the four directions? Is one direction very weakly represented, or not represented at all?

The representation seems remarkably uniform for all four directions. This uniformity tends to support the hypothesis. However, the sample is small, and a much larger number of readings would be needed to judge the hypothesis at an acceptable level of confidence.

EX. 16-C

To increase the sample size, carry out the same procedure for diagonal sets of directions: NE-SW and NW-SE. For this you will need a 45-degree triangle, which you can make by cutting a square piece of card along the diagonal. Starting in an upper corner of the map, sweep the diagonal edge across the map. Locate, mark, and tally as before. Flip the diagonal and continue. Enter data in the table above. Complete the sums at right and bottom.

(2) Evaluate the data of Set 2, as in Question 1. Compare the two data sets. Has your level of confidence been strenghtened or weakened?

The data of Set 2 show greater variation between directions than for Set 1. The total number of occurrences is substantially larger, as well. The hypothesis, although weaker, remains tenable. No strongly preferred direction has emerged with sufficient strength to discredit the hypothesis.

(3) In carrying out your observations, did you find it difficult to make a decision as to when to select a segment, and when to pass it up, for a segment that was close to the reference direction? Did you feel an urge to favor or disfavor individual selections in order to get the final results you were hoping for? Could you be acting under the influence of an unconscious bias? If so, what recommendation can you make to reduce the effect of personal bias in carrying out the sampling?

[Various different answers and suggestions can be anticipated. Perhaps most students will admit to a feeling of pressure of bias for or against the hypothesis. One suggestion might be to combine the results of all members of the class. Another, to devise an automated computerized scanning system that follows the rules with total impartiality.]

Part B   Textures of Drainage Patterns

We can look at the concept of self-similarity within a dendritic drainage system from a somewhat different angle, but with fractal geometry fully involved. Study the four panels in Figure E. All show the dendritic drainage pattern. In the first three, find first-order basins outlined by a dashed line. Color these areas in a bright color. The three panels can be thought of as parts of a single fractal system, like that shown in Figure C, because all three map squares have the same area (1 mi^2), but each in turn contains a greater number of replications of the same basic pattern. Panel 4, showing a region of badlands, has no channel system drawn, because the details are too fine to be shown, but you can actually see the channels on the air photo of Figure F. Figure G is a special map of a small area in the same area of badlands. All channels on the map were carefully surveyed and measured in the field. The map is 0.1 mi in width, with an area of 0.01 sq mi (1/100 mi^2).

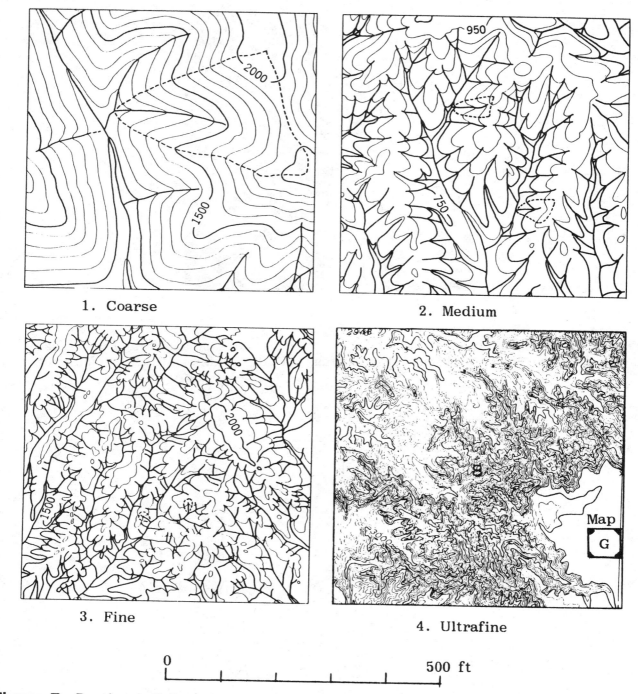

1. Coarse

2. Medium

3. Fine

4. Ultrafine

0 ⊢——————————————⊣ 500 ft

Figure E Portions of four U.S. Geological Survey topographic contour maps, illustrating a wide range of textures. Each square is one mile wide. (a) Driftwood, PA. (b) Nashville, IN. (c) Little Tujunga, CA. (d) Cuny Table West, SD.

Think of these four examples as resembling a set of woven fabrics, all using the same design of interwoven fibers consisting of a vertical set (warp) and horizontal set (woof). Imagine first a wicker floor mat of wide strips of plant fiber, such as bamboo or rattan; call that "coarse texture." Next, think of burlap cloth, which is not as coarse, and call it "medium texture." Go on to a cotton cloth, such as muslin, with much closer threads, and call it "fine texture." Lastly, think of a sheer, delicate fabric, such as voile, made of threads too small to see with the unaided eye; call that "ultrafine texture." We have applied these same names to the four scales of dendritic drainage shown in Figure E.

Figure F Vertical air photo of an area of one square mile of the Big Badlands of South Dakota. North is toward the bottom. (U.S. Department of Agriculture.)

Figure G Portion of a contour topographic map of a small drainage system in the Badlands National Monument, SD. The square covers a width of 0.1 mi; an area of 0.01 mi^2. (From Kenneth G. Smith, 1958, <u>Bull</u>., <u>Geological Soc</u>. <u>of Amer</u>., vol. 69, p. 999, Figure 14.)

(4) To get a rough idea of the ratios of length scales of the fingertip tributaries (first-order channels) in this series of maps estimate an average length for a typical first-order channel on each map. Enter these "ball-park" estimates in the table below. Give the ratio of each length to the next longer length. What is the ratio of the first to the fourth?

Texture:	Length (ft):	Ratios:
(a) Coarse	2,000	
		3.3 /1
(b) Medium	600	
		3.0 /1
(c) Fine	200	
		4.0 /1
(d) Ultrafine	50	

Ratio of (a) Coarse to (d) Ultrafine: 40   to 1

Texture is known to be related to several geological and environmental factors. Two are particularly strong influences: (1) Resistance of the bedrock to erosion. Massive, well-cemented sandstone, quartzite, and unweathered granite tend to produce coarse texture, whereas soft easily-eroded shale or clay and deep soft regolith produce fine to ultrafine texture. (2) Density of the natural plant cover. Forests with dense, closed leaf canopies and a thick ground cover of plant debris (humus) tend to produce coarse texture. Open woodland, brush and scrub, and desert grassland, which have much bare soil exposed, tend to give fine texture. Thus climate becomes a closely related influence.

(5) For each of the areas shown, suggest which factors may be important in determining the texture. We give you the location and bedrock condition. It's up to you to infer the climate and its natural vegetation type. Consult text Chapters 8 and 9, with further help from Chapter 20, if necessary.

(a) Coarse. North-central Pennsylvania, Appalachian Plateau Province (lat. 41°N, long. 78°W). Massive, hard sandstones and conglomerates of Pennsylvanian age.

The bedrock favors coarse texture. The moist continental climate supports dense forests of deciduous and/or coniferous trees, which protect the soil surface and reinforce the geologic effect.

(b) Medium. South-central Indiana, Interior Low Plateaus (lat. 39°N, long. 86°W). Shales and thinbedded sandstones of Mississippian age.

Weakness of the strata would favor fine texture, but the moist continental climate with forest or prairie grassland would counteract that effect to yield a medium texture.

(c) Fine. Southern foothills of San Gabriel Mountains, Los Angeles County (lat. 34°18'N, long. 118°23'W). Loosely consolidated sandstone and shale of Pliocene age.

Weakness of the bedrock would favor fine texture, and this would be reinforced by the semiarid climate, which typically has a partly open brush cover (chaparral), favoring easy erosion.

(d) Ultrafine. Badlands National Monument, South Dakota (lat. 44°N, long. 102°W). Brule Formation, clays and ash beds.

The extremely weak clay is not only easily eroded, but produces a large amount of surface runoff during local rainstorms. Vegetation is absent, so intense erosion of small rills is severe, giving ultrafine texture.

Name _____   Date _____

_____   _____

Exercise 16-D Domes and Their Hogbacks
          [Text p. 357-59, Figures 16.13, 16.14,
             16.15, 16.16, 16.17.]*

Sedimentary domes are not common geologic structures, but some are found in the Middle Rockies and bordering parts of the Great Plains. Perfectly circular domes, such as the Sundance Dome shown in text Figure 16.13, are an extreme rarity. Perhaps it follows that the drawings of a circular dome, text Figure 16.14, are idealized beyond the norm.

Before proceeding further, we should think about how circular and elliptical sedimentary domes are formed and what lies beneath them. Two quite different origins are recognized, with excellent evidence for both. One is the <u>laccolithic dome</u> formed by an igneous intrusion between sedimentary formations. It is a relative of the igneous sill, described on text p. 244-45 and illustrated in Figure 11.6. Instead of spreading widely, the magma pushes up the overlying strata into a circular dome. The igneous mass itself is called a <u>laccolith</u>. An excellent example of a laccolithic dome is Navajo Mountain, rising from the Rainbow Plateau in southern Arizona. It lies close to the famed Rainbow Bridge near Lake Powell. Figure A is a cross section of Navajo Mountain, showing where the laccolith is supposed to be. Actually, igneous rock is not exposed, because erosion hasn't yet cut through the massive sandstone layers that form the dome. Perhaps Sundance Dome is also of the laccolithic type; it would be hard to tell from what shows at the surface.

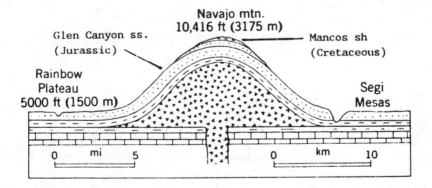

Figure A Idealized cross section of Navajo Mountain, showing the inferred laccolithic core. (Based on data of H.E. Gregory.)

*Modern Physical Geography, 3rd Ed., p. 313-15, Figures 18.21-18.24.

Larger sedimentary domes, such as the Black Hills Dome, represent uplifts of the crust deep beneath the surface and they are interpreted as tectonic in origin. Figure B compares tectonic and laccolithic domes. The upper diagram shows that ancient crustal rock, Precambrian in age, lies under the strata and has been elevated along with the strata. Recent seismic studies show that several such domes and arches of the central Rocky Mountain region were pushed up by overthrust faults that rise steeply under the dome. In some cases, the thrust plane cuts through to the surface. This kind of thrusting would explain domes with steep dips on the eastern side, but gentle dips on the western.

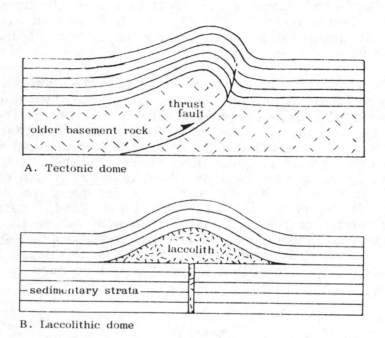

A. Tectonic dome

B. Laccolithic dome

Figure B Comparison of a tectonic dome with a laccolithic dome. (A.N. Strahler.)

Figure C is a topographic contour map of an imaginary sedimentary dome of the tectonic variety. It's a sort of cross between the circular dome diagrammed in text Figure 16.14 and the large elongate Black Hills dome of Figure 16.17. Given the help of these two text figures, you should be able to interpret the contour map without difficulty.

(1) Hogbacks are mentioned on text p. 357 and illustrated in Figures 16.14, 16.15, and 16.17. Find two good examples on the topographic map and label them with the letter **H**.

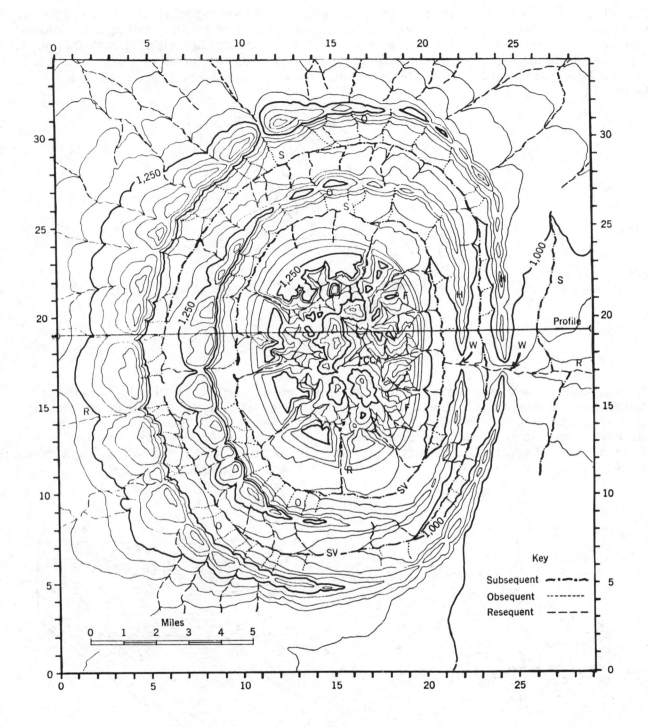

Figure C Synthetic topographic contour map of an ideal mountainous dome. (A.N. Strahler.)

(2) Flatirons are not mentioned in your text; they are special landforms related to hogbacks. Where a steeply upturned hard formation is cut through by closely spaced streams, the hard layer is shaped into a triangular slab, point-up, as in the sketch below. Find flatirons in in the lower diagram of text Figure 16.14. Locate two good examples of flatirons on the map and label them with the letter **F**.

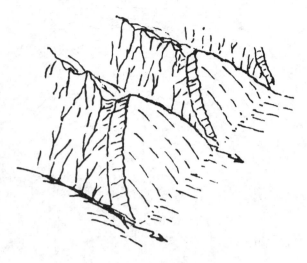

Stream and valley development on dipping strata. Figure D ties together the landforms of horizontal strata, gently-dipping strata (such as those of coastal plains), and domes. One group of landforms grades into the other on this diagram, as the dip changes from zero at the rear to 45° or more on the front face of the diagram. Subsequent streams (S) occupy long, narrow valleys of weak shale. These are separated by cuestas (Cu) and hogbacks (H).

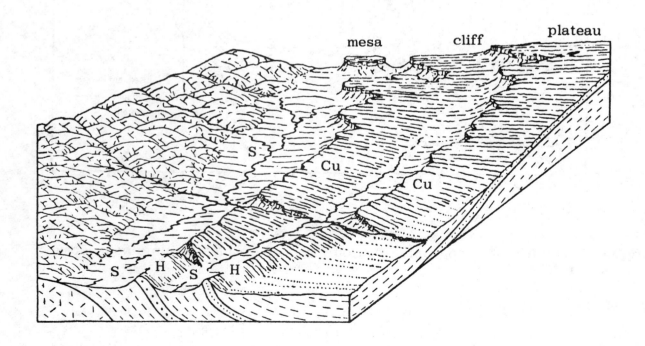

Figure D Schematic diagram of landforms transitional from horizontal strata to steeply-dippping strata. (After W.M. Davis.)

(3) On the topographic map (Figure C) locate two subsequent valleys and label them **SV**. Draw the subsequent streams in these valleys, using a red pen or crayon; label them **S**.

Resequent and obsequent streams. Figure E shows that a subsequent stream is fed by two opposing sets of short streams. One set, called <u>resequent streams</u> (R), flows with the direction of dip. The other set, called <u>obsequent streams</u> (O), flows opposite to the direction of dip. The diagram also shows consequent streams (C), flowing down the slope of the original land surface (right). (See text p. 355 and Figure 16.9 for explanation.) Notice that the resequent streams flow in the same direction as the consequent streams, but they are controlled by the dip of a resistant rock layer and represent a younger generation of streams.

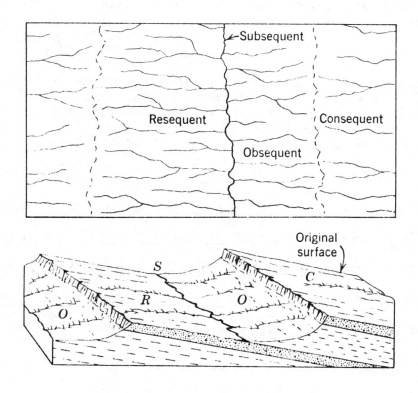

Figure E Stream types on eroded dipping strata. (A.N. Strahler.)

(4) On the map, draw all the resequent streams in blue lines; the obsequent streams in green lines. Label examples with the letters **R** and **O**, respectively, in both subsequent valleys. Include and label as resequent streams those that flow radially outward from the outer ridge. Add a color key in the lower right corner of the map.

(5) Examine the central crystalline area of the dome as shown by contours on the map; label it **CCA**. What drainage pattern is developed here? If this were an arid region, what distinctive landforms might you expect to find here? Draw in black line the stream channels of the central crystalline area.

The drainage pattern is dendritic. In an arid climate, there would be buttes and

mesas here, bounded by steep cliffs.

(6) Construct a topographic profile from 0-9 to 29-19 on Map C, using the blank graph provided (Graph A). Add to the profile a geologic structure section consisting of layers of sandstone and shale conforming to the the hogbacks, flatirons, and subsequent valleys. Show crystalline rocks (granite pattern) in the core of the dome. (Using tape, replace the completed profile in its original place on the page.)

(7) Assuming that the sandstone formations that make the two hogback ridges on the map (Figure C) are of uniform thickness over this entire region, explain why the ridges are broad and cuestalike on the west side of the dome, but narrow and straight-crested on the east side of the dome. Does your answer explain why the major streams drain out through the east side of the dome?

The dome is asymmetrical. Dip is steep on the east flank; gentle on the west

flank. Steep dip gives narrow hogbacks; low dip gives broad cuestas. In the

early stages of dome erosion, consequent streams flowing down the steeper

eastern side of the dome would have cut through the resistant formations first.

The first such stream to succeed would have sent out subsequent branches first,

taking control of the entire drainage system.

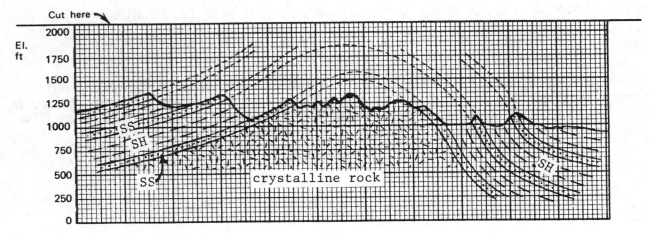

Graph A

Exercise 16-E Mountain Ridges on Folded Strata
           [Text p. 360-62, Figures 16.20, 16.21,
           16.22, 16.23, 16.24; p. 56-57, Figure 2.28;
           p. 50, Figure 2.21.]*

Foreland folds were introduced in Chapter 13, illustrated by the Zagros
Mountains of Iran and the Jura Mountains of the northern European Alps. Keep
in mind that in both examples the folding is geologically very recent--Late
Cenozoic--and resulted from the collision between the Africa plate and the
Eurasian plate. Text Figure 16.9, Block A, illustrates the Zagros-type folds, in
which most of the mountains are anticlines. They are in the class of initial
landforms--landforms produced directly by tectonic or volcanic processes. We now
turn to mountains developed by deep erosion (denudation) of very ancient fold
structures; they belong to the class of sequential landforms. (See text p. 287.)

Deeply eroded anticlines and synclines form striking zigzag ridges in the Ridge
and Valley Province (or Folded Appalachians), a narrow belt that extends from
Pennsylvania southward through Maryland and Virginia and ends in Alabama.
Here, strata of Paleozoic age were thrown into foreland folds in the continental
collision that took place in late Carboniferous (Permian) time, about 240 million
years ago. Since then, the regional history has been largely one of denudation,
exposing deeper and deeper parts of the fold structures. Today's topography is
illustrated in text Figure 16.19, Block C. Note especially that synclines as well
as anticlines form the high ridges, and some of the valleys lie along the axes of
anticlines.

Part A   Zigzag Ridges of Pennsylvania

For our first exercise, we use the side-looking airborne radar (SLAR) image of
zigzag ridges of the Ridge and Valley Province near Hollidaysburg, Pennsylvania,
text Figure 2.28, p. 57. Find this area on the Landsat image, text Figure 16.20,
p. 361. Note that the location of the SLAR image is along the northwestern edge
of the fold belt and includes a small portion of the Allegheny Plateau (upper-left
corner), where the strata are nearly horizontal.

(1) Cover the SLAR image with a tracing sheet. Secure it firmly with masking
tape along the top and trace the rectangular outline of the area. Using pencil
(for easy correction), trace the sharp crest of the narrow ridge that enters the
area at top center of the frame and exits near the lower left corner. (We will
refer to this feature as the West Ridge.) Do the same for a similar ridge that lies
farther east--designated the East Ridge.

---

*Modern Physical Geography, 3rd Ed., p. 315-17, Figures 18.27-18.31, Plate I-4.

Figure A reproduces text Figure 16.22 with added lines and labels. A dashed line is drawn along the axis of the syncline and another along the axis of the anticline. Arrowheads at the far ends of these two lines show that the direction of <u>plunge</u> of the folds is away from the observer. Study the mountain that ends in a curved cliff at the point of crossing the synclinal axis; this feature is a <u>synclinal nose</u>. Following the same cliff toward the right to the point where it reaches the anticlinal axis, we observe that it doubles back on itself, enclosing an <u>anticlinal cove</u>. Away from this summit the mountain slopes gently down to a point and disappears; this feature is an <u>anticlinal nose</u>.

(2) Describe the two kinds of "noses" and explain how their contrasting topographic forms are related to geology.

A single cliff formed by a resistant sandstone layer is continuous along the zigzag ridge. The synclinal and anticlinal "noses" point in opposite directions; i.e., a "zig" alternates with a "zag". The synclinal nose has its curved cliff facing in the up-plunge direction. The anticlinal nose tapers down in the direction of plunge, while the cove opens out in the up-plunge direction.

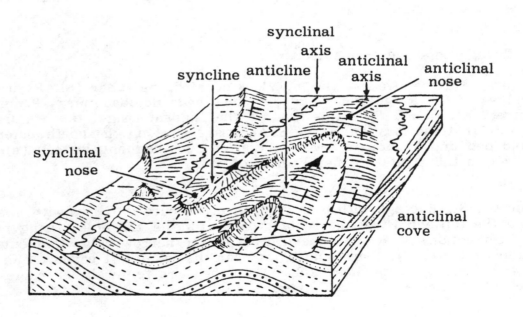

Figure A Landforms on eroded plunging folds. (After E. Raisz.)

(3) On the tracing overlay of the SLAR image, locate, mark and label the following features:

(a) Two synclinal noses and two anticlinal noses along the West Ridge and one of each of the same on the East Ridge. (Use "S-nose" and "A-nose" for labels.)

(b) Along both ridges, place numerous dip symbols to show the direction of dip. Place the symbol on the down-dip side of the ridge crest. The dip symbol is a broad T, as follows:

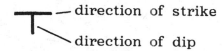

             —direction of strike

             \direction of dip

(c) Draw a dashed line along the axis of each syncline and each anticline. Label "synclinal axis" and "anticlinal axis," respectively.

(d) Identify a synclinal valley and an anticlinal valley. Label **SV** and **AV**.

(4) Near the center of the map, the West Ridge appears to be broken and offset. This feature is caused by a fault cutting diagonally across the ridge. Draw a solid line where you think the fault lies. Label **F**.

Remove the tracing sheet and attach it over the space provided on p. 473.

Part B Plunging Folds Shown by Topographic Contours

Figure B is a synthetic topographic map, not representing any real area, but designed to illustrate the typical landforms of the Appalachian Ridge and Valley Province. For guidance in interpreting the map, Figure C is an idealized geological cross section that shows both the landforms and structures of a series of deeply eroded open folds. A sandstone formation (ss) and a conglomerate formation (cgl) are the ridge-formers; the thick shale formations above and below are the valley-formers.

At this point we introduce two new terms, not found in your textbook. A ridge in which the dip of the hard layer is in one direction only is called a homoclinal ridge (HR). It is identical in definition and form to the hogback, a ridge developed in a deeply eroded a dome (Exercise 16-D). A valley in which the strata dip in one direction only is a homoclinal valley (HV). Thus we have a set of six terms in all to describe the ridges and valleys.

(5) Identify and label each ridge and valley on the map, using the initials given in the key to Figure C. Enter numerous strike-and-dip symbols on the ridges. Show the direction of plunge of fold axes by long arrows. Label the arrows **A** for anticline; **S** for syncline.

(6) Draw lines to show a complete drainage system for the map area. Extend the streams shown on the map as needed to complete the minor branches. Use the following colors: subsequent streams (S) in red; obsequent streams (O), green; resequent streams (R), blue. Label examples of each with their code letters. (See Exercise 16-D for definitions.)

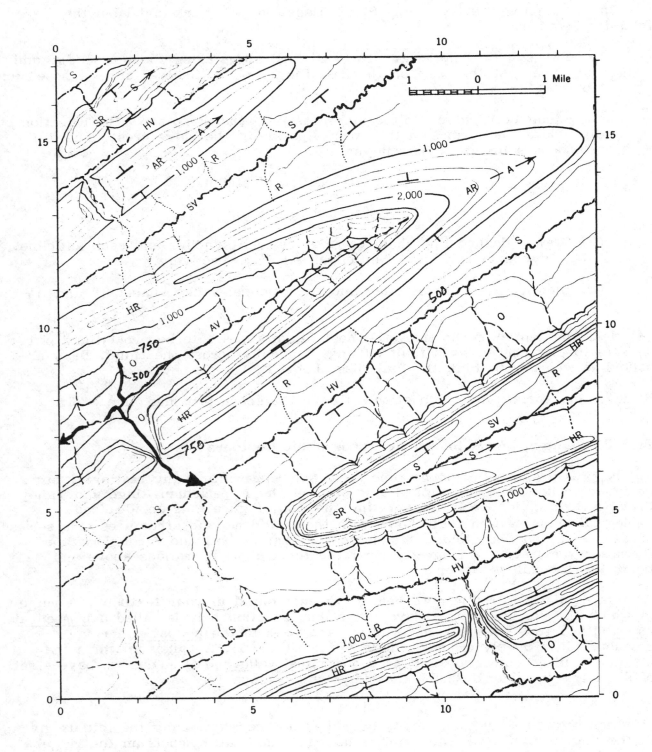

Figure B Synthetic topographic contour map of deeply eroded plunging folds. (A.N. Strahler.)

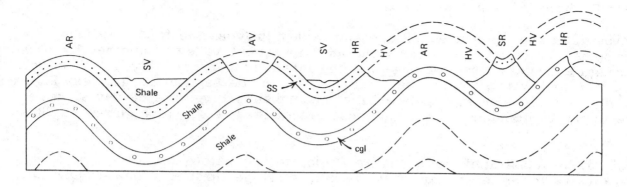

AR	Anticlinal ridge or mountain	AV	Anticlinal valley
SR	Synclinal ridge or mountain	SV	Synclinal valley
HR	Homoclinal ridge or mountain	HV	Homoclinal valley

Figure C Schematic cross section of relationship of ridges to fold structures. (A.N. Strahler.)

(7) What name is given to the kind of drainage pattern shown on this map? Describe the pattern in terms of the types of streams that compose it. Compare it with the drainage pattern of a deeply eroded dome (text Figure 16.16).

This is a trellis drainage pattern. The larger elements are subsequent streams following the strike of the main valleys between ridges. Obsequent and resequent streams are short and enter the main subsequent streams at nearly right angles. The pattern is similar in composition to the annular + radial pattern of a dissected dome, which can be described as a trellis pattern bent into a circular form.

(8) Construct a topographic profile from the upper left-hand corner of the map to the lower right-hand corner. Use the blank profile graph provided (Graph A). Draw in a complete geologic structure section consisting of sandstone and shale formations in agreement with your map interpretations already made and labeled. After completion, return the profile to its former position on the page and secure it with tape.

Part C Watergaps and Windgaps

Throughout the Appalachian Ridge and Valley Province we find numerous examples of streams that cross from one subsequent valley to another by passing through a deep, narrow watergap in the intervening ridge. If you look closely, you can spot watergaps on the SLAR image. On the Landsat image, text Figure 16.20, several major watergaps of the Susquehanna River are clearly shown in the upper right-hand corner and these are diagrammed in text Figure 16.23.

(9) On the topographic map, Figure B, locate two watergaps. Give grid coordinates for each and identify the kind of ridge through which each is cut. Describe in detail the topographic form of each gap.

At 11-2, a watergap cuts through a homoclinal ridge with north dip. The gap is V-shaped, with its narrowest point at the north end and flaring out to the south. At 1-13, a watergap cuts through an anticlinal ridge. This gap is constricted at both north and south ends, widening out in the middle. On both sides of the gap, a cliff forms a perfect arch, and the two arches are perfectly opposed, because the anticline is symmetrical in the dip of its two flanks.

Throughout the Appalachian ridges we occasionally run across what looks like a watergap in a ridge, but on closer look has no stream occupying the gap. A still closer look may show that the floor of the gap is considerably higher in elevation than that of the valley floor on either side. Such a feature was named a windgap by early settlers to the region. They often took advantage of a windgap to accommodate an easy trail or road across a ridge that otherwise was a formidable barrier to travel.

Windgaps are correctly interpreted as former watergaps in those cases where no fault zone is present to explain the gap. The abandonment of the gap by the stream which formerly occupied it and carved it is explained by stream capture, a process by which the headwater drainage basin of one stream, the captor stream, is gradually extended headward toward the trunk of another stream that is situated at a higher elevation. Figure D shows the situation before the capture. Eventually, the flow of the higher trunk stream is diverted into the favored captor, and this leaves as its victim a beheaded stream with only a trickle of flow in its former channel below the point of capture (Figure D, after). Our diagrams show that the captor stream is a tributary of a large river, with a profile at a low elevation and capable of easily eroding its watergaps to keep its profile low. The victim, a small stream, has difficulty maintaining its course across the hard rock barrier of its watergap; its elevation remains high. It easily succumbs to the predator stream, which has only a short distance to flow on weak rock to meet the main river. The key to this process is difference in elevation of the two streams in the divide area that separates them.

(10) Our imaginary topographic map, Figure B, shows a windgap at 2-7. Using a special color (such as a yellow highlighter) show the streams as they were before the capture. Which stream was the captor? Why was it successful? State your evidence.

The captor stream is the stream that leaves the map at 0-7. Evidence is that the elevation of the captor channel is below 250 ft, whereas the valley bottom on the south side of the windgap is above 500 ft. Evidently, a large stream lies off the map not far to the west.

Figure E, a portion of a U.S. Geological Survey topographic map, shows a great anticlinal ridge, Wills Mountain, in the vicinity of Cumberland, Maryland. Wills Creek maintains a watergap--The Narrows--across the anticline, below which it joins the North Branch of the Potomac River. A windgap lies about $2\frac{1}{2}$ mi southwest of The Narrows in a section of Wills Mountain known locally as Haystack Mountain; through it runs Braddock Road and a new highway.

(11) Which stream formerly flowed through the windgap? By what captor stream was it diverted?

Braddock Run formerly ran through the windgap. It was captured by a small tributary to Wills Creek.

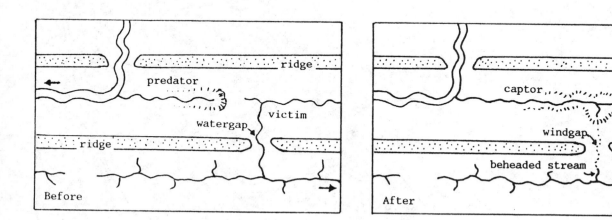

Figure D Schematic maps of stream capture, leaving a windgap. (A.N. Strahler.)

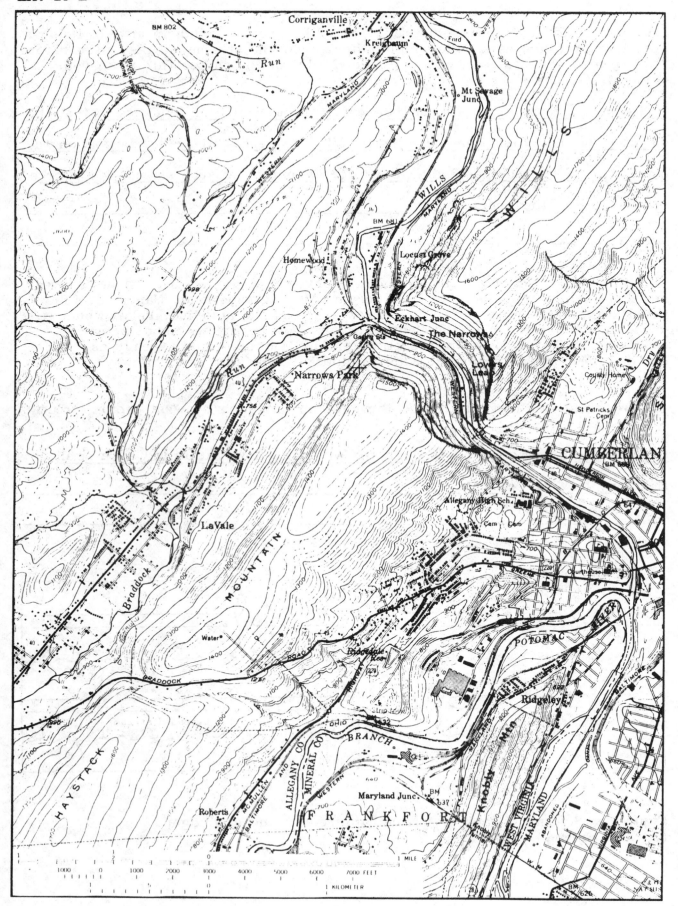

Figure E Portion of the Cumberland, MD-PA-WV, Quadrangle. U.S. Geological Survey.

(12) About how much higher is the windgap floor than Braddock Run (northwest side of the mountain)? How much higher than the North Branch Potomac River (southeast side)? What do these figures suggest as to the age of the stream capture that left the windgap.

About 340 ft above Braddock Run. About 620 ft above the North Branch. The capture must have occurred a long time in the past, on the order of several or many tens of thousands of years ago, or more than 100,000 years ago. (Extremely difficult to set a figure.)

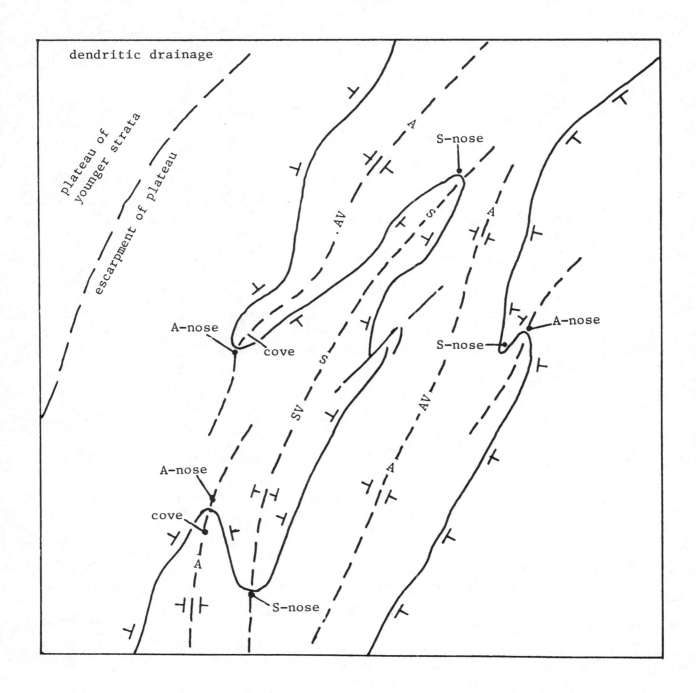

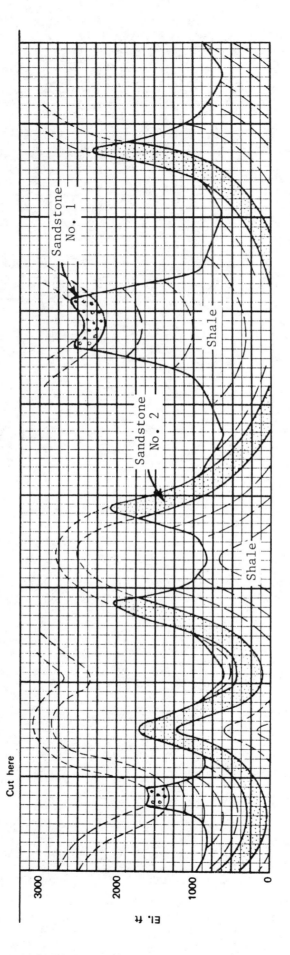

Graph A

Exercise 16-F   Stone Mountain--A Granite Monadnock
    [Text p. 364-66, Figure 16.31, 16.32, 16.33.]*

Granite exposed in domes seems to excite and attract sculptors, especially those who work on a grand scale, using pneumatic jack hammers and explosives. In the granite of Mount Rushmore of the Black Hills uplift, Gutzon Borglum carved those great busts of four American presidents, visible from as far away as 60 mi (100 km). Borglum died in 1941 and the work was completed by his son, Lincoln. Years earlier, in Georgia, the Daughters of the Confederacy had commissioned the carving of equestrian statues of Robert E. Lee, Stonewall Jackson, and Jefferson Davis in the steep granite face of Stone Mountain, near Atlanta. Borglum designed and partially completed the work, but in 1924 resigned in a tiff with his sponsors and destroyed his plans. Two other sculptors worked on it, and despite long delays, it was dedicated in 1970. The State of Georgia now owns Stone Mountain and has transformed it into a state park.

As your text explains, Stone mountain is a monolith of granite, what remains of an ancient granite pluton intruded into metamorphic rocks of the Piedmont region. Throughout most of the Cenozoic Era, this region has been undergoing deudation, and most of it has been reduced to a rolling upland, called the Piedmont Peneplain. Numerous monadnocks rise above that level; many of them are of quartzite and have elongate, ridgelike forms. Stone Mountain is perhaps unique in the Piedmont in terms of its compact outline and smoothly rounded form, but there's a similar one in the Blue Ridge near Asheville, North Carolina. Other prominent domes compete for attention in other lands. Rio de Janiero has its granite Sugar Loaf, while Australian tourism extolls Ayres Rock, a prominent sandstone mass in the desert of Northern Territory. But Pasadena, California, has its own little-league contender in Eagle Rock, a small but prominent dome of massive conglomerate. Actually, Australia has some fine granite domes resembling Stone Mountain, but they are on the Eyre Peninsula of South Australia. Splendid granite domes also occur in the Nubian Desert of North Africa. And what about the Yosemite Domes? We'll come to them soon enough.

We investigate Stone Mountain with the help of a photograph and two topographic contour maps. The old oblique air photograph, Figure A, shows the steep northeast face of the mountain, hidden from view in the color view of text Figure 16.33. Details of the dome are shown on the large-scale map, Figure B, while the relationship of the dome to the surrounding Piedmont upland is better seen in the small-scale map, Figure C.

*Modern Physical Geography, 3rd Ed., p. 319-21 , Figures 18.39-18.43.

Figure A   Taken in the 1920 by the U.S. Army Air Corps, this photo of Stone Mountain shows the scenery before development of the area as state park.

(1) In what important way or ways does Stone Mountain differ geologically from the Yosemite domes of Exercise 14-A?

(a) Most important is that Stone Mountain is formed of granite and is surrounded by other (metamorphic) rock that is more susceptible to denudation, whereas the Yosemite domes and their surrounding rock mass are one and the same body of granitic rock. (b) The Yosemite domes show massive exfoliation shells, absent on Stone Mountain. [It has been suggested that Stone Mountain formerly had exfoliation shells, but that these have long since disintegrated.]

(2) Locate on the map, Figure B, the ground point above which each of the two photos was taken. From each point, draw an arrow across the dome to show the direction of view. Label your arrows "Fig. 16.33" and "Fig. A."

(3) What is the nature and cause of the parallel dark streaks or lines on the flanks of Stone Mountain?

These lines mark the courses of rills of rainwater that flow down the flanks of the dome. Actual rill channels have been worn into the rock in a few places. Elsewhere, these lines are sites of algal growth and/or mineral oxides that produce a dark colored surface layer.

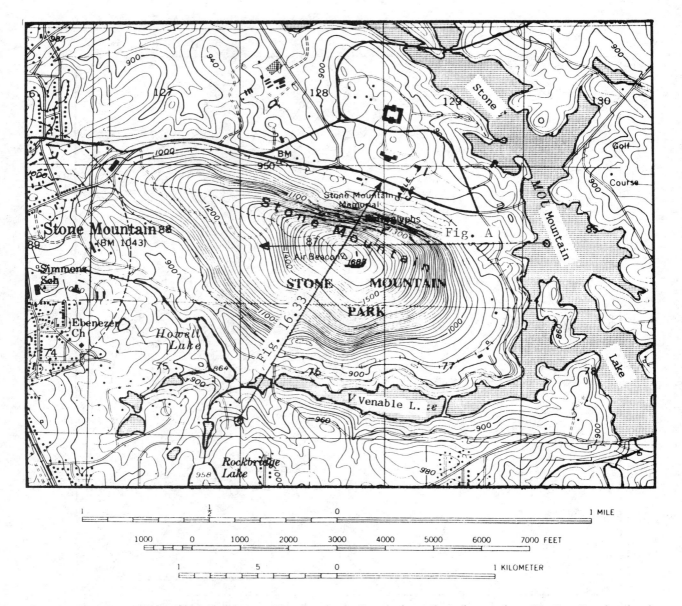

Figure B Portion of the Stone Mountain, Georgia, Quadrangle. U.S. Geological Survey, 1954. Scale 1:24,000. Contour interval 20 ft.

(4) Using the small-scale map of Figure C, construct a topographic profile through the summit of Stone Mountain from 0.5-1.5 to 12-11. Use the blank graph (Graph A). When the profile is completed, replace it and secure with tape.

(a) Below the profile, draw the outlines of the resistant granite stock that lies beneath Stone Mountain. Label it "granite."

(b) Using a color pen or pencil, draw a horizontal line across the profile at the approximate level of the Piedmont Peneplain, which can be taken as between 1000 and 1050 ft.

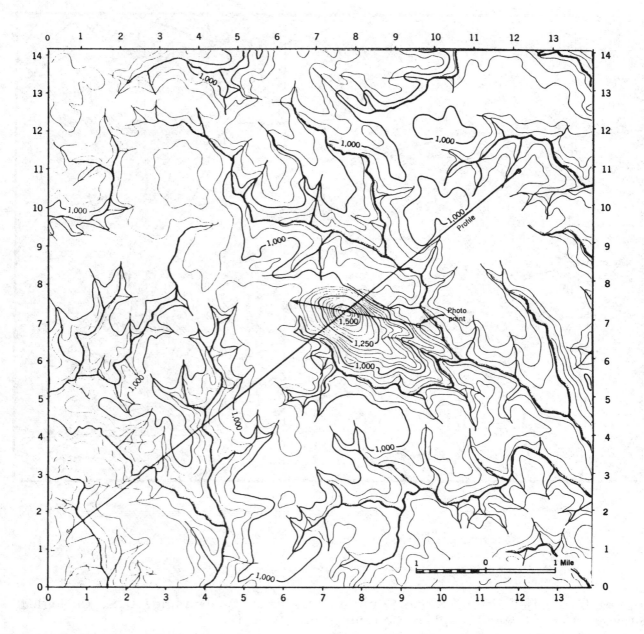

Figure C Portion of the Atlanta, Georgia, Quadrangle, U.S. Geological Survey. Scale 1:125,000. Contour interval 50 ft. (Map scale has been enlarged to about 1:87,000.)

The Piedmont Peneplain reached its full development many millions of years ago, and was then raised during a broad crustal uplift that affected the entire eastern part of North America. Streams on the peneplain became rejuvenated and trenched their valleys below the peneplain level, which is now represented only in the higher summits of the area surrounding Stone Mountain.

(5) Using blue pencil or pen, draw a complete drainage pattern on map of Figure C. Show streams wherever indicated by a V-indentation of the contour lines. What kind of drainage pattern is shown? Is it similar to that shown in text Figure 16.31? Is it similar to or different from the dendritic patterns shown in Exercise 16-B, Figure B, and in Exercise 16-C, Figure D? What general statement can you derive from your comparison?

A dendritic pattern is present. Yes, it resembles that of the Idaho batholith, shown in text Figure 16.31. The pattern differs in no important respect from the dendritic patterns in exercises 16-B and 16-C. The general conclusion is that dendritic patterns occur on both horizontal strata and crystalline (igneous or metamorphic) rocks that are uniform throughout in their textures.

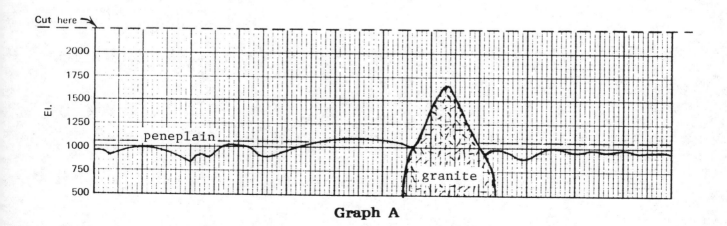

Graph A

Exercise 16-G   A Volcanic Neck and Its Radial Dikes
[Text p. 367-68, Figures 16.34 and 16.37.]*

In an arid climate, volcanic necks that were one surrounded by thick shale formations now rise as striking isolated peaks, surrounded by plains. They are locally called "buttes," but there's a world of difference between their geology and that of the sandstone buttes, such as the Mitten Buttes of Monument Valley (Exercise 16-A). While there are quite a few volcanic necks in the Colorado Plateau region of northeastern Arizona and northwestern New Mexico, Ship Rock certainly reigns supreme.

We investigate Ship Rock and its radial dikes by means of two photographs and a contour topographic map. Figure A is an oblique air photo taken many years ago by Robert E. Spence, a noted documentary photographer of western scenes, particularly in southern California. The color photograph in your textbook, Figure 16.37, shows Ship Rock from a different angle. It was taken by Dr. John S. Shelton, for many years a professor of geology at Pomona College. A pilot and small-plane owner, Shelton produced hundreds of superb landscape photos in both color and black-and-white. Many other fine Shelton photographs adorn your textbook. The best of his collection can be seen in his popular textbook, Geology Illustrated, published in 1966.

Figure C shows comparison cross-sections of a volcanic neck, such as Ship Rock, and a sedimentary butte, such as the Mitten Buttes of Monument Valley. Notice that the massive igneous rock of the volcanic neck has a set of small, irregular vertical dikes. Rough vertical joints are also present, and weathering develops sharp points of varying heights.

(1) On the topographic map, Figure B, find the points above which the plane was located when each of the two photos was taken. Draw a line from each to show the center line of the photo. Label the lines "Fig. 16.37" and "Fig. A."

(2) Determine the contour interval used on the map. Estimate the summit elevation of Ship Rock and its height above the surrounding plain. How high is the dike at 6.7-1.2?

Contour interval: __50__ ft.

Summit elevation: __6400__ ft, approx.

Height above plain: __6400__ minus __5400__ equals __1000__ ft.

Height of dike: __200__ ft.

---

*Modern Physical Geography, 3rd Ed., p. 321-33, Figure 18.44, Plate I-1 .

Figure A Oblique air photograph of Ship Rock, New Mexico, and its radial dikes. (Robert E. Spence.)

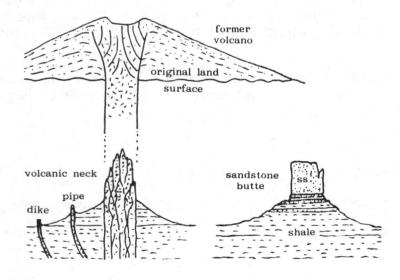

Figure C Sketch comparing a volcanic neck with a sandstone butte. (A.N. Strahler.)

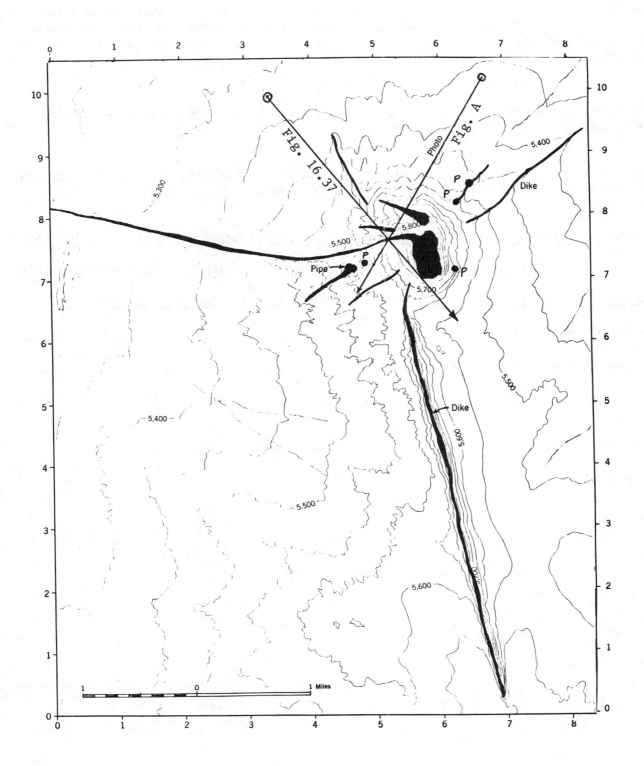

Figure B Portion of Ship Rock, New Mexico, Quadrangle, U.S. Geological Survey. Scale 1:62,500.

(3) On the map, color in red all exposed volcanic rock. The dikes should be shown as thin red lines. In addition to the two great dikes, at least two small dikes are indicated by sharply pointed contour lines. Label the volcanic neck and the dikes.

(4) With the aid of the photo, Figure A, locate three small pipes of volcanic rock. Label these and color them red.

(5) Contours near 5.0-6.0 are highly crenulated (crinkled). What is the meaning of this contour form? What landform type is present. What kind of rock is indicated?

These contours indicate badlands; that is, stream-eroded clays or shales in which stream channels are closely spaced and texture is ultrafine. The region is one of semiarid (steppe) climate; the badlands are largely barren of vegetation and therefore easily eroded.

Name _____   Date _____

_____   _____

Exercise 17-A   The Ria Coast of Brittany--A Youthful Shoreline
[Text p. 378-81, Figure 17.19, 17.20, 17.21.]*

The ria coast, listed first in your textbook under coastlines of submergence, gets it name from the northeastern coast of Spain, where numerous capes project out into the Atlantic and between them lie narrow, branching bays. The title <u>Ria</u> is used for these bays--Ria de Betanzos and Ria de Muros y Noya, for example. In Spanish <u>ria</u> means "estuary."

A ria coast must meet certain requirements. First, it is a partially drowned (submerged) land surface made up of erosional landforms. Tectonic and volcanic landforms are excluded. This means we are dealing with land surfaces eroded by (a) fluvial processes (streams, along with mass wasting, and weathering) or (b) glaciers, either alpine or continental. For ria coasts, the choice is (a), so we expect to find river valleys in a partially drowned condition. A close look at the bottom topography of the bays is necessary in order to identify valleys in which streams formerly flowed.

Our exercise is based on a topographic contour map of the coastal region around the port of Brest, France. It lies on the peninsula of Brittany, which projects westward into the Atlantic Ocean. Similar in many ways to the Ria coast of Spain, the Brittany coast is deeply embayed, and it is modified strongly by wave erosion only where the shore is exposed to waves of the open Atlantic Ocean. The coast was never modified by Pleistocene glaciers or ice sheets.

The contour interval of the map is 20 m on land. Submarine contours are shown as dashed lines for depths of 1, 5, 10, 20, 40, and 50 m. A stippled pattern shows areas that lie between the high-water line (bold line) and the low-water line (fine line). Marine cliffs of rock are shown by lumpy (bumpy) projections along the high-water line.

(1) In red pencil or pen, mark on the map all parts of the shoreline where a marine cliff is well developed. In green, shade all probable sand or shingle beaches. ("Shingle" means well rounded pebbles or cobbles.) Label three examples of beaches; three of marine cliffs. Find and label a good example of a pocket beach on the outer shoreline.

(2) Using a pencil, draw lines on the map to give a reconstruction of the drowned stream system in the Brest harbor, or estuary. Reconstruct branches that connect with the mouths of small streams that enter side branches of the estuary. Carry the lines inland, up the tributary valleys.

(3) Locate two meander bends of the former stream system. (Give grid coordinates.)

Meanders: __34-18__   and   __38-14__

---

*Modern Physical Geography, 3rd Ed., p. 334-35 , Figures 19.29, 19.30.

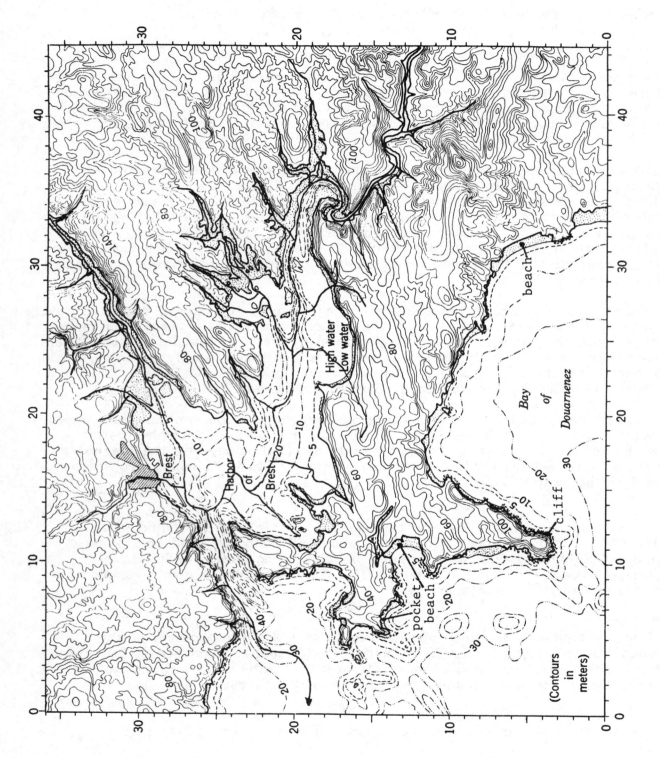

Figure A Portion of the Brest, France, topographic map, Sheet 21. Original scale 1:200,000.

(Contours in meters)

(4) Why are there few prominent cliffs along the shoreline of the Harbor of Brest? Give the grid coordinates of cliffs that are indicated on the map within the harbor.

The harbor has a long, narrow mouth, in which large waves of the open ocean are absorbed. Cliffing is indicated on two narrow promontories, located at 14-20 and 19-22.

(5) Study the peninsula ending at 11-4. Which side seems to have undergone the greater amount of marine erosion? What is the topographic evidence? Is this configuration what you would expect, knowing that the open Atlantic Ocean lies to the west, whereas a bay 15 km wide lies to the east?

The cliff on the east side appears higher and steeper than that on the west side. (Note that the 100-m contour is closer to the shoreline on the east side than on the west.) The west side should be the steeper, if exposure to wave attack were the sole factor. Two possibilities can be considered: (a) Two shoals, indicated by the the 10-m contour, may reduce the force of wave attack from the west. (b) The peninsula may consist of a resistant, layer-like rock formation that dips westward and thus controls the height of the cliff.

(6) The stippled zone between high and low water levels is alternately exposed and inundated as the tide falls and rises. In the heads of bays near 30-20, what type of sediment deposit lies in the stippled zone?

These are probably tidal mud flats or tidal marshes. They consist of fine sediment (silt and clay) including much organic matter, and possibly a layer of peat.

Name _____ Date _____

_____   _____

Exercise 17-B Baymouth Bars and a Recurved Spit
[Text p. 373-74, Figures 17.8, 17.9;
p. 379-80, Figure 17.20.]*

As a ria coast is attacked by storm waves, promontories and headlands are cut back rapidly. At first there is little excess sand to form beaches, but at a later stage beaches appear in a wide variety of forms. Chances are that you have visited some of these kinds of beaches without knowing how they got there.

(1) Figure 17.20 of your textbook shows these depositional features, but doesn't name them. We reproduce this illustration at Figure A. Using the list below, enter the code letter of each landform at the place it is shown on the diagram. Most of these landforms aren't defined in your textbook, so you will need to make guesses. To help with the identification we offer sketch maps of a few, in turn-of-the-century cartographic styling (Figure B).

T	Tombolo	CH	Cliffed headland	BHB	Bayhead beach
S	Spit	DT	Double tombolo	BSB	Bayside beach
RS	Recurved spit	HB	Headland beach	BHD	Bayhead delta
CS	Complex spit	BMB	Baymouth bar	L	Lagoon
CT	Complex tombolo	MBB	Midbay bar	I	Inlet
LB	Looped bar	CB	Cuspate bar	CP	Cuspate delta

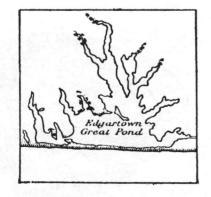

Baymouth bar sealing off a freshwater pond on Marthas Vineyard, MA.

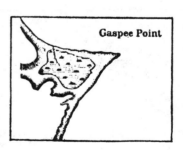

Cuspate bar, enclosing a marsh, near Providence, RI.

---

*Modern Physical Geography, 3rd Ed., p. 328-29, Figure 19.15-19.17 .

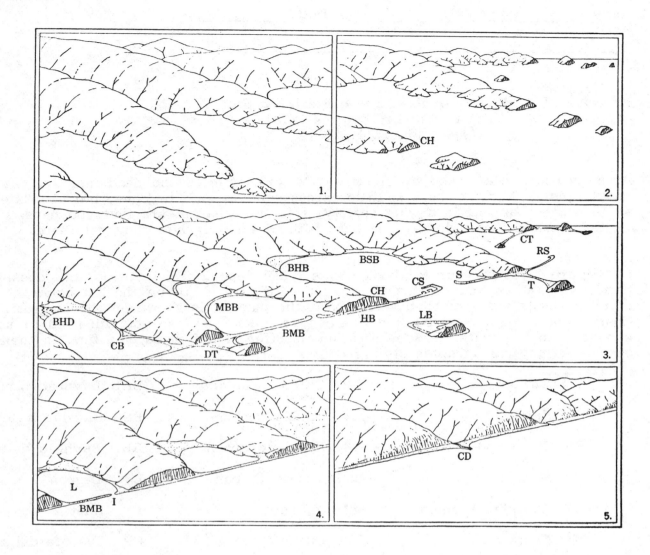

Figure A Stages in the evolution of a ria coastline. (A.N. Strahler.)

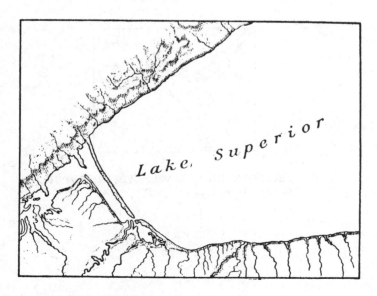

Bayhead bar near Duluth, MN.

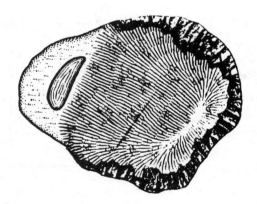

Looped bar on the lee side of Shapka Island, AK.

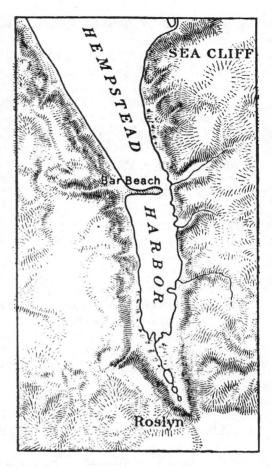

Monte Argentario, Italy, a former island now tied to the mainland by a double tombolo. The town of Orbetello is on an earlier, incompletely formed tombolo.

Bar Beach in Hempstead Harbor, Long Island, NY, is a midbay bar.

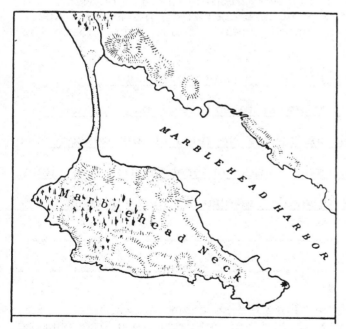

Marblehead Neck, MA, a former island connected to the mainland by a single tombolo.

Figure B Seven sketch-maps showing a variety of deposi- tional coastal landforms. (From D.W. Johnson, Shore Processes and Shoreline Development, Copyright 1919 by John Wiley & Sons, New York. Used by permission.)

A Shoreline of Submergence with Baymouth Bars

Figure C is a topographic contour map of a section of the north shoreline of Lake Ontario. The topography here was developed in preglacial time by fluvial denudation that carved stream valleys. The area was then heavily scoured by the Erie Lobe of the Pleistocene ice sheet, leaving behind glacial till. Lake sediments were added in a late glacial stage, when Lake Ontario was larger than it is today. Waves and currents, reworking and shifting the cover of glacial and lake deposits have formed two sand bars that now separate Yeo Lake and Spence Lake from the open water. Sandbanks Provincial Park now occupies the coastal zone here.

(2) Which of the two sand bars is a baymouth bar, and which a midbay bar? Label them accordingly.

(3) At 12.5-6.0, the sand bar has long narrow ridges, parallel with the shoreline. Near 7-8 the sand bar has several very small hill summits. How do you interpret these features? What caused them?

The parallel ridges may be beach ridges, thrown up by storm waves and left behind as the bar was widened by progradation. They may now bear coastal dunes. The small hill summits near 7-8 may be sand dunes, parts of a large dune ridge with its steep slip-face overriding the shoreline of Yeo Lake.

(4) Describe the form of the narrow outlet channel through the bar between Spence Lake and Athol Bay. Explain in terms of beach drifting processes. Are tidal currents responsible for keeping this channel clear? If not, what keeps the channel clear?

The outlet channel runs diagonally across the bar from north to south. Apparently, beach drifting of sand from northwest to southeast has gradually moved the mouth of the outlet toward the southeast. No tides of any consequence occur on the Great Lakes. Probably, the excess water of precipitation and runoff accumulating in Spence Lake and flowing through the channel like a stream is able to keep the passage clear.

(5) Using a color pencil or pen, redraw the shoreline as a smooth, simple shoreline of the future, after the headlands have been cut back to a line passing through 0.0-14.0, 6.0-11.0, 10.0-8.0, and 19.0-0.0. Redraw the sand bars in their new positions.

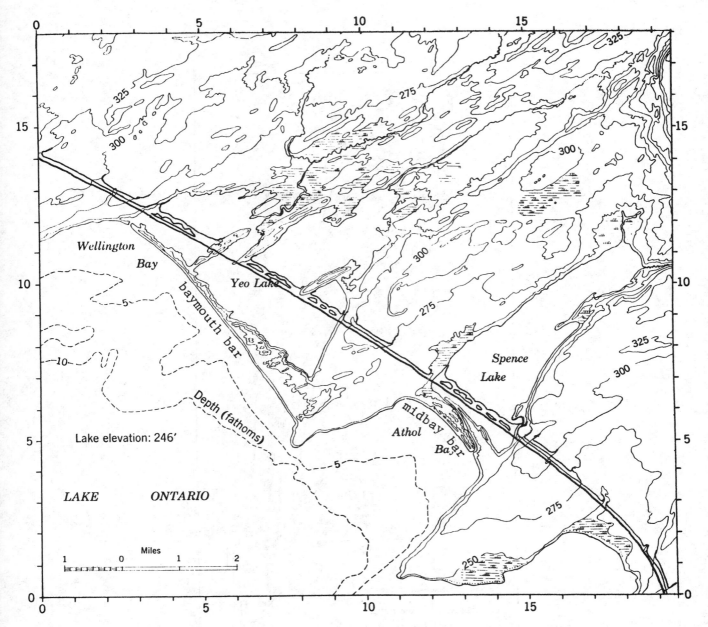

Figure C Portion of Wellington, Ontario, topographic sheet. Geological Survey of Canada. Original scale 1:63,360.

Sandy Hook--A Large Complex Spit

Sandy Hook, shown in the topographic contour map, Figure D, is a large spit extending north from the Navesink Highlands of New Jersey into the Lower Bay of New York Harbor. The basic structure of the spit has been formed of sand carried northward along the New Jersey coast by shore-drifting processes. Numerous old beach ridges show stages in the growth of the spit. These beach ridges have been modified by wind action and bear coastal dune forms. The tip of Sandy Hook is said to have grown about one mile since 1764; one-half mile since 1865. Formerly a military property, used for coastal artillery, Sandy Hook is today a State Park open to the public.

(6) Using a color pencil or pen, sketch a series of curved lines to show a succession of beach ridges produced during the growth of the spit. Use the contour patterns as a guide.

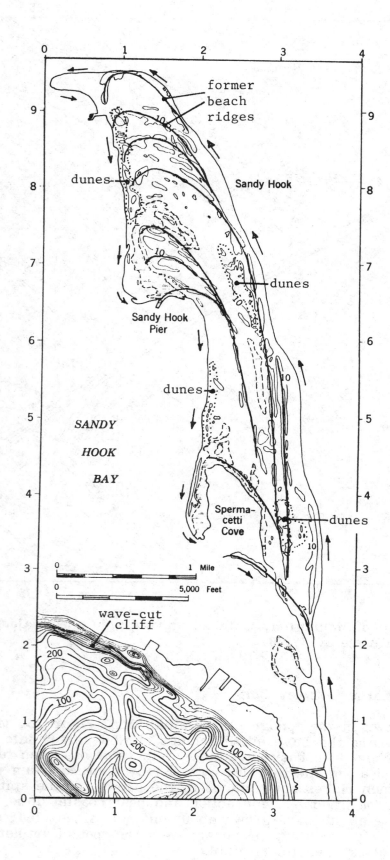

Figure D Portion of Navesink, NJ, topographic sheet. State of New Jersey, Department of Conservation and Development. Original scale 1:24,000.

(7) Draw many short arrows on the water, close to the shoreline, to show the dominant direction of beach drifting on both the outer (eastern) and inner (western) sides of the spit.

(8) Using a color different from that in Question 6, color at least three areas on the spit in which you might expect to find sand dunes resting upon the beach deposits. Label these as "Dunes."

(9) What is the origin of the marsh near 1.4-7.6?

The marsh lies in low ground between successive beach ridges; i.e., in a swale, or two swales. Sand spits don't grow uniformly, but rather in spurts, producing a succession of ridges separated by swales.

(10) What is the origin of Spermacetti Cove?

Shore drift of sand southward along the shore of Sandy Hook Bay built a small, secondary spit pointing south, that now encloses a small body of water.

E (Note correction.)

Figure *D* is a map of the Sandy Hook area in which earlier stages in the evolution of this coast have been reconstructed. It was presented in 1919 by Professor Douglas Johnson of Columbia University, then a leading authority on coastal geomorphology. The hypothetical original shoreline (1) is intended to show the coastline as it was several thousand years ago, when melting of the great Pleistocene ice sheets had caused a rise of sea level, drowning stream valleys of the Atlantic Coastal Plain.

(11) Follow with a pencil point, starting each time at the bottom of the map and tracing it to the northernmost limit, each of the numbered shorelines in order from 1 through 6. What landscape feature that exists today lies along the line of stage 2? Find this feature on the topographic contour map and describe its form. Label it on the map. How high is the feature above sea level?

Wave erosion of the bedrock mass called "Highlands of Navesink" produced a high marine cliff or scarp, shown on the topographic map by closely crowded contours. The cliff is about 200 feet high. [Notice that at 1.2-1.8 a small hill at the scarp base suggests the presence of a landslide mass resulting from oversteepening of the cliff. The bedrock here consists of weakly consolidated strata of Upper Cretaceous age.]

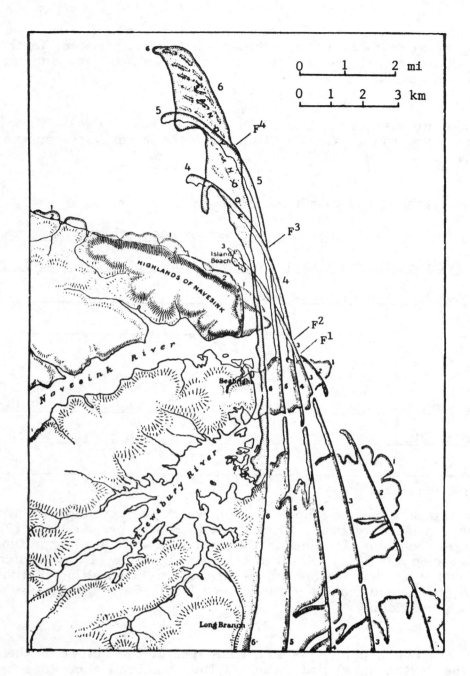

Figure E Sketch map of the northern coast of New Jersey, showing the development of Sandy Hook. (Same source as Figure B. Copyright © 1919 by John Wiley & Sons, New York. Used by permission.)

(12) In Figure E, the letter "F" appears four times, given as F^1, F^2, F^3, and F^4, in succession from south to north. "F" stands for "fulcrum." Deduce from the map what meaning the author intended for this term. How do the coastal-forming processes south of each fulcrum point differ from those north of the same point?

North of the fulcrum, progradation is in progress, building a spit northward into the open water. As the spit grows, it is shifted toward the east. South of the fulcrum, retrogradation is occurring, destroying the land and pushing the cliffed shoreline westward. As retrogradation occurs south of the fulcrum, the baymouth bars and the stem of the spit are also pushed westward.

_____   _____

Exercise 17-C   Barrier Island Coasts
         [Text p. 380-83, Figures 17.22, 17.24, 17.25.]*

Barrier island coasts are well represented along the passive eastern continental margin of North America, with its partly submerged coastal plain extending from New Jersey to Georgia, and around the entire Gulf of Mexico from the Florida panhandle to Tampico, Mexico.

The Texas Barrier Island Coast

(1) We begin the exercise by a close inspection of the Landsat image of a portion of the Texas barrier island coast, Figure 17.24. Fasten a sheet of tracing paper over this illustration and secure it with tape. Mark the corners of the rectangle. Find and label the features listed below. Use the suggested abbreviations. Use arrows as needed. Consult the map, Figure 17.25, for locations. Attach the finished tracing sheet to the blank rectangle provided on p. 506.

CB	Corpus Christi Bay
CC	Corpus Christi, city
NR	Nueces River (entering CB from west)
AP	Aransas Pass
SC	Ship channel to Aransas Pass
MI	Mustang Island (just south of Aransas Pass)
IF	Irrigated farmlands west and south of CC
PI	Padre Island
BB	Baffin Bay (deep bay with branching arms)
DV	Drowned valleys of BB.
LM	Laguna Madre
SD	Sand dunes (along inside shore of PI)
CL	Clouds
KR	King Ranch (red area near bottom)

Barrier Islands of the Delmarva Peninsula

Figure A is a topographic contour map of a section of the barrier island coast in Virginia. The area shown is in the narrow southern end of the Delmarva Peninsula that lies between Chesapeake Bay and the open Atlantic and ends in Cape Charles. Not far to the north along this coast is Assateague National Seashore and the community of Chincoteague, familiar to many TV viewers as the locality where a herd of wild horses has its home.

Metomkin Island and Cedar Island are two segments of the barrier island, separated by Metomkin Inlet. Large areas of open lagoon--Metomkin Bay and Floyds Bay--compete with salt marsh, shown in a distinctive pattern laced by a network of sinuous tidal creeks. The mainland can be taken to begin along a line coinciding with the 10-ft contour, which defines the outer limit of flat upland surfaces bearing the local designation of "necks."

---

*Modern Physical Geography, 3rd Ed., p. 335-37, Figures 19.31-33, Plate J-2.

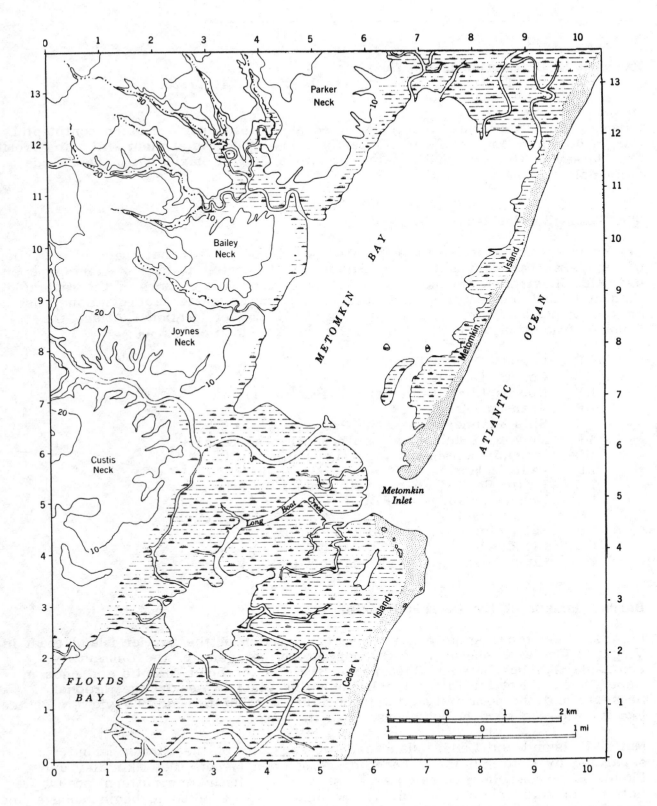

Figure A Portion of Accomac, VA, Quadrangle. U.S. Geological Survey. Original scale: 1:62,500. Contour interval 10 ft.

(2) Note that the four necks--Parker, Bailey, Joynes, and Custis--are sharply defined on the east by the 10-foot contour. This suggests the presence of a more-or-less straight scarp sloping down to tidewater level. The landward limit of the salt marsh is a nearly straight line paralleling that contour. Can you suggest an origin and history for this scarp?

The scarp (if correctly interpreted as such) must have been cut by wave action prior to the building of the barrier beach. The erosion would have occurred after the sea level had risen rapidly to its present level, following the rapid wasting of the continental ice sheets. [More likely, it dates back to an earlier interglacial epoch.]

(3) Tidal marsh and tidal streams extend inland into the branching stream valleys between Parker Neck and Bailey Neck and between Custis Neck and Joynes Neck. How can these features be explained in terms of glacial/interglacial changes of sea level. Reconstruct the events of valley erosion and valley filling.

During the maximum of the Wisconsinan stage of glaciation, sea level stood much lower than today, allowing streams to deepen their valleys, which were also extended far eastward across what was then exposed seafloor of the continental shelf. Post-glacial rise of sea level was accompanied by up-building of a barrier island. At the same time, the deepened stream valleys were invaded by tidal waters. Deposition of sediment in the tidal zone followed.

(4) At Metomkin inlet, there is an offset in the straight outer shorelines of Cedar Island and Metomkin Island, the former being farther seaward. Can you explain this offset in terms of shore drifting processes?

The sand supply seems to be from the south, with a persistent net shore drift northward. Under such conditions, the updrift barrier is built more rapidly seaward, causing an offset. [In many cases, elsewhere, the updrift barrier beach overlaps the downdrift barrier beach.]

(5) Notice that the width of tidal zone (distance from 10-ft contour to the barrier island) decreases from south to north. Give two possible explanations for this widening.

(a) The seaward slope of the former seafloor on which the barrier beach is built

may increase in steepness toward the north. (b) The supply of sediment for

building the barrier island may lie to the south. Diminishing sand supply toward

the north would allow wave action to push the barrier beach closer to the

mainland. [This explanation seems to fit with the answer to Question 4.]

The Barrier Beaches of Long Island, New York

The south shore of Long Island is yet another fine example of a barrier-island coast. It excels as an example of the shaping of tidal inlets and their relationship to the direction of prevailing shore drift of sand. Figure B shows the principle. Shore drift from east to west allows the updrift end of the barrier to grow rapidly, overlapping the downdrift end, which is narrowed and eroded. The inlet thus migrates downdrift.

Figure C is a simplified map of the western part of Long Island. Notice that Fire Island overlaps Jones Beach, and that Rockaway Beach overlaps Coney Island. However, Jones Beach seems to align quite well with Long Beach across Jones Inlet.

(6) Can you offer an explanation of the lack of overlapping of the barrier beach at Jones Inlet?

(a) Jones inlet may be of recent origin, breached by storm overwash and

remaining open, but with too little elapsed time to result in an offset. (b)

Because of the large proportion of salt marsh in this section of the coast, tidal

flow through Jones Inlet may be weak. If so, tidal currents would not interfere

seriously with shore drift past the inlet. Fire Island inlet may be very old,

allowing it to have migrated far westward, gradually acquiring a large offset and

overlap. Tidal current flow through Fire Island Inlet should be powerful, because

of the large open water area of Great South Bay.

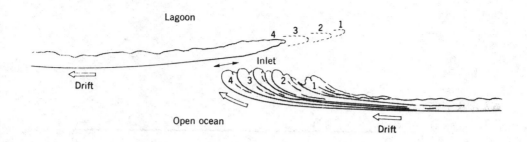

Figure B Schematic map of migration of an inlet, with offsetting and overlap. (A.N. Strahler.)

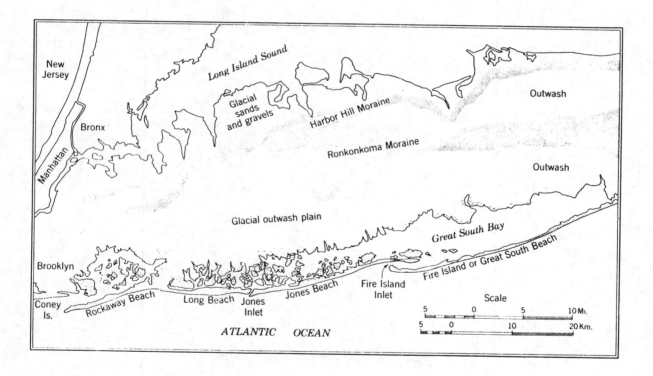

Figure C Coastal features and moraines of western Long Island, New York. (Based on maps of the U.S. Geological Survey.)

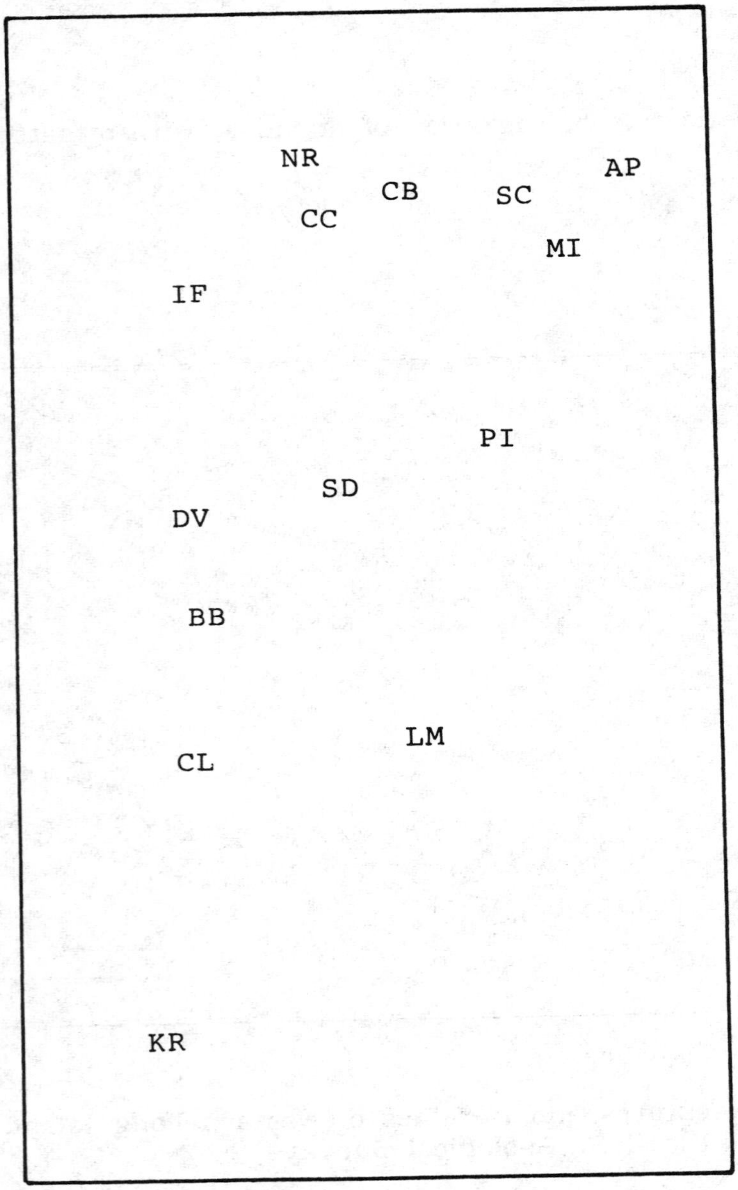

NR

CB   SC   AP

CC

MI

IF

PI

SD

DV

BB

LM

CL

KR

Name _____   Date _____

_____   _____

Exercise 17-D   The Ocean Tide
             [Text p. 376-77, Figure 17.15.]*

For hundreds of millions of years, the ocean tide has played a major role in shaping coastal landforms. In our modern era of high technology, the ocean tide sometimes plays a new and different role. If it should just happen that a great oil tanker with full load, on its way out of the Port of Valdez, Alaska, were to run aground on a submerged rock just after high water of a spring tide, there might be a delay of full lunar cycle of 29 days before the vessel could be floated off that rock. During that period, leaking oil could spread through the branching estuaries of the harbor, carried alternately landward and seaward until the oil reached the shores of the entire estuary system of Prince William Sound. Let's hope that it never happens--but of, course, it did.

What causes the tidal rise and fall of the oceans? The moon is in control of the earthly tidal cycle. Figure A shows an imaginary uniform ocean layer over a perfectly spherical solid earth. The force of the moon's gravitational attraction is greatest at point A, closest to the moon; least at C, farthest from the moon; intermediate in strength at B. This diminishing force from A to C causes the tide-raising force, shown by small arrows in Figure B. The ocean water is drawn toward two tidal centers, A and C, where the water tends to accumulate and its level rises to a summit. Thus we have two centers of high water level (at A and C). At the same time, the water subsides to lower level along a great circle passing through the poles. This global "girdle" of low water separates the two centers of high water.

Next, we bring into play the fact that the earth is continually rotating on its axis, so the two bulges and the low girdle continually sweep around the solid earth. An individual at a particular fixed location in low latitudes will observe a rise of ocean level twice each day, alternating with a decline to low level. These are the "high waters" and "low waters," respectively.

Part A   The Semi-Daily Tidal Cycle

The tidal rhythm explained above gives the kind of tide curve shown in text Figure 17.15. It is called the semi-daily tidal cycle. Now, the lunar day--one earth turn with respect to the moon--is about 25 hours (25^h) of mean solar time (clock time). We can anticipate or predict, then, that two successive high waters will occur $12\frac{1}{2}^h$ apart; and same for successive low waters. Does observation confirm this prediction? To find out, we need to plot the data of an actual tidal cycle.

*Modern Physical Geography, 3rd Ed., p. 332, Figure 19.25.

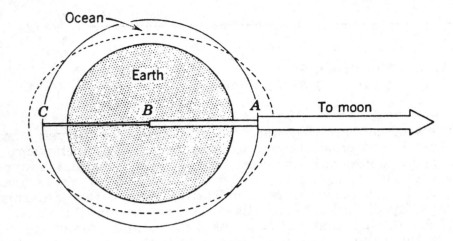

Figure A Gravitation as the tide-producing force. (A.N. Strahler.)

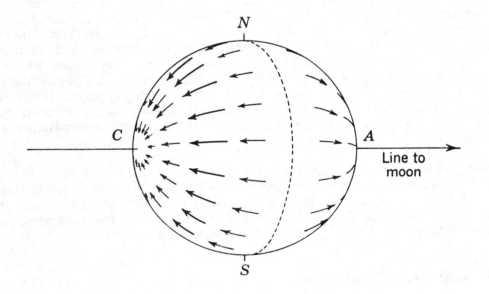

Figure B Ocean waters tend to move toward two centers, A and C. (A.N. Strahler.)

(1) Use the data of Table A, showing hourly observations of staff gauge height for San Francisco harbor during a 24-hour period. Plot the data as points on the blank graph (Graph A). Connect the points with a smooth curve.

Table A San Francisco, California

Hour	Height (Feet)	Hour	Height (Feet)
12 midnight	0.0	1 P.M.	−0.3
1 A.M.	−1.5	2 P.M.	−1.5
2 A.M.	−2.3	3 P.M.	−2.4
3 A.M.	−2.6	4 P.M.	−2.5
4 A.M.	−2.3	5 P.M.	−1.9
5 A.M.	−1.5	6 P.M.	−1.0
6 A.M.	−0.3	7 P.M.	0.1
7 A.M.	0.8	8 P.M.	1.2
8 A.M.	2.0	9 P.M.	2.0
9 A.M.	2.6	10 P.M.	2.3
10 A.M.	2.8	11 P.M.	2.2
11 A.M.	2.3	12 midnight	1.2
12 noon	1.2		

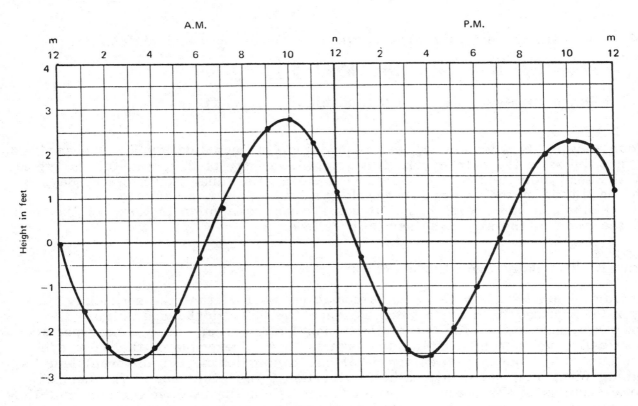

Graph A

(2) Our data cover most of two semi-daily tidal cycles, each of about $12\frac{1}{2}^{h}$ duration. Check your plotted tide curve to see if each cycle lasts for $12\frac{1}{2}^{h}$. Actually, the true value of the average semi-daily cycle is more accurately given as $12^{h} 25^{m}$ (12.42^{h}). Measure the elapsed time between successive low waters and between successive high waters. Calculate the average of the two values.

Between successive low waters: <u>12</u> hrs <u>35</u> min

Between successive high waters: <u>12</u> hrs <u>15</u> min

Average: <u>12</u> hrs <u>25</u> min

Comment on your data. Do they agree with the more exact value given above? Both readings differ from the accurate value, one being shorter and the other longer, but the average comes close to that value.

---

(3) Determine as closely as you can the heights of the two high waters and the two low waters. Give to the nearest one-tenth foot.

First high water   +2.8   ft   First low water   -2.6   ft

Second high water   +2.3   ft   Second low water   -2.5   ft

(4) What were the height ranges between successive low and high waters?

Range from first low to first high: 5.4   ft

Range from second low to second high: 4.8   ft

Your answers to Question 4 give the amplitude of each tidal cycle. The amplitude is not the same for both cycles in a single lunar day of 25^h, and this is typical of the semi-daily tide curve. Only twice during the lunar month of $29\frac{1}{2}$ days is the amplitude of the two cycles closely matched. This is a topic we must leave unexplained in our exercise, but it has a clear explanation in terms of the path of the moon in the sky, whether high or low.

Part B The Daily Tidal Cycle

At various places along the world's coastlines we find that the semi-daily tidal cycle is replaced by a cycle twice as long--the daily tidal cycle. Figure C compares these cycles for two stations: Portland, Maine, shows a near-perfect semi-daily cycle; Manila, Philippine Islands, a good example of the daily cycle. Sections of the ocean behave somewhat like tuning forks. For different lengths, tuning forks respond in sympathetic vibration to different sound frequencies. Some sections of the ocean just won't "vibrate" on the two-cycle input, but they vibrate nicely on the daily rhythm. We have just such an example in the tidal cycle for St. Michael, Alaska, a port on the coast of the Bering Sea at about latitude 64°N, just south of the Seward Peninsula.

(5) Plot the data of Table B on Graph B. Connect the points with a smooth curve. Notice that this graph spans two days, or 48 hours, which is twice the time covered in Graph A. The table data are for 2-hour intervals and each graph unit spans 2 hours.

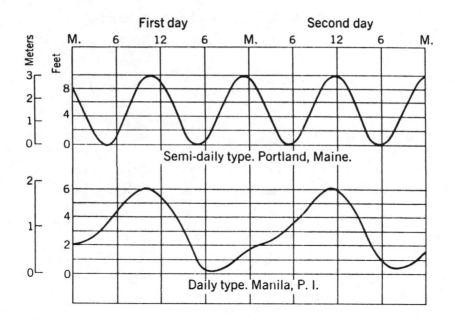

Figure C Examples of two kinds of tide curves. (A.N. Strahler.)

(6) What interval of time elapsed between successive high waters? between successive low waters? (Read to nearest 10 minutes.)

Time of first high water: __04:00__

Time of second high water: __04:30__

Elapsed time: __24^{h}30^m__

Time of first low water: __17:10__

Time of second low water: __18:40__

Elapsed time: __25^{h}30^m__

How do your readings compare with a value twice that of the semi-daily cycle, given in Question 2?

Compared with the correct value of 24^{h}50^m, the elapsed time for the high waters is shorter, but only by 20^m. For the low waters the observed value is 40^m too long. The average of the two, 25^{h}00^m, is only 10^m too long. [Measurement error may be important here, since the point of inflection of the curve is difficult to locate.]

Table B St. Michael, Alaska

First Day		Second Day	
Hour	Height (Feet)	Hour	Height (Feet)
0	1.0	0	1.1
2	2.0	2	2.3
4	2.3	4	3.0
6	1.7	6	2.9
8	0.4	8	1.9
10	−0.5	10	0.4
12 noon	−1.2	12 noon	−0.5
14	−1.6	14	−1.3
16	−1.8	16	−1.8
18	−1.8	18	−2.2
20	−1.6	20	−2.2
22	−0.5	22	−1.8
24	1.1	24	−0.5

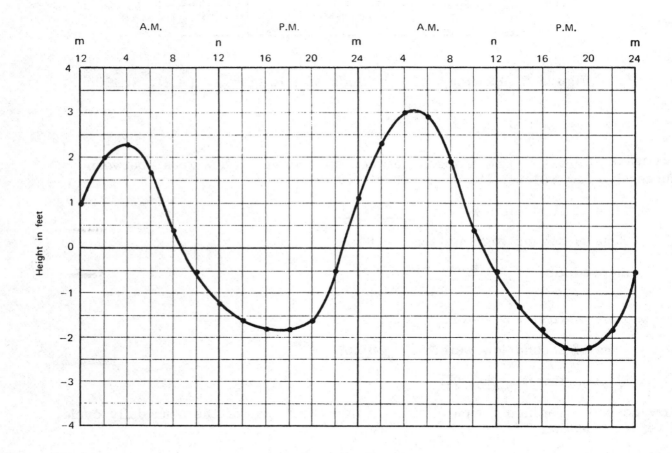

Graph B

(7) Compare the two cycles--semi-daily and daily--as to the uniformity and symmetry of the curves. Look first at Figure A, then at Graphs A and B. Describe your impression of these curves.

The peaks and valleys of the semi-daily curves are highly uniform and symmetrical, even when the amplitude changes. The daily curve tends to have sharply formed and symmetrical peaks, but the valleys are definitely more open, or U-shaped. [This asymmetry can be traced to the supressed effect of the daily cycle. Note the small daily inflection of the trough of the Manila curve.]

Part C Neap and Spring Tides

Those who fish or sail the coastal waters of North America are well aware of yet another tidal cycle that affects their marine activities. They will talk about "spring tides" and "neap tides," much to the obfuscation of their landlubber friends. Most of the latter would, of course, know that spring tides occur in the spring of the year (dead wrong!), but "neap" would ring no such bell. So we must back up to basics and look again at the tide-producing force and what controls it.

The sun's gravitational pull on the earth is much stronger than that of the moon, but it takes second place to the moon in its ability to raise tides of the oceans. Why this should be so is a question we can only hint at here. Look at Figure A and replace the word "moon" with "sun." Our sun is 93 million mi distant (average value); the moon a mere 240,000 mi (0.24 million miles). These facts lead to the conclusion that with the sun on the same reference line, its gravitational attraction at points A, B, and C differs only very slightly, as compared with the equivalent differences with the moon on the line. So, while the sun holds our earth in orbit, it yields to the moon on the control of the tidal cycle.

The sun's tidal influence is to increase or decrease the tide-raising force. When moon and sun are in the same line that force is stronger; when the two bodies exert their pull at right angles to each other, that force is diminished. Figure D shows these relationships; it pictures the moon's orbit as a circle, with the sun far off to the left. Conjunction and opposition are the two in-line positions; quadrature, occuring twice each month puts the moon's line at right angles to that of the sun. (The strange term "syzygy" describes both or either of the in-line positions in one word.)

Figure E shows how the high and low waters of the tide are affected by the moon's position, or phase, in relation to the sun. Spring tides, occurring close to the phases of full moon (opposition) and new moon (conjunction), have exceptionally large ranges. Neap tides, occuring near phases of first quarter and third quarter (quadrature), have exceptionally low ranges. Spring tides are about 20 percent greater than average; neap tides about 20 percent less.

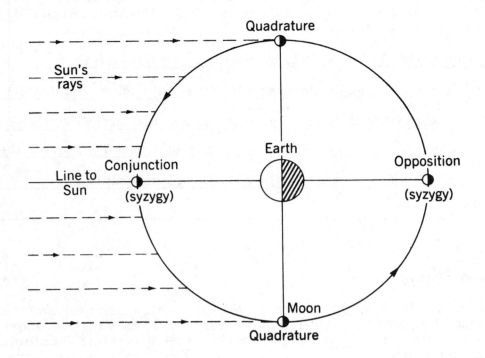

Figure D Key positions in the moon's orbit. (A.N. Strahler.)

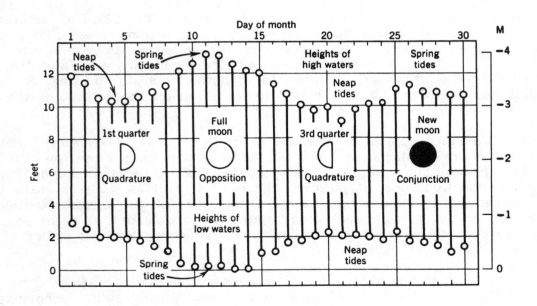

Figure E An example of spring and neap tides in Boston harbor. (A.N. Strahler.)

(8) From Figure E, determine the following values:

Greatest range of spring tides at full moon:

Height at high water: __13.3__ ft

Height at low water: __0.2__ ft

Range: __13.1__ ft

Least range of neap tides at first quarter:

Height at high water: __10.2__ ft

Height at low water : __1.9__ ft

Range: __8.3__ ft

Average of spring of and neap ranges: __10.9__ ft

Percentage difference of spring and neap from average: __24__ %

Name _____ Date _____

Exercise 17-E   Crescent Dunes and Sand Seas
[Text p. 392-94, Figures 17.39, 17.40., 17.41.]*

For sheer vastness, the North African-Arabian deserts as a group rank first for their huge seas of bare dune sand, over which the sharp dune crests continually shift with the wind. Until satellite imagery became available, North America wasn't world-class in this field, but we now find excellent examples of a wide range of desert dunes in the great Sonoran Desert (Gran Desierto) of northwestern Mexico and the adjacent southwestern United States. Satellite images also reveal new details of vast inland dune fields in the Kalahari Desert of southwestern Africa and deserts in central and western Australia.

In this exercise we investigate first those active dunes whose crests lie transverse to the wind--crescent dunes and transverse dunes of sand seas. To this we add, briefly, a look at the great star dunes.

Part A   Crescent Dunes, or Barchans

The isolated crescent dune achieves remarkable perfection of symmetry, but can vary considerably in outline and in the proportions of slip face to windward slope. An alternative name, <u>barchan</u>, is preferred by many writers. (Pronounce it 'bar-can'.) The variant spelling, <u>barkhan</u>, gives a clue to its origin in central Asia. Figure A, showing a side view of barchans along the Columbia River in Oregon, was taken about a century ago by the distinguished American geologist, G.K. Gilbert, and remains unexcelled. Notice that Dr. John Shelton's oblique air photo of a barchan in Peru (text Figure 17.40), shows an almost identical form. That the dune lies upon a rather rough ground surface underlines a salient point about barchans: they must travel downwind. More barchans are shown in Figure B (lower-right corner). Although solitary by nature, barchans also like to link arms as they travel. The little ones travel faster than the large ones, and it is said that the little ones catch up with the big ones, with which they merge.

(1) Using a color pencil or crayon, color several of the barchans in Figure B.

(2) In Figures A and B, and text Figure 17.40, examine closely the desert surface surrounding the barchans. What material composes this floor? Why does it remain unmoved by the same winds that are capable of driving sand over the dune surface? (Clue on text p. 391.)

<u>The surface appears to have a cover of scattered pebbles and/or shrubs. The pebble cover would form a desert pavement, resisting wind erosion. [The pebbles and large sand grains of the pavement allow dune-sized sand grains to travel easily in saltation.]</u>

*Modern Physical Geography, 3rd Ed., p. 334-35, Figures 20.4-20.7, Plate J.4

Figure A Crescentic (barchan) dunes near Biggs, OR. (G.K. Gilbert, U.S. Geological Survey.)

Your text, p. 392, describes the manner in which sand grains travel over sand surfaces by leaping and rebounding. The scientific term for this activity is saltation, from the Latin saltare, to leap or dance. Figure C shows a ground plan and longitudinal profile of a typical barchan. Saltation carries sand grains up the windward slope to the dune crest, from which the grains fall through the air to land on the slip face. Enclosed by the slip face is a sheltered stagnant air zone upon which the slip face is encroaching. From each of two horns, sand in saltation streams downwind from the dune in a spray zone.

(3) On Figure C label all of the terms named in the above paragraph. Use arrows as needed to point to the features named.

(4) Complete the circular outline of the barchan to make a complete, symmetrical geometrical figure. What geometrical term describes this outline? Label it on the drawing.

    Outline figure: _____ellipse_____

The contour topographic map, Figure D, shows barchans of a large dune field near Moses Lake, Washington.

(5) Using a color pencil or pen, draw in the basal outline of at least twelve well-developed barchans shown on the map. Place a large arrow on the map to show the prevailing wind direction.

(6) Choosing the three highest, best-formed barchans, measure the width, W, (horn-to-horn, across the wind) of each, in feet. Calculate the height, H, of each in feet. Write these figures directly beside the dune. (Note: The grid on the map border is scaled in 1000-yd units.)

Figure B Sand sea of transverse dunes between Yuma, AZ, and Calexico, CA. (Robert E. Spence.)

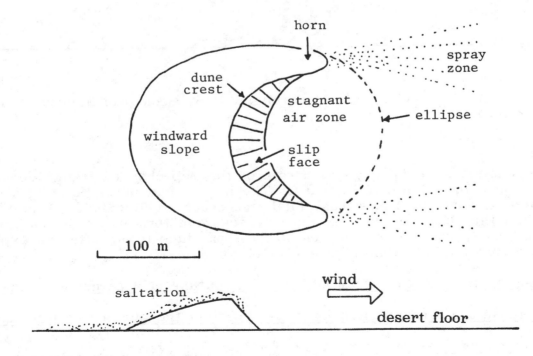

Figure C Sketch map and profile of an idealized barchan. (A.N. Strahler.)

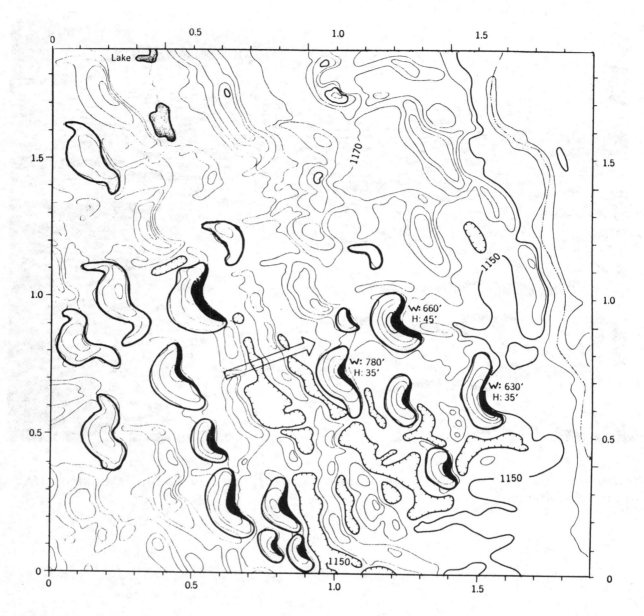

Figure D Portion of Sieler, WA, Quadrangle. U.S. Geological Survey. Original scale: 1:24,000.

(7) Barchans close to the western margin of the map show a strange departure from the ideal form shown in Figure C. The horns, instead of pointing straight down-wind, are curved backwards. Find an example at 0.25-1.00. Draw the basal line around this dune to show the special backswept form of the horns. Find and outline in color three other examples of the same dune form. Offer an explanation for this feature. (Hint: text Figure 17.43.)

The sweeping back of the horns suggests the outline of a parabolic dune.

Perhaps these are a mixed (hybrid) type. Possibly the pure crescentic form is in

process of changeover to the parabolic form; or vice versa.

(8) Study the two small dunes located at 0.60-1.20 and 1.10-1.15. Outline their basal plan. Does their form provide evidence for your explanation given in Question 7.

These are clearly parabolic dunes. They lack the characteristic slip face of the

barchan.

(9) What evidence is there on the map that the ground-water table lies close to the surface? What influence might this condition have on dune development?

Lakes at 0.4-1.6 suggest that a water table lie close to the surface. The closed

depressions in the south-central part of the map may hold marshes with

vegetation that would trap blowing sand. The effect might be to favor parabolic

dunes.

Part B Transverse Dunes of a Sand Sea

Figure E is a contour topographic map of the same transverse dunes seen in the oblique air photo, Figure B, and in text Figure 17.41. This field lies between Yuma, Arizona, and Calexico, California. Millions of travelers have seen these dunes from Interstate Highway 8, paralleling the All American Canal (seen in the distance in Figure B) that brings water from the Colorado River to California. Locally, this dune field goes by the name of the Sand Hills.

(10) Compare the two air photos. By looking very closely, you can find the identical dune features on each. How do these photos differ in direction of view and the dune features they emphasize. (Get your directions from the map, Figure E.

Figure B is looking westward, directly at the slip faces of the dunes. Figure

17.41 looks northeastward, along the dune crests, with the slip faces in deep

shadow. The same barchan dunes are shown in both, with one sinuous dune

ridge of connected barchans distinctly visible in both photos.

(11) Using color pencil or crayon, shade the slip faces of the dunes. Show direction of prevailing wind with a bold arrow.

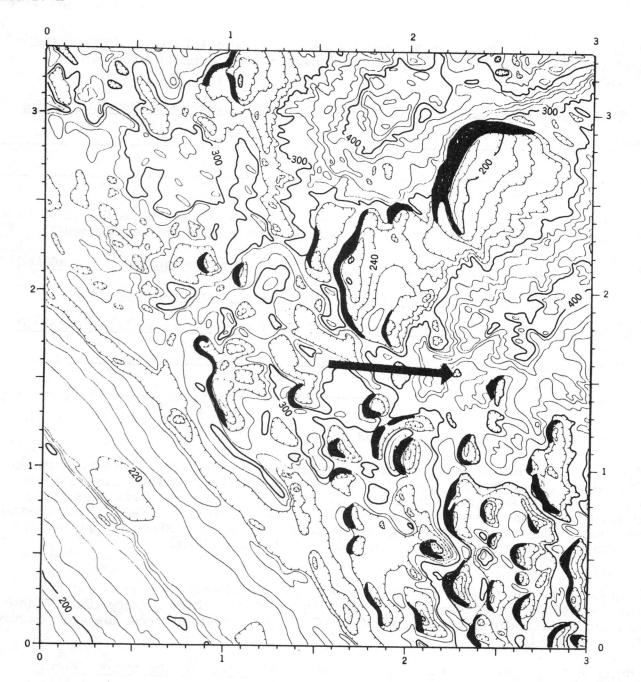

Figure E Portion of Glamis Southeast, CA, Quadrangle. Original scale 1:24,000.

(12) Special credit question. In the southwestern part of the map, contours run in parallel alignment, trending northwest to southeast. Do these contours represent dune ridges? Offer an explanation of this topography and cite supporting evidence.

The extreme parallelism and uniformity of contours seems to exclude dune forms, especially since the ridges shown trend obliquely to the prevailing wind. An alternate hypothesis is that the long, low ridges are beach ridges of a former lake or bay. Uniformity of the crest elevations suggests a water-plane as a controlling factor.

Part C   Star Dunes

Study the rows of star dunes shown in text Figure 17.42. Similar rows of star dunes are found in the Sonoran Desert. Figure F shows sketches of a star dune with six points, another with three points and a "pinwheel" look.

(13) Your text, p. 394, states that isolated star dunes in parts of the Sahara Desert "remain fixed in position and have served for centuries as landmarks for desert travelers." Assuming that the statement is accurate, what might you infer as to wind conditions necessary for the formation of the star shape and its fixed position?

One single dominant prevailing wind must be ruled out in favor of two, three, or more important wind directions, each with its own season. When one of these seasonal winds moves the dune crest in one direction, at another season the sand movement would be reversed and the crest would be restored to its earlier position. Competing wind directions might also be explained by reversing directions accompanying moving cyclones and weather fronts.

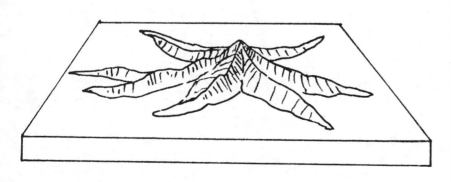 

Figure F   Sketches of star dunes. (A.N. Strahler.)

_____   _____

Exercise 17-F   Blowout Dunes and Hairpin Dunes
          [Text p. 394-97, Figures 17.43. 17.44, 17-45.]*

Provincetown, Massachusetts, is a famous fishing port on Cape Cod, thriving since colonial days. The Mayflower Pilgrims landed here in 1620, stayed a month or so, then sailed west across the bay to disembark at Plymouth Rock. Provincetown is safely nestled in a quiet harbor within the curved fist of Cape Cod, and well protected from vicious Atlantic storms attacking from the north and east. It was these same gale-force winds that brought waves of sand into the streets of Provincetown.

Figure A is a sketch map the northern tip of Cape Cod, known as the Provincelands. Wavelike dune ridges cover most of the area, with summits rising over 80 ft (25 m) above sea level. Although sand movement was previously kept in check by a native forest of pitch pines, the trees were cut away by the first settlers. By about 1725, the invasion of Provincetown by sand began to be a serious menace. Steep dune slopes began to back up against houses, and driving sand frosted window panes. Sand drifts, accumulating in the streets, had to be removed constantly by use of horsedrawn carts. When the naturalist Henry David Thoreau visited Provincetown in 1849, he wrote:

> The sand is the great enemy here. . . . The houses were formerly built on piles, in order that the driving sand might pass under them. . . . There was a school-house just under the hill on which we sat, filled with sand up to the tops of the desks, and of course, the master and scholars had fled. Perhaps they had imprudently left the windows open one day, or neglected to mend a broken pane.[1]

Eventually, the Provincetown dunes were pretty well stablized and remain that way today, but slip faces of dunes near the north shore continue to transgress the blacktop of main roads.

---

[1]From Henry D. Thoreau, <u>Cape Cod</u>, Copyright © 1961 by Thomas Y. Crowell Co., New York, p. 174-176.

Part A   Blowout Dunes of the Provincelands

One special feature of the Provincelands that relates to this exercise is the occurrence of several nicely developed parabolic blowout dunes. Their location is shown in Figure A, directly north of Pilgrim Lake. Figure B is a contour topographic map showing these dunes. Figure C, a ground photo, shows the inner floor of the dune directly north of High Head. You are standing on one dune ridge, looking downwind (about south-southeast) at the enclosing dune ridge. Plantings of dune grass cover the blowout floor and also the crest of the dune, but the windward slopes, up which the sand travels in saltation, remain bare.

---

*Modern Physical Geography, 3rd Ed., p. 345-47, Figures 20.8, 20.9, Plate J.4.

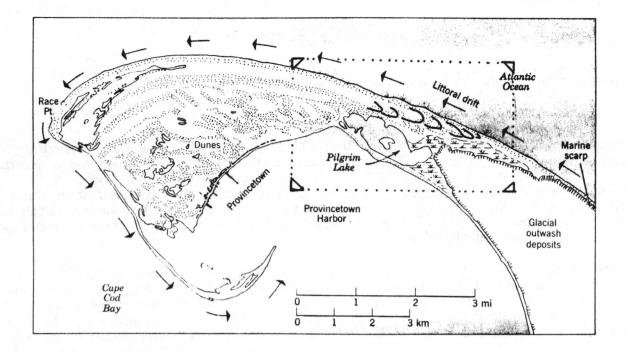

Figure A Sketch map of the Provincelands of Cape Cod, as it was in 1887. Dunes show as dot patterns. (A.N. Strahler.)

(1) Using a color pencil or crayon, shade the ridges of parabolic blowout dunes shown by contours on the topographic map, Figure B. In another color, shade the flat floors within the same dunes. (They are delineated by hachured depression contours.)

(2) Using the graphic scale on this map, give an approximate average width of these dunes, measured at right angles to wind direction. About how high do the dune summits rise above the level of the floors?

    Width: __1,500__ ft;   __0.5__ km

    Height: __40-45__ ft;   __12-14__ m

(3) The wind arrow on the map shows northwest winds as dominant for the development of these parabolic dunes. Under what meteorological conditions would strong northwest winds occur? Why aren't the legendary howling nor'easters (northeast storm winds) involved in shaping these dunes?

Northwest winds would follow the passage of a strong cold front in a deep wave

cyclone. Typically, these are drying winds. Nor'easters are typically accompanied

by drenching rains that wet the sand surface, suppressing its rapid surface

movement by saltation.

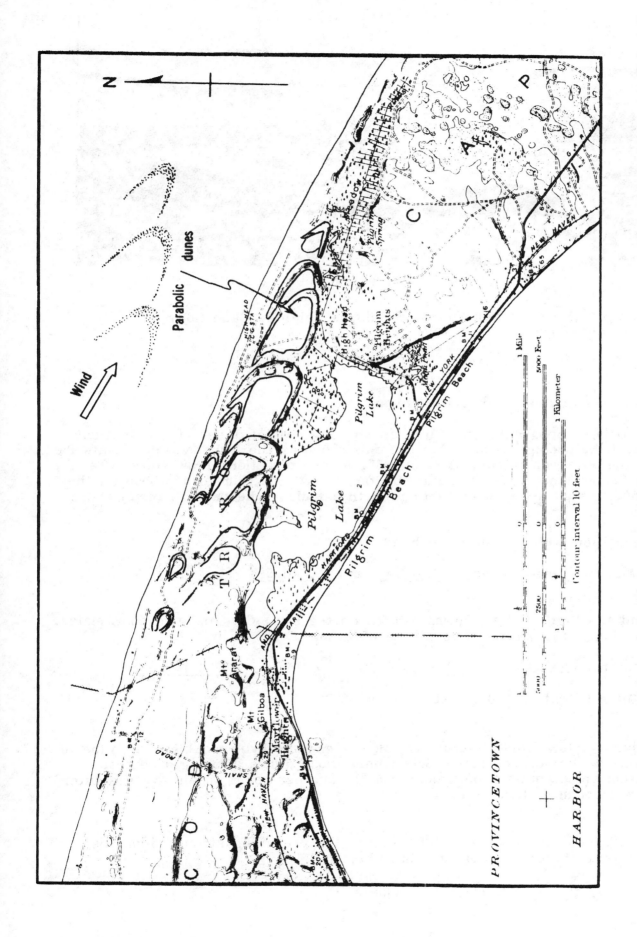

Figure B Topographic map of the Truro region of Cape Cod. U.S. Geological Survey. Original scale 1:24,000.

Figure C A parabolic dune north of Pilgrim Heights, Cape Cod. (A.N. Strahler.)

Part B Coastal Dunes of Lake Michigan

Figure D is a topographic contour map showing coastal dunes of the southern coast of Lake Michigan, set aside for public use as the Indiana Dunes State Park. Notice that the map is rotated about 45°, as the direction arrow shows. Text Figure 17.44 shows the advancing slip face of a dune, similar to those on this map, but located a few miles farther northeast along the Lake Michigan shore.

(4) About how wide is this dune belt?

Width: __600__ yds; __0.3-0.4__ mi; __0.55__ km.

(5) Find the highest dune summit shown on the map. Give its grid coordinates. Give its height with respect to a base elevation of 600 ft.

Summit elevation: __730+__ ft   Grid coord.: __0.53-2.18__

Dune height: __130+__ ft

(6) Using a color pencil or crayon, color each well-developed blow-out dune to emphasize its horseshoe shape. Select only those dunes showing a central depression enclosed by a prominent rim. In another color, emphasize the outer slip face of each of these dunes.

(7) Draw a bold arrow on the map to show the prevailing strong winds, as inferred from the dune forms you have selected.

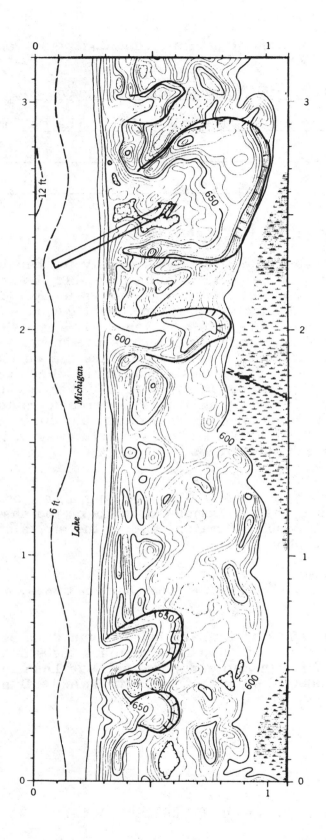

Figure D Portion of Dune Acres, IN, Quadrangle. U.S. Geological Survey. Original scale 1:24,000.

(8) What type and distribution of natural vegetation would you expect to find over this coastal dune belt?

Foredunes along the beach and blowout dunes may have a light cover of beach grasses and sbrubs. Steep slip faces will be bare. The flat marsh area will have a forest of deciduous trees.

Part C Hairpin Dunes of the California Coast

Figure E is an oblique air photo of a famous locality for hairpin dunes--the shore of San Luis Obispo Bay on the central California coast. It was taken in 1937, when the dunes were not as yet heavily impacted as today by human activity, such as mining of sand, denudation of natural plant cover, and the churning of sand surfaces by powerful ORVs (off-road vehicles). The main line of the Southern Pacific Railroad cuts right through one particularly long and beautifully-shaped hairpin dune. Several small lakes can be identified as located within a hairpin dune, or landward of the end of the dune. Groves of eucalyptus trees, imported from Australia and intended as a lumber resource, lie landward of the dunes (rectangular plots). Figure F is a topographic contour map of the same area. Text Figure 17.45 is another air view of hairpin dunes in the same general area; it shows clearly the contrast between the scrub vegetation of the fixed hairpin dunes and the cultivated fields.

(9) Cover the photograph, Figure E, with a tracing sheet, as in previous exercises. When the assignment is completed, transfer the sheet to the space provided. Using a color pencil or crayon, outline the crests of all hairpin dunes you can identify on the photo.

(10) On the Map, Figure F, outline the same hairpin dunes, as in Question 9.

(11) On both the photo and map find the long hairpin dune across which the railroad cuts a trench. Note that Mud Lake occupies the seaward end of the floor of this dune. Label it on both map and photo as "Mud Lake Dune." Identify and label other dunes as follows: White L. Dune, Big Twin L. Dune, Pipeline L. Dune.

(12) Measure the length and width of Mud Lake Dune.

Length:	2,000 yds	1,700	m
Width:	500 yds	450	m

(13) What explanation can you offer for the presence of permanent lakes in this semiarid climate?

These are water-table lakes, occupying depressions that lie below the level of the ground water table.

(14) Bolsa Chica Lake is triangular in outline. What geomorphic features are responsible for this shape?

On two sides (N and S) lie ridges of two adjacent hairpin dunes, meeting in a V. On the southeast side is an embankment shown by three parallel contours. It appears to be the scarp of a terrace at 100 ft elevation. This may be a marine terrace.

(15) On the photos, active (live) dunes of bare sand are present over large areas. Describe and name the kind or kinds of dunes present, in terms of their shapes. What symbol is used on the map to show these live dunes?

Those closest to the camera are transverse dunes whose crests trend north-south. The slip faces are on the landward side, which is the direction of their advance. A few dunes (lower-left in photo, Figure E) resemble barchans (crescent dunes) that merge with each other. On the map, the live dunes are shown by a stipple pattern of fine dots arranged in circular arcs.

(16) On the map, Figure F, along a line parallel with the shore (1.6-0.0 to 1.6-5.0) is a broad dune ridge enclosed by the 100-ft contour. This ridge appears on the photo to bear active transverse dunes. Shoreward of the ridge is a line of depressions conspicuous by their scrub vegetation. Between these depressions and the beach is a narrow belt of live dunes. Can you think of an explanation for the broad dune ridge that might involve changes in the elevation of the land with respect to sea level?

The ridge may represent a barrier beach of sand produced when the sea level was higher than today. Then, the coast experienced a tectonic uplift of some 80 feet. What was the former lagoon, landward of the barrier, became a low plain. The barrier beach became a sand source for the hairpin dunes, which advanced

Figure E Oblique air photo of coastal dunes along the shoreline of San Luis Obispo Bay, CA. (Robert E. Spence.)

across the plain. They became stablilized. Live transverse dunes then took over and largely obliterated the hairpin dunes.

Figure F Portion of Oceano, CA, Quadrangle. U.S. Geological Survey. Original scale 1:24,000.

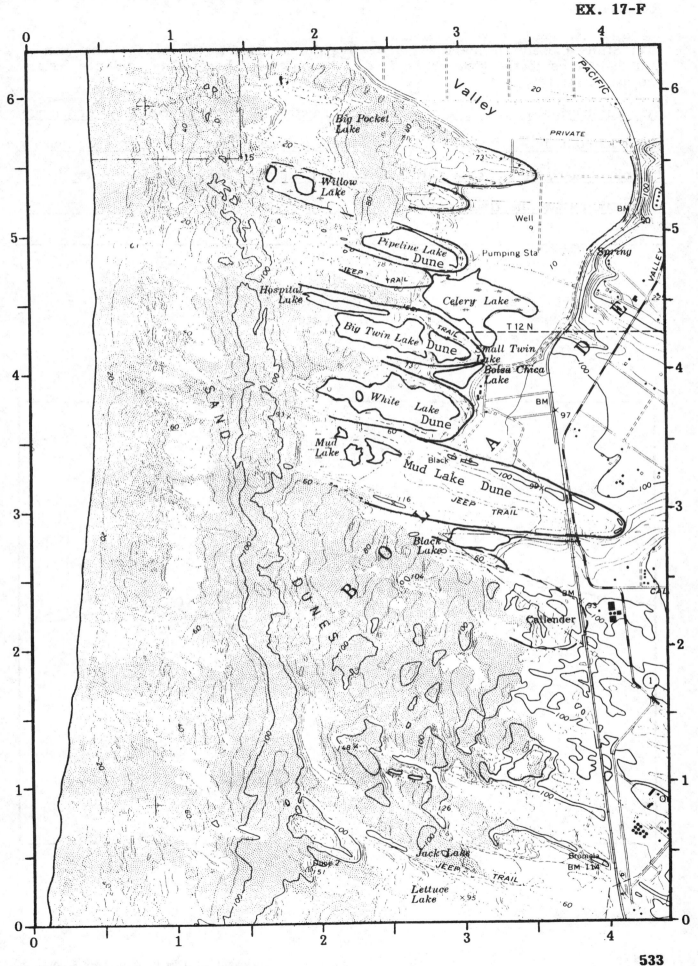

(17) Study closely the narrow belt of live dunes close to the beach. What can you deduce as to the kind or kinds of dunes present here? Cite evidence from the map contours.

<u>Straight contours, trending in the same direction as the side ridges of the large</u>

<u>fixed hairpin dunes farther inland, indicate a similar kind of dune ridge, namely</u>

<u>hairpin dunes, or elongate parabolic blowout dunes. These are of a newer</u>

<u>generation than the fixed dunes.</u>

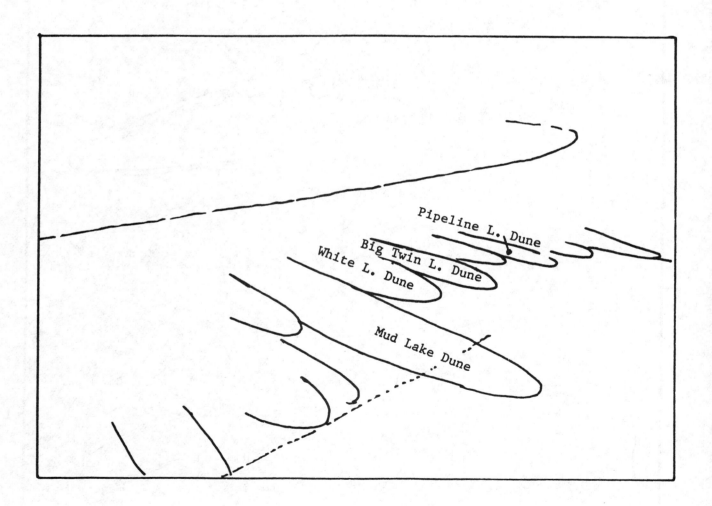

Exercise 18-A   Living Glaciers of Alaska and British Columbia
    [Text p. 402-405, Figures 18.1, 18.2, 18.3.]*

North America has its share of large alpine glaciers. They flourish in a great arc around the Gulf of Alaska: in northern British Columbia, the southwest corner of the Yukon Territory, and across southern Alaska. Here, the combination of lofty mountains ranges and a strong onshore flow of moist marine air guarantees the two essential requirements of alpine glaciers: high mountains and abundant precipitation.

Glaciers of the Alaska Range

Glaciers of the Alaska Range show clearly in text Figure 18.3. This is a good time to read again the legend beneath the satellite image, along with the text describing glacier flow and glacier regimen.

(1) Fasten a tracing sheet over text Figure 18.3. Trace in black the outlines of all the glaciers you can identify, including their upper branches.

(2) At least three of the glaciers show a marked widening near the terminus. Identify these with the letter **W**. Explain why this widening should occur.

Widening occurs because the glacier has pushed out over a nearly flat plain and is no longer confined between rock walls. The ice is free to spread laterally, tending to develop a "paw-shaped" outline. [This feature is referred to as a glacier foot.]

---

*Modern Physical Geography, 3rd Ed., p. 351-54, Figures 21.1, 21.2; Pl. K.1.

(3) Identify those glaciers that clearly show evidence of having experienced strong surging motion. Place the letter "S" over each of these. Does the sinuous (wrinkled) pattern of the medial moraines (moraines carried along on the ice surface) continue upstream to the higher parts of the glacier? Describe and explain what you see. Use the space below to draw a medial moraine to illustrate its form. (Attach the tracing sheet to the space provided.)

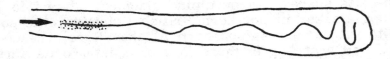

Sinuosity is greatest near the glacier terminus, and lessens upstream, finally dying out completely. This indicates that there is a piling-up effect in which the amount of compression increases downstream. This happens because ice flow meets with increasing internal and frictional resistance.

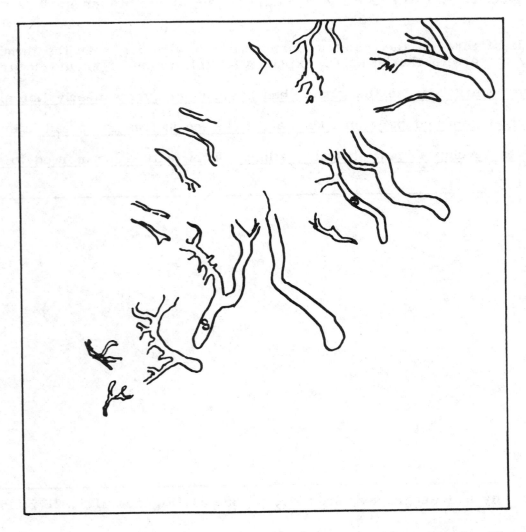

(4) Examine the glaciers on the northwest side of the Alaska Range; i.e., those flowing toward the northwest. Although they are partly obscured by clouds, you should be able to draw partial outlines. What important difference do you note between these glaciers and those that flow southeast from their common divide? Explain.

Glaciers flowing northwest are shorter and narrower. A plausible explanation is that orographic precipitation is much greater on the eastern side of the range, which is windward to moist maritime air masses. Arctic air masses approaching from the west would have a low moisture content, producing much less snowfall on that side of the divide. [The ice-drainage divide has probably shifted westward as a result.]

Glaciers of the Chugash Mountains

Figure A is a contour topographic map of glaciers in the Chugash Mountains, a coastal range at the head of the Gulf of Alaska. Port cities of Valdez and Cordova are located near here. Valdez, at the head of Prince William Sound, is the terminal point of the Alaska Pipeline. Cordova lies southwest of the map area. The major topographic feature of reference is the Copper River. Its valley lies just off the map to the east, and we see a small piece of its estuary and floodplain in the upper-right corner of the map.

On the original U.S. Geological Survey map, contours on glacial ice are printed in blue, making an easy contrast with the brown contours of the land. A dashed line separates the two kinds of areas.

To give you a better understanding of the glaciated terrain of our map, we include a remarkable air photograph of a similar landscape (Figure B). It shows the great Eagle Glacier and its source icefield in the Coast Mountains of northern British Columbia, near the International Boundary with Alaska. Called a trimetrogon photograph, it looks both down and sidewise. When viewing it, imagine that you are in a window seat in an airplane. You can look almost straight down, and also into the far distance. The strange, almost grotesque object you see in the lower right corner of the photo is a huge cirque, containing a small relict glacier. The photo gives a beautiful display of medial moraines. Smooth white glacier surfaces in the distance are firn fields in the zone of accumulation.

(5) Using a color pen or pencil (blue), trace on the topographic map the dashed line that outlines Heney Glacier and the unnamed glaciers that lie east of it on the map. You may wish to complete the tracing over the entire map, or at least over that part including the main tributary glacier system to Heney Glacier.

(6) Using a different color (green or red), trace the drainage divide that surrounds Heney Glacier and the two glaciers to the east of it. This requires careful concentration. A pencil line, lightly drawn, is recommended first, so as to permit easy correction.

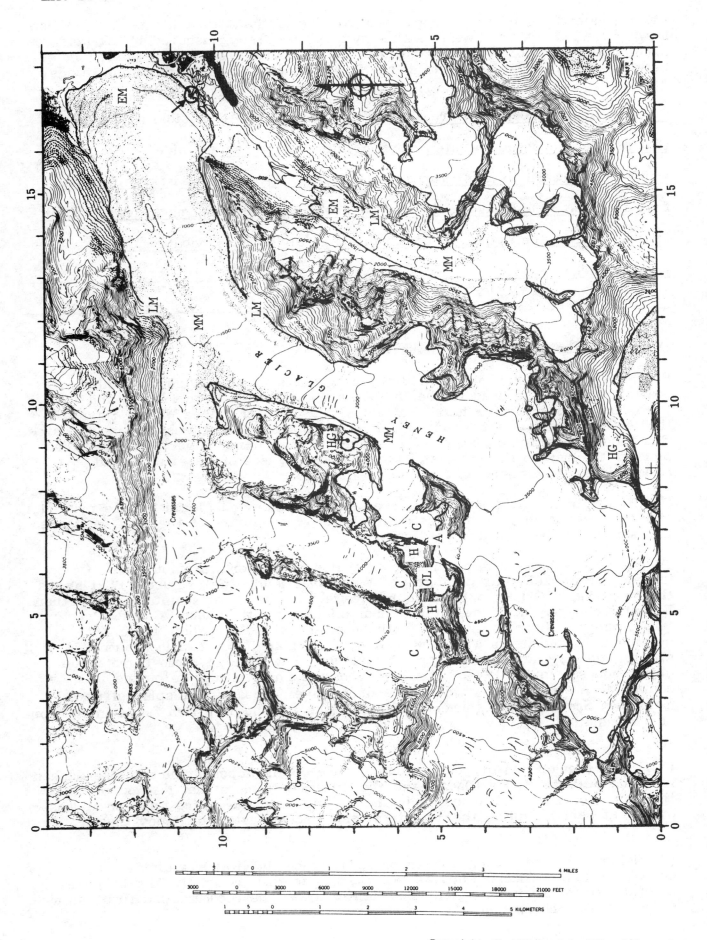

Figure B Eagle Glacier, British Columbia. (USAF Trimetrogon photo. Canada Department of Mines, No. 99471.)

(7) Three exceptionally fine large cirques lie at the upper end of Heney Glacier, between 2.0-1.5 and 4.0-4.0. Label these cirques with the letter **C**. Label three other cirques that lie along a line from 3.0-6.0 and 6.0-6.0. These belong to north-flowing glaciers tributary to the western branch of Heney Glacier.

(8) Along the divide between the two sets of cirques of Question 7, find and label good examples of a <u>horn</u> (**H**), an <u>arête</u> (**A**), and a <u>col</u> (**CL**). (An arête is the sharp, knifelike divide between two cirques; a col is a low point on an arête).

Figure A Portion of the Cordova (D-3), AK, Quadrangle. U.S. Geological Survey. Original scale: 1:63,360.

(9) Study the divide between 11-2 and 17-3 that separates the north-flowing glacier from cirques that face south. Why are the north-facing cirques full of glacial firn and ice, whereas the south-facing cirques are empty?

<u>The north-facing cirque walls are in shadow during much of the day in summer,</u>

<u>allowing firn to accumulate. South-facing cirque walls are exposed to direct</u>

<u>sunlight, melting the winter snowfall and preventing firn accumulation.</u>

(10) Follow Heney Glacier downvalley from its cirques to its terminus, noting the spacing of the contours that cross the glacier. Look for changes in gradient. What is the significance of closely crowded contours and associated crevasses (short lines)? Give specific locations by grid coordinates. Explain. (Refer to text Figure 18.2.)

<u>Crowding of contours and numerous crevasses occur conspicuously at two places:</u>

<u>4.5-2.5 and 10.5-7.5. These suggest the presence of an ice-fall over a rock</u>

<u>step.</u>

(11) Medial moraines are illustrated and labeled in text Figure 18.5B. Two of them are beautifully shown in the text photo, Figure 18.1. Find medial moraines on the map near 9.0-6.0 and 13.0-5.0. Label these **MM** and color them to show their full extent and source points.

(12) Lateral moraines are illustrated and labeled in text figures 18.2 and 18.5B. Find two good examples on the map. Give grid coordinates and label them **LM**. Color them, same as for the medial moraines.

Grid coordinates: <u>12.0-9.0</u>   and   <u>14.0-6.0</u>

(13) An end moraine is illustrated and labeled in text Figure 18.2. Find two good examples on the map. Give grid coordinates, label them **EM**, and color them, same as for the other kinds of moraines.

Grid coordinates: <u>15.0-7.0</u>   and   <u>17.0-12.0</u>

(14) Find a good example of a small glacier, occupying a cirque, but not connected to a branch or trunk of a glacier system. Give grid coordinates. Label it **HG**, for "hanging glacier."

Grid coordinates: <u>9.0-1.0</u>

(15) Locate precisely a point at the end of the Heney Glacier where a subglacial meltwater stream emerges from an ice tunnel. Give grid coordinates. Circle it with a color line.

Grid coordinates: <u>17.2-10.5</u>

Name _____   Date _____

_____   _____

Exercise 18-B   Glacial Landforms of the Rocky Mountains
[Text p. 405-409, Figures 18.5, 18.7.]*

The Middle and Southern Rockies provide many fine examples of landforms of alpine glaciation fully exposed by disappearance of the glaciers that carved them. Two examples that we study in this exercise are from this region; one from the Bighorn Mountains of north-central Wyoming, the other from the Uinta Mountains of northern Utah.

Part A   Cirques, Tarns, and Troughs of the Bighorn Mountains

For this exercise we use another of those remarkably precise block diagrams, drawn by Erwin Raisz, (Figure A) to give a vivid picture of the terrain shown on the contour topographic map, Figure B.

(1) Determine the contour interval of the map. Measure precisely the width of the area it covers.

    Contour interval: ___200___ ft.

    Width of map area: ___11,000___ yds;   ___6.25___ mi.

(2) This is a good contour map on which to practice the art of plastic shading to bring out the visual effect of surface relief. Using a very soft lead pencil, apply shading to steep slopes that face south, southeast, and east. Make the shading darkest where the slope is steepest, as in the headwalls of the cirques. Make a sharp edge on the dark shading to reveal the sharp break between the gently sloping mountain summit upland and the cirque headwall.

Each small lake in a cirque and its trough occupies a rock basin, scoured out by the glacier, although in some cases a moraine may form a dam that holds a lake. Such rock-basin lakes are sometimes called tarns, an ancient Scandinavian term. Figure C shows a splendid rock-basin lake in the Northern Rockies of Glacier National Park. It occupies a cirque with steep walls rising more than 4,000 ft to the surrounding horns and aretes. Chains of tarns, such as those on your map, are sometimes called "paternoster lakes" from their resemblance to beads of a rosary.

Troughs of the Bighorn Mountains look very much like the one shown in text Figure 18.7, which lies in the nearby Beartooth Mountains, a geologically similar uplift with a crystalline rock core into which the glacial features are carved. Notice that the the upland on either side of the trough is a rolling, plateau-like surface. Both regions are described as being in a "youthful" stage of alpine glaciation, as shown in the right-rear corner of Block C of text Figure 18.5.

---

*Modern Physical Geography, 3rd Ed., p. 354-57, Figures 21.4-21.6; Pl. K.2, K.3.

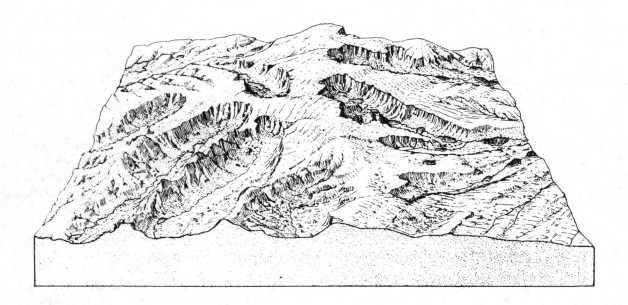

Figure A Block diagram of landforms of the Bighorn Mountains included the topographic map, Figure B. (Drawn by Erwin Raisz.)

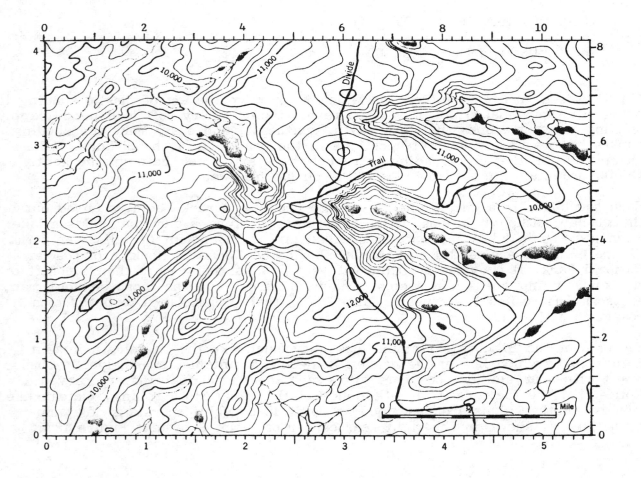

Figure B Portion of the Cloud Peak, WY, Quadrangle. U.S. Geological Survey. Original scale: 1:125,000.

Figure C Lake Ellen Wilson in Glacier National Park, Montana. (Douglas Johnson.)

(3) Using a color pencil or pen, trace on the map (Figure B) the main drainage divide of the mountain range. Label it "Divide." Explain why cirques heading on the east side of the range have eaten back closer to the main divide than those heading on the west side of the range. You may need to speculate as to the cause. (Hint: Think about (a) winds and (b) insolation.)

In middle latitudes, glacial activity is more intense on the east side of a north-south mountain range. This effect occurs for two reasons, both relating to snow accumulation: (a) Prevailing westerly winds blow snow off the upland to fall on lee, east-facing slopes. (b) Maximum daily air temperature occurs in the afternoon hours, when west-facing slopes are also under maximum intensity of solar rays; thus promoting faster melting of snow.

(4) Using a color pencil, lay out on the map the shortest possible route for a trail across the mountain range from 5.5-2.3 to 0.0-1.5 in such a way that the gradient nowhere exceeds 1,600 ft per mi. Avoid all cirques and troughs.

Part B   Cirques, Arêtes, and Horns of the Uinta Mountains

The example of glaciated mountains in the Unita range, shown in Figures D, E, and F is strikingly different from that in the Bighorn Range. Like the Bighorn range, the Uinta range is a great arch, or anticline, in which sedimentary strata were uplifted in the Laramide Revolution that closed the Mesozoic Era. The top of the arch is flattened. Whereas the Bighorn arch has been eroded to the point that a large crystalline rock core is exposed, the Uinta arch still retains strata that pass over the summit. These strata are quartzite formations of Precambrian age. Weaker formations within the quartzite layers allowed high, steep cliffs of quartzite to retreat as the cirques enlarged, much as the Redwall and Coconino cliffs have retreated in the walls of Grand Canyon. Figure F is a very old photograph showing the high, steep quartzite cliffs surrounding a cirque with its rock-basin lake. Notice the succession of cliffs and ledges. (This cirque is just off the top of your map.)

(5) Write the word <u>cirque</u> on several good examples on the map. Label examples of <u>arêtes</u>, <u>horns</u>, and <u>cols</u>.

(6) Color blue all areas on the map that you would suppose were covered by glacial ice at a stage of maximum glaciation. Add numerous arrows to show directions of ice movement. In another color, draw in all medial moraines that you might expect to have existed on the glaciers.

(7) Summarize the major differences between the Uinta glacial landforms and those of the Bighorn range.

The Uinta range is in an advanced stage of glaciation in which the cirques have met back-to-back in narrow arêtes and sharply peaked horns. No extensive upland plateau surface remains, as it does in the Bighorn range. Furthermore, the control exerted by the flat-lying quartzite strata has exaggerated the height and steepness of the cirque walls.

Figure D Block diagram of landforms of the Uinta Mountains included in the topographic map, Figure B. (Drawn by Erwin Raisz.)

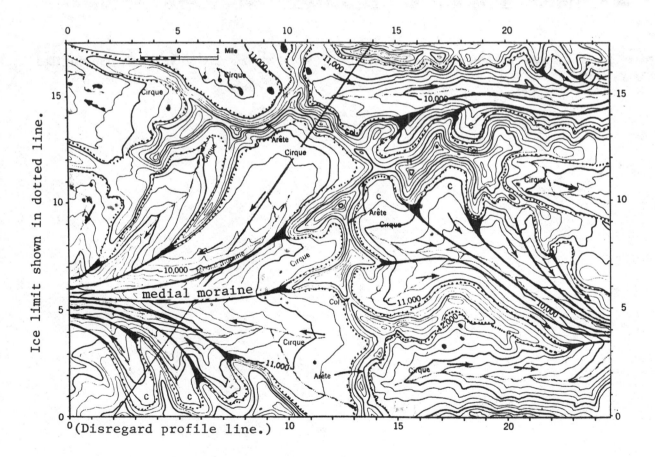

Figure E Portion of the Hayden Peak, UT, Quadrangle. U.S. Geological Survey. Original scale 1:125,000. North is to the right.

Figure F A cirque in the Uinta Mountains. (Wallace W. Atwood, U.S. Geological Survey, 1909.

Name _____   Date _____

_____   _____

Exercise 18-C   A Terminal Moraine and Its Lake
        [Text p. 408-409, Figures 18.2, 18.8.]*

The debris that an alpine glacier transports to its terminus accumulates in depositional landforms of two kinds. It may simply be dropped down on the melting ice surface to accumulate as a terminal moraine (end moraine), or it may be picked up and carried by braided meltwater streams to form a valley-train alluvial sheet that steadily raises the level of the valley.

Text Figure 18.8 shows a massive terminal moraine of a former glacier that originated high on the steep eastern face of the Sierra Nevada of California. Looking closely, you can see that the moraine ridge at the left, which runs back into the glacial trough as a lateral moraine, is double-crested. The inner ridge is easily traced on a U-turn across the valley mouth and up the opposite side. Evidently the recession of the ice front was temporarily halted to produce the secondary moraine ridge.

A terminal moraine of rather unique form is shown on the topographic contour map, Figure A. The region is the Southern Rockies of Colorado. Cirques heading on the east side of the Continental Divide (1.0-2.0) fed a trunk glacier that flowed due east in what is now the valley of Lake Fork, a well-scoured U-shaped glacial trough. Upon emerging from this trough, the glacier advanced over the level floodplain of the upper Arkansas River (15-5). Here, the glacier spread out into an expanded-foot shape, semicircular in outline. For a long time it held this position, its end moraine building continuously to form a great circular barrier. Retreat must have been rapid, because there is little sign of moraine in its trough floor or walls.

(1) About how high is the moraine ridge?   How wide?

        Height: _200_   ft.   Width: _0.9_   mi.

(2) Does Turquoise Lake have an outlet? If so, give its location.

Lake outlet is to the south, at 11.5-0.5. "Gaging Station" is located on the outlet

stream._____

---

*Modern Physical Geography, 3rd Ed., p. 356, Figure 21.5.

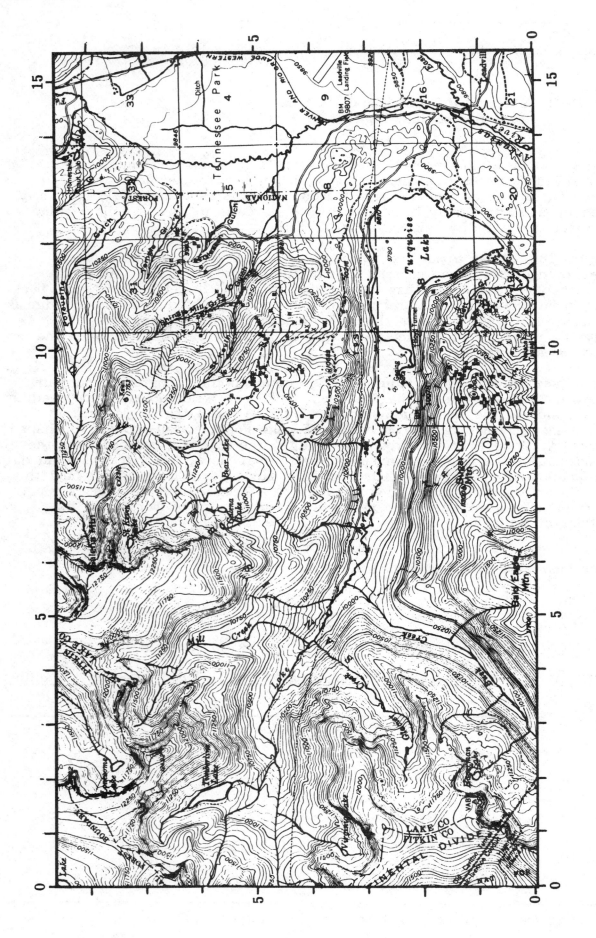

Figure A Portion of Holy Cross, CO, Quadrangle. U.S. Geological Survey. Original scale 1:62,500.

(3) The surface of the end moraine appears irregular and bumpy with what appear to be a few small surface depressions. How do you explain these features?

The surface irregularities are probably the result of melting of ice blocks within moraine left behind as the glacier gradually wasted back. Collapse of the overyling glacial till produced pits and hollows.

(4) What is the meaning of the numerous small squares and Y-shaped symbols scattered over the mountain slopes to the north and south of Turquoise Lake? Make a guess. Guess as to what kind of bedrock underlies most of this area.

These are symbols for mine shafts and mine pits. These are probably mines or prospect pits for a metallic ore, such as lead, silver, tungsten, or gold. (Note such names as "Galena Mountain" and "Leadville.") The bedrock is probably very old igneous or metamorphic rock with a complex geologic history. No indications of layering or bedding that would be a sign of sedimentary strata or lava beds. [Geologic map of Colorado shows both granite and schist/gneiss, Precambrian age, at this location.]

Name _____ Date _____

_____ _____

Exercise 18-D Moraines and Outwash Plains
         [Text p. 415-18, Figure 18.21.]*

Over much of the northcentral and northeastern United States, and extending far north into Canada, landforms shaped by the great Wisconsinan ice sheets and the meltwaters streaming from them are today scarcely touched by weathering and mass wasing and the eroding forces of runoff. But humans have heavily impacted these landforms in and around our urban areas--excavating them and regrading them for highways and shopping malls. Try to find a good exposure of a glacial delta to visit on a field trip and you will soon get the point. The beautifully washed and graded sand of the delta is now elsewhere, bound in concrete. Outwash plains, also deposits of sand and gravel, are mostly concealed under houses, lawns, and pavements of a thousand housing developments.

The total complex of landforms found near what was formerly the terminal or recessional belt of an ice sheet is laid out in the two block diagrams of text Figure 18.21. To understand how these landforms take expression in topographic contours, we use a map of an area along the Wisconsin-Minnesota boundary.

(1) Study the map, Figure A, alongside Diagram B of text Figure 18.21. Look closely at the lower left end of the block and the cross section it bears. Match the various kinds of landforms with their locations on the map. Label examples each of the following on the map:

         Moraine   Kettle lake
         Knob-and-kettle   Ground moraine
         Outwash plain   Drumlin
         Ice-block depression   Basal glacial till

(2) Compare the surface elevation of the outwash plain close to the moraine with that of the surface just north of the moraine. Explain the relationship in elevations of these two surfaces.

Elevation of the outwash plain is over 1,200 ft (perhaps as high as 1,230 or

1,240 ft), while the surface north of the moraine is between 1,000 and 1,050 ft.

This is a difference of about 200 ft. The outwash plain was being built while the

ice occupied the area to the north. Thickness of the outwash can be estimated at

about 200 ft.

_____

_____

*Modern Physical Geography, 3rd Ed., p. 363-67, Figure 21.19.

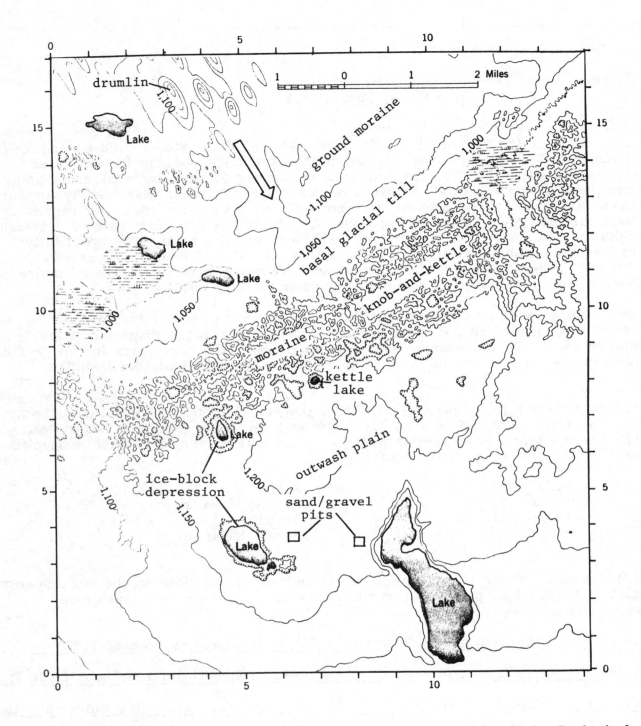

Figure A Portion of the St. Croix Dalles, WI-MN, Quadrangle. U.S. Geological Survey. Original scale 1:62,500.

(3) How can you explain the presence of ice-block depressions (kettles) in the outwash plain lying well south of the stagnant ice margin, where the moraine was being deposited?

The ice must have previously advanced beyond the position of the moraine, and then wasted back rapidly. Either this was a minor advance and retreat, or the retreat was part of the general ice recession from its terminal position far to the south.

(4) Draw a bold arrow on the northern part of the map to show the direction of ice movement. What two kinds of evidence suggest the flow direction? Are they in agreement?

Flow direction is given quite clearly by the long axes of the drumlins at 4-16. This direction is also at right angles to trend of the moraine, which is an independent indicator of direction.

(5) What is the origin of the basins occupied by lakes and marshes north of the moraine?

These are either rock basins scoured by the ice, or depressions resulting from uneven deposition of basal glacial till, or a combination of both. Note the absence of steep scarps surrounding these basins.

(6) Study the group of small hills and depressions lying between 0-14 and 4-14. How do you interpret them in terms of features of glaciation?

They appear to be part of a weakly developed morainal ridge. It would have been formed during a brief pause in the ice-front recession.

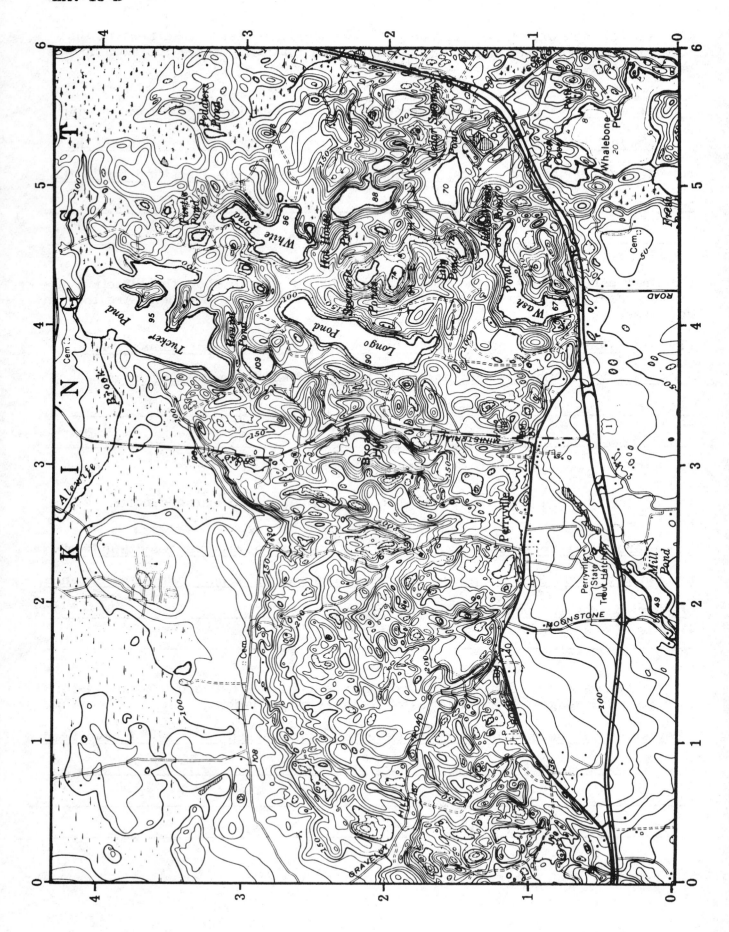

(7) If you planned to excavate sand and gravel for commercial use in the production of concrete, where on this map would you locate the pit? Indicate by one or more squares labeled "sand/gravel pit." Explain your choice.

Thick sand and gravel layers would be found beneath the outwash plain. The texture would probably become finer with increasing distance southward, away from the former ice margin.

Figure B is a topographic map showing a belt of knob-and kettle moraine in southern Rhode Island. The map scale is 1:24,000. The contour interval is 10 ft. Bedrock here consists of intrusive igneous rocks of late Paleozoic age. Low hills north and south of the moraine are formed of heavily abraded bedrock masked by a layer of basal glacial till.

(8) A striking feature of the moraine belt is that lakes (ponds) are almost completely absent in the western half of the map, but numerous and large in the eastern half. Offer an explanation for this concentration of lakes. Take into consideration that the ocean shoreline (Long Island Sound) lies only a mile away to the south of the map border, and that a large estuary lies within a mile of the eastern map boundary. The water body in the extreme southeast corner of the map is a salt-water bay.

The lakes may reflect the level of the water table. The eastern depressions have lower floors than those in the west. Thus the bottoms of the depressions in the east are well below the water table level; those in the west lie well above it. Note that water levels in the ponds are highest at the north, declining quite persistently toward the south and southeast.

Figure B Portion of the Kingston, RI, Quadrangle. U.S. Geological Survey.

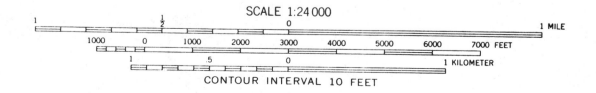

SCALE 1:24 000

CONTOUR INTERVAL 10 FEET

Name _____   Date _____

_____   _____

Exercise 18-E   Eskers and Drumlins
[Text p. 417-18, Figures 18.21, 18.23, 18.24.]*

This exercise brings together two very odd landforms of unlike origins--eskers and drumlins--although both originate beneath an ice sheet. Eskers are stream deposits in subglacial tunnels in stagnant ice; drumlins are till deposits plastered over the subglacial floor by moving ice. It stands to reason that eskers and drumlins cannot be formed in the same place at the same time, but it would be possible for an esker to run right over a drumlin, if the esker formed after the moving ice had become completely stagnant.

Eskers

The word "esker" is said to be of Gaelic-Irish origin, and you will find it in their literature spelled also as eskir, eskar, excar, and eiscir. Sweden has many fine eskers, which the Swedes call "osars." A more descriptive name used in New England is "horseback," and indeed, the crest of an esker would provide a fine horse trail well above adjacent bug-ridden swamps and muskeg--all the better because the crest of an esker is typically free of forest trees. Canada must have more eskers than any other nation; they are found from Labrador on the east to Great Bear Lake on the west, forming fanlike patterns that show clearly the flow directions within the great ice sheets.

Figure A is an oblique air photo of an esker in the Canadian shield. It is an old photo, but they don't come much better. It shows the heavily glaciated Precambrian crystalline rocks with their numerous ice-scoured basins. Notice that the esker crosses the lake basins as well as the hill summits between them, almost as if dropped at random across the terrain. Actually, eskers form on or close to the rock floor and the water in the ice tunnel can flow uphill as well as downhill, depositing gravel as it goes. Eskers show thinning and thickening of their height and width, looking a bit like a python after a big meal. The esker crest in this photo appears highly reflective of light, because of the scarcity of trees on the crest.

Map A in Figure B is a contour topographic map of an esker in Maine, known locally as the Enfield Horseback. The contour interval is 20 ft.

(1) Estimate the width of the esker just north of the Passadumkeag River. How high is it above the surrounding plain?

   Width: __300__ to __400__ yds. Height: __60__ to __80__ ft.

_____

*Modern Physical Geography, 3rd Ed., p. 365, Figures 21.19, 21.23, 21.24; Pl. K.4.

Figure A An esker near Boyd Lake, Canada. (Canada Dept. of Mines, Geological Survey, Atlas 126.)

(2) Why does the esker crest rise and fall in elevation along its length?

The thickness of gravel deposit on the ice tunnel floor would vary from point to

point.

_____

(3) Explain the rise in elevation of the esker crest in the vicinity of 3.3-6.0.

At this point, the esker appears to have been deposited over a hill of bedrock or

till, similar to that shown near 4.4-3.0.

_____

(4) Explain the very small closed depressions at 2.3-13.9.

These are small ice-block depressions. Small bodies of ice were probably buried

under the esker gravel and subsequently melted.

_____

Drumlins

The term <u>drumlin</u> is from Ireland, where it is used to describe various kinds of rounded hills, and has been adopted internationally as a scientific term for a subglacial deposit of till shaped by the ice flow into a streamlined form. County Down in northeast Ireland contains nearly 4,000 bonafide drumlins. The drumlin in text Figure 18.24 shows what is often considered the ideal drumlin form. (Read the figure caption.) New York claims the largest American drumlin field, with some 10,000 being counted along the south shore of Lake Ontario. To the north of that lake, the province of Ontario has four important drumlin fields. Drumlins of a great field in east-central Wisconsin number about 5,000, and there are many in New England and Nova Scotia, as well.

From the viewpoint of American history, however, Boston's drumlins have the highest distinction. Two that lie close to the historic heart of the city are Breeds Hill and Bunker Hill. In May of 1775, colonial militia had laid seige to the British forces in Boston. General Howe then arrived with reinforcements with the intention of breaking the seige, and with rumored plans of taking the heights of Charlestown. To counter this threat, the Continental commander, William Prescott, was sent to fortify Bunker Hill. Instead, he chose the other drumlin, Breeds Hill, as his position and there in June fought a furious and bloody battle before being forced to withdraw. But, as history would have it, this was the Battle of Bunker Hill.

Boston Harbor has many fine drumlins, several of which are shown in Map B of Figure B. They are more rotund in outline than the classic examples from farther inland, and you will need to explain that feature.

(5) Measure the length, width, and height of the drumlin at 1.2-2.7.

Length: <u>2,100</u> ft. Width: <u>1,200</u> ft. Height: <u>110</u> ft.

(6) Using the long axes of the drumlins as an directional indicator of ice movements, compare the directions indicated by the drumlins in the larger map (Nantasket Beach, Telegraph Hill, etc.) and those on Paddocks Island. Paddocks Island actually lies only a short distance due west of Telegraph Hill. Offer an explanation of the differences in direction you observe.

<u>The first group shows a direction close to west-to-east (azimuth 90°). Of those on Paddocks Island, the southernmost also trends west-east, but the northeastern two are oriented northwest-to-southeast (azimuth 135°). Ice flow might have been locally deflected in direction by hard-rock knobs on the bedrock.</u>

(7) Drumlins of western New York State are shown in Map C of Figure B. How many drumlins are shown on this map?

Number of drumlins: <u>65</u> [count varies]

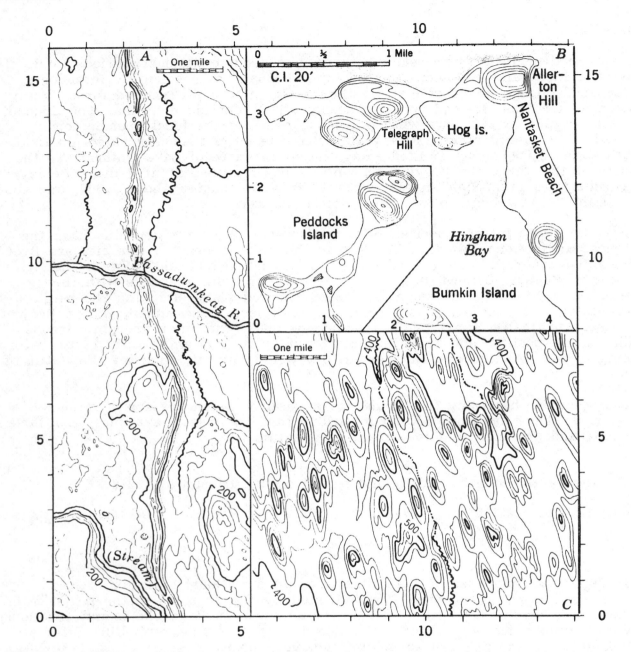

Figure B Portions of topographic maps of the U.S. Geological Survey. (A) Passadumkeag, ME, Quadrangle. Original scale 1:62,500. (B) Hull, MA, Quadrangle. Original scale 1:31,680. (C) Weedsport, NY, Quadrangle. Original scale 1:62,500.

(8) Estimate the direction (azimuth) of ice flow indicated by this group of drumlins.

Azimuth: __165__ °

(9) The drumlins of Map B are somewhat differently shaped than those on Map C. Describe and explain this difference.

Drumlins of Map B are more rotund (stubby); those of Map C are elongate, with

long tails. The greater elongation may reflect more rapid ice movement combined

Figure C Vertical air photograph of drumlins and till grooves near Carp Lake, northern British Columbia. (U.S. Air Force.)

with more nearly uniform direction, because of a smoother bedrock surface

beneath the ice.

_____

_____

Figure C is a vertical air photograph of drumlins near Carp Lake in northern British Columbia. The area shown is about 4 mi wide. The drumlin forms are side by side with long narrow grooves and ridges, also formed in glacial till. Some of the drumlins show narrow grooves on their surfaces. In some areas of Canada the land surface consists entirely of straight, parallel grooves with intervening narrow ridges, and is called a "fluted till surface."

(10) Using a color pen, outline several well-shaped drumlins. Draw lines to emphasize a set of well-developed parallel grooves.

Exercise 19-A   Soil Textures
            [Text p. 427-28, Figures 9.3, 9.4, 9.5.]*

Did you ever wonder what makes those "Idaho" baking potatoes so wonderfully smooth and uniformly shaped? Careful plant breeding by agronomists, perhaps, or do the growers just select the best ones for market? Could the secret lie in a remarkable kind of soil in which they are grown? Wheat and potatoes are important agricultural exports of the Palouse region that lies in a part of the Columbia Plateau Province where Washington, Oregon, and Idaho meet.

Looking up the soil order and suborder of the Palouse region on the soil map of the United States (see Exercise 19-C, Map A), we learn that these soils are Xerolls, the suborder of Mollisols that is characteristic of the Mediterranean type of soil-water budget. These are naturally fertile soils with a loose texture, and that in itself could at least partly explain the fine potatoes. Xerolls, however, cover a much larger area in that part of the U.S. than just the potato region. Further research uncovers descriptions of two local soils: Palouse Soils and Walla Walla Soils. The U.S. Department of Agriculture text states that the parent matter of these soils is "largely loess (floury wind-blown material," and that "silt loam is the predominant texture."

Probing further, we consult a map of the aeolian deposits (dune sand and loess) of the United States. Over the Palouse region there is indeed a layer of loess; it covers 100 percent of the surface. It is Wisconsinan in age and ranges in thickness from 4 to 16 feet. Typical American loess of that age consists of about 90 percent silt and about 5 percent each of sand (mostly very fine sand) and clay. On the uplands, this silt fell directly from the turbid air of dust storms, while torrential winter rains washed some of it into the valley bottoms, making the Walla Walla Soils. Here, where irrigation is feasible during the dry, hot summers, the remarkable potatoes are grown. The tubers are free to grow to perfection in the soft silt that surrounds them. And don't forget the huge, sweet Walla Walla onions!

Other soil textures can be unfavorable to agriculture, and even make the soil next-to-impossible to cultivate. Vertisols, rich in clays that swell and become sticky upon being soaked with water, also harden, shrink, and crack upon drying out. Over the centuries, vertisols of Africa and India have resisted the primitive plows drawn by humans and animals, yet their natural soil fertility is extremely high.

Modern soil science is rigorous in its precise measurements of the physical and chemical properties of soils. Our exercise illustrates how this quantitative approach is applied to the textures of parent materials of the soil.

---

*Modern Physical Geography, 3rd Ed., p. 382, Figures 22.2, 22.3.

Soil Texture Grades

Soil texture grades as defined by the U.S. Department of Agriculture are given in Table A. This information agrees with that of text Figure 19.3, but contains additional subdivisions of the sand and gravel grades.

(1) On the blank graph below, locate the size limits of each grade. Draw a vertical line at each grade limit from bottom to top of the graph. At the top of the graph, label each vertical line with the diameter in millimeters. Then label each size grade at the top of the graph, using the following letter code. (These are listed in order from finest to coarsest, plotting from left to right on the graph scale.)

Code	Grade name
C	Clay
S	Silt
VFS	Very find sand
FS	Fine sand
MS	Medium sand
CS	Coarse sand
FG	Fine gravel
CG	Coarse gravel and larger

In the soil science laboratory, samples are passed through a series of sieves in order to measure the percentage of each grade. Particles larger than 2.0 mm are not not included in determining the texture class of the soil. Keep this in mind as we turn later to the subject of texture classes.

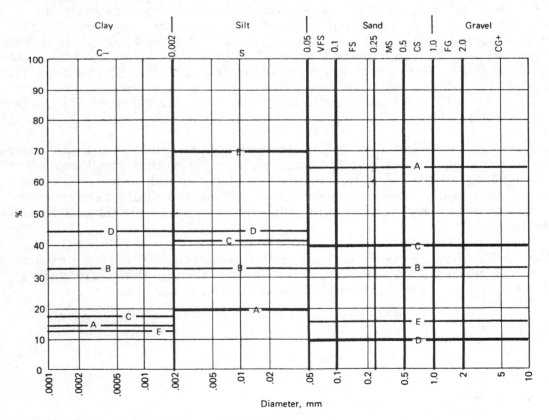

Diameter, mm

Graph A

Table A   Soil Texture Grades

Name of grade	Diameter (mm)
Coarse gravel	Above 2.0
Fine gravel	1.0-2.0
Coarse sand	0.5-1.0
Medium sand	0.25-0.5
Fine sand	0.1-0.25
Very fine sand	0.05-0.1
Silt	0.002-0.05
Clay	Below 0.002 (2.0 microns)
Colloidal Clay	Below 0.0001 (0.01 micron.

[Data source: U. S. Department of Agriculture.]

(2) Examine the horizontal scale (x-axis) used on the graph. What name is applied to a scale of this kind?

Logarithmic scale (or constant-ratio scale).

(3) Fine sand, medium sand, coarse sand, and fine gravel are separated by the following diameters: 0.25, 0.5, 1.0, and 2.0. What mathematical term is applied to a succession of numbers related in this manner? What ratio of increase holds constant in this number sucession?

This is a geometrical progression. The ratio or increase is 2, meaning that the

values are doubled successively.

(4) Does a zero-value point exist on the horizontal scale (x-axis), toward the left, assuming that we extended the graph as far as necessary in that direction? How far can the scale extend to the right? Explain.

There is no zero point. The scale can be extended to without limit in both

directions, with the limit approaching infinity toward the right.

(5) Refer back to Exercise 11-D, Part B, The Wentworth Scale of Particle Grades, Table A. Compare the USDA boundaries with those of the Wentworth scale. (If you have already completed Exercise 11-D, you may adapt the answer to Question 4 here.)

Both systems use a scale of doubling values, but limits differ between some of the grades. The upper limit for sand is the same for both, namely 1.0 mm, but larger grains are given the name of "pebbles" in the W scale. The silt/sand boundary on the W scale is 0.0625, rather than 0.05. The W scale silt/clay boundary is set at 0.004 mm, whereas it is 0.002 in the USDA system. Colloids on the W scale set in below 0.24 microns, but below 0.01 microns in the USDA scale.

(We will return to Graph A to insert additional data.)

Soil Texture Classes

Soil texture classes make use of the scale of texture grades, combining the percentages of three grades: sand, silt, clay. Text Figure 19.4 shows the components of five of the classes, using pie-diagrams. We reproduce this illustration here as Figure A.

Figure B enlarges on this concept to cover all texture classes used by the USDA. It is a triangular graph that simultaneously accommodates three sets of numbers. All possible combinations of three percentages, ranging from 0 to 100 can be plotted as points. We have plotted the five examples of Figure A as labeled points. Check each of these points to see that the percentages are correctly plotted.

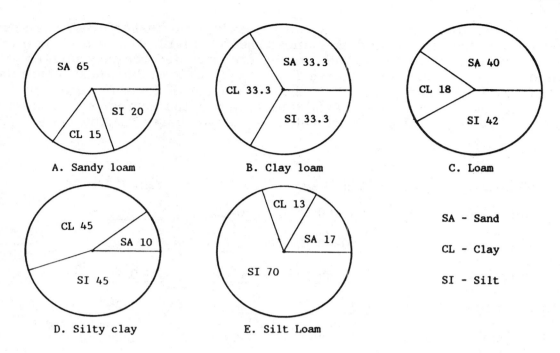

Figure A Pie-diagrams of five examples of soil texture classes.

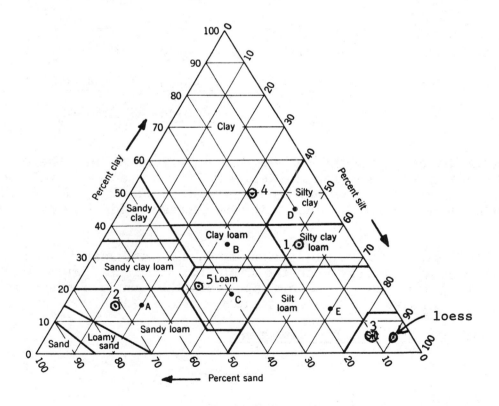

Figure B Soil texture diagram. (From U.S. Department of Agriculture, Soil Survey Staff, 1975, <u>Soil Taxonomy</u>, Handbook No. 436, p. 471, Figure 38.)

(6) Altogether, there are 12 soil texture classes, and these fill all the space on the triangular diagram. Boundaries and names of all 12 classes are shown on the diagram. In the table below, blank spaces are provided for percentages of sand, silt, and clay for each class. Estimate as closely as possible these percentages from the graph. Be sure you follow the correct set of lines for each grade. It will help to hold a clear plastic straightedge along the correct set of lines for each grade, moving it from lowest value to highest value for the class.

Class name	Sand (%)	Silt (%)	Clay (%)
Sand	85-100	0-15	0-10
Loamy sand	70-90	0-30	0-15
Sandy loam	43-85	0-50	0-20
Sandy clay loam	45-80	0-28	20-35
Loam (example)	43-52	28-50	7-27
Silt loam	0-50	50-87	0-27
Silt	0-20	80-100	0-12
Silty clay loam	0-20	40-72	27-40
Clay loam	20-45	15-52	27-40
Sandy clay	45-65	0-20	36-55
Silty clay	0-20	40-60	40-60
Clay	0-45	0-40	40-100

(7) Using the data from the pie-diagrams in Figure A, plot the percentages of each texture grade on Graph A. Represent the percentage by a horizontal line drawn across the zone covered by the grade. If possible, use a different color for each sample. Label the lines with the corresponding letters A, B, C, etc.

(8) Table B gives the percentages of sand, silt, and clay in each of five soil samples. Locate each sample on the triangular diagram, Figure B. Make a dot, surrounded by a small circle, and label with the sample number. In the spaces below, write the name of the soil texture class.

Sample 1   Silty clay loam

Sample 2   Sandy loam

Sample 3   Silt

Sample 4   Clay

Sample 5   Loam

Table B Texture Classes of Five Samples

Sample No.	Sand		Silt		Clay	
	(%)	(arc)	(%)	(arc)	(%)	(arc)
1	15	54°	51	184°	34	122°
2	72	260°	14	50°	14	50°
3	10	36°	85	306°	5	18°
4	18	65°	32	115°	50	180°
5	47	169°	32	115°	21	76°

(9) Using the blank circles below, construct a pie-diagram of each of the above five samples. Note that beside each percentage in Table B is the arc, in degrees, for each sector in the pie. Use a protractor. Start at the horizontal zero reference radius in the 3-o'clock position, shown on each circle. The sector for sand is laid off anticlockwise from that reference line; the silt sector is laid off clockwise from that line. The clay sector will automatically fall between the sand and silt sectors, but you should measure it with the protractor as a check against error.

(10) Refer back to the third paragraph of this exercise, where the composition of a typical American loess soil is given. Find and label this point on the triangular diagram, Figure B. Name the texture class to which it belongs.

Texture class: Silt

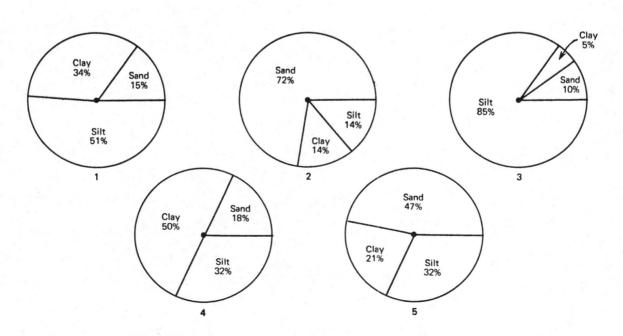

Name _____   Date _____

_____   _____

Exercise 19-B   Soil Suborders and Soil-Water Budgets
         [Text p. 161-62, 431-33, 446, 449-50, Figure 7.13.]*

In presenting the soil-forming processes we emphasized the role of climate in forming distinctive soil horizons (text p. 431-33). For example, a large soil-water surplus promotes eluviation--the carrying down of ions and colloids from an upper soil layer to a lower one, where illuviation takes place (see text Figure 19.9). A large soil-water surplus also results in decalcification. On the other hand, a large annual soil-water deficit allows calcium carbonate to accumulate--the process of calcification. Our conclusion: The soil-water balance is a key to soil development.

Modern soil science goes far beyond using only descriptive words in describing soil-forming processes--it moves into quantitative science that requires field and laboratory measurements of the soil features it investigates. In this exercise we take that challenge to "go quantitative" about soils. Fortunately, this will involve no more than the use of the soil-water budget that you have already mastered in the earlier chapters on climates. The challenge of this exercise is to use those soil-water budgets as a means of defining soil suborders in a hard, quantitative manner.

Two of the ten soil orders have been singled out in your text descriptions as each having important suborders related to differences in climate:

Order	Suborders
Alfisols	Boralfs, Udalfs, Ustalfs, Xeralfs
Mollisols	Borolls, Udolls, Ustolls, Xerolls

Soil scientists recognize five <u>soil-water</u> regimes:

Regime name	Prefix	Latin root	Meaning
Aridic	Aridi-	<u>aridus</u>, dry	arid
Ustic	Ust-	<u>ustus</u>, burnt	seasonally dry
Xeric	Xer-	<u>xeros</u>, dry	dry summer, wet winter
Udic	Ud-	<u>udus</u>, humid	moist
Perudic	Ud-	(same)	very moist
Aquic	Aqu-	<u>aqua</u>,	water saturated

_____

*Modern Physical Geography, 3rd Ed., p. 389.

Besides these five soil-water regimes, a sixth regime based on temperature is required. It is a boreal regime, with the prefix bor, from the Greek boreas, for "northern." The connotation is the same as in our climate system--Boreal forest climate (11)--in which soil water is solidly frozen for several consecutive months.

Names of suborders combine a prefix with a formative element in the name of the order. For the Alfisols, the formative element is the syllable alf, so we have Aqualfs, Ustalfs, Xeralfs, Udalfs, and Boralfs. For the Mollisols, the formative element is oll, giving us Aquolls, Ustolls, Xerolls, Udolls, and Borolls. (There are no aridic Alfisols or Mollisols.) We will not need to be concerned with the aquic regime, because it refers to soils of continuously wet places (bogs, swamps). Don't worry about pronunciation. These terms were all coined in the 1950s and most aren't in even the latest dictionaries.

Figures A and B put these suborders into a schematic form, which may help to visualize the climatic relationships. They use a graph on which the soil-water regimes are put on the vertical scale, the temperature regimes on the horizontal scale. Temperature is also accounted for in computing the soil-water balance (through its influence over potential evapotranspiration). The diagrams should be considered largely pictorial, rather than quantitative.

Table A lists the soil orders and suborders within each regime (aquic excluded), along with their latitudinal ranges and the climate types and subtypes in which each regime may be found.

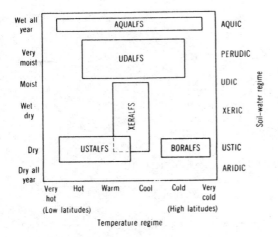

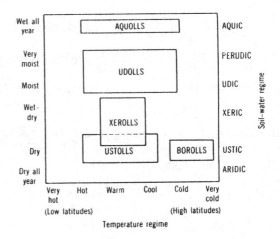

Figure A Suborders of the Alfisols.   **Figure B** Suborders of the Mollisols.

Table A Soil-Water Regimes and Soil Orders

Regime	Soil order	Soil suborder	Latitude range	Climate type and subtype	Base status
ARIDIC	Aridisols		45°N-50°S	4d, 5d, 5sd, 9s, 9sd, 9d	High
	Mollisols	Borolls	45°-55°N	9s, 11s	High
	Alfisols	Boralfs	50°-65°N	11s, 11sh	Moderate
USTIC	Mollisols	Ustolls	20°-50°N,S	4s, 5s, 9s	High
	Alfisols	Ustalfs	35°N-35°S	3, 4s, 9s	Moderate
	(Oxisols)	(Ustox)	10°N-20°S	2, 3	Low
XERIC	Alfisols	Xeralfs	30°-55°N,S	7	Moderate
	Mollisols	Xerolls	40°-55°N	9s	High
UDIC	Alfisols	Udalfs	25°-55°N,S	8h, 10h	Moderate
	Mollisols	Udolls	40°-45°N, 30°-40°S	6h, 10h	High
	Ultisols	(Udults)	40°N-40°S	1, 2, 6h, 10h	Low
	Oxisols		10°N-10°S	1, 2	Low
	Spodosols		40°-70°N	11h, 11sh	Low
PERUDIC	Spodosols		45°-55°N	11p	Low
	Oxisols		10°N-10°S	1, 2	Low

As a basis for illustrating the soil-water regimes, we present a typical soil-water budget for each in Figure C, graphs 1 through 5. You may wish to refer back to text Chapter 7 and to Exericse 7-D, which explain how the soil-water budget is calculated and what the symbols mean. Our examples closely resemble those used with each climate type in text Chapters 8 and 9 and Exercises 8-D and 9-D. We will be referring back to text illustrations of the representative soil-water budgets.

For the five soil-water budgets of Figure C, we use as a reference base two maps of the United States and southern Canada found in Exercise 19-C: Figure A, Soil orders; Figure B, Climate types and subtypes. From here on, we refer to these two maps as "the US soils map" and "the US climate map."

Aridic Regime

The aridic soil-water regime is that of the desert and semidesert climate subtypes and spans the full latitudinal range from low latitude through middle latitudes. Our example is from El Paso, Texas, situated on the Rio Grande. Find the city on both the US soils map and the US climate map. Mark the spot with a small circle and label it "El Paso."

(1) Identify and name the soil order and climate type for El Paso.

Soil order: <u>Aridisols</u>

Climate type: <u>Dry subtropical, desert subtype (5d)</u>

(2) Describe the temperature regime of El Paso in terms of its seasonal range. How is this temperature regime reflected in the annual cycle of potential evapotranspiration (Ep)?

<u>The annual temperature range is very large, about 24C°, with hot summers and</u>

<u>cool winters. The cycle of Ep is similarly one of strong annual range, from very</u>

<u>small Ep in winter to very large in summer. Total Ep is large (96 cm).</u>

(3) Describe the annual cycle of precipitation (P) at El Paso and its relationship to the cycle of Ep. How large is the annual shortage (D)? Is there a water surplus (R)?

<u>Precipitation, annual total 22 cm, is small in all months, but with a small peak in</u>

<u>midsummer. Ten months show a shortage, and the other two months are about</u>

<u>balanced. The result is a very large annual shortage (74 cm) and no surplus.</u>

574

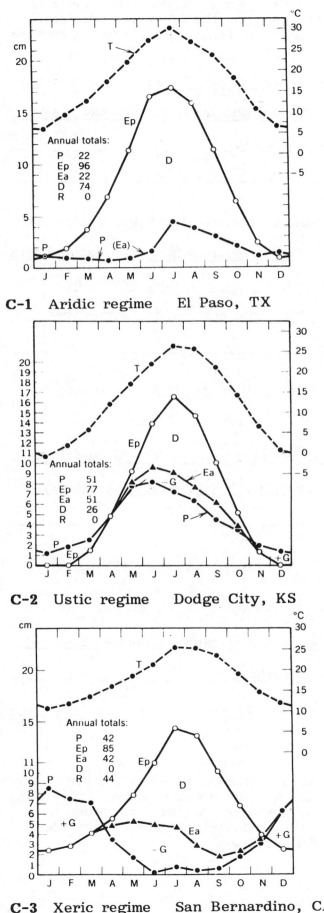

C-1 Aridic regime   El Paso, TX

C-2 Ustic regime   Dodge City, KS

C-3 Xeric regime   San Bernardino, CA

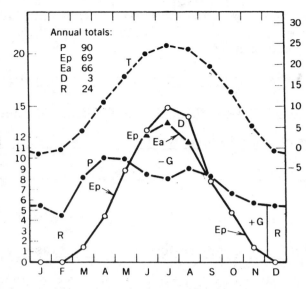

C-4 Udic regime   Urbana, IL

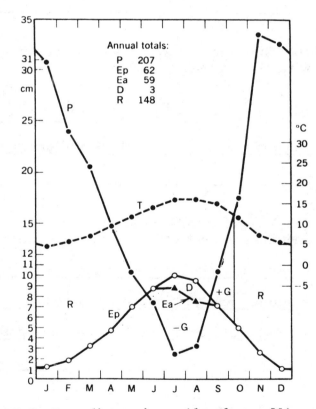

C-5 Perudic regime   Aberdeen, WA

Figure C Soil-water budgets of each of five soil-water regimes.

(4) In text Chapter 9, find a station of similar soil-water regime. Give name of station and climate type. Locate and label the place on the US soils and climate maps. Compare its water budget data with that of El Paso. Is the same soil order present here?

Parker, AZ, (Figure 9.2). Very similar soil-water budget. Parker has a greater Ep maximum and annual range, and a greater total shortage. Parker has a small recharge in two winter months. Both have Aridisols.

(5) How does the aridic regime affect soil-forming processes and the soil profile?

Extreme calcification occurs because of year-around evaporation potential and lack of soil-water storage. A massive caliche layer would be expected below the surface horizon. In poorly drained areas, salts would accumulate (salinization).

(6) Find another aridic soil-water budget. It is in text Chapter 8. Compare it with the American examples.

Khartoum, Sudan, Figure 8.18. This station has about double the Ep and D of the American stations. It is uniformly hot through the year with no cool season.

(7) Is agriculture possible under the aridic soil-water regime? Is fertility of Aridisols a problem?

Crop cultivation is not possible without irrigation. Well drained Aridisols of suitable (loamy) texture are extremely fertile when irrigated. Example, Imperial Valley, California. Salinization is a potential problem.

Ustic Regime

The ustic soil-water regime in midlatitudes is intermediate between the aridic and udic regimes. Although found in a dry climate with no water surplus, soil water is available from precipitation during a growing season, so that cultivation of cereals (dry farming of wheat) is possible without irrigation. The midlatitude ustic regime is illustrated by Dodge City, KS, Figure C-2.

(8) Locate and label Dodge City on the US soils and climate maps. With what soil order, suborder, and climate type is this area associated? In what sense is this regime a transition (intermediate) between the aridic and udic soil-water regimes? (Examine Figure C-4.)

Soil order: <u>Mollisols</u>   Suborder: <u>Ustolls</u>

Climate type and subtype: <u>Dry midlatitude, semiarid subtype (9s)</u>

<u>It has a substantial summer shortage (D) that is intermediate between that of the</u>

<u>aridic and ustic regimes. It has a short winter season with monthly surpluses in</u>

<u>four consecutive months, but lacks the surplus of the udic regime.</u>

(9) Find in text Chapter 9 a similar soil-water budget. How do the two differ? Is there a difference in the soil order and suborder?

<u>Medicine Hat, Alberta, Figure 9.21, is very similar in form, but values for P and</u>

<u>Ep are much smaller in the Alberta example because of its much more northerly</u>

<u>latitude and colder winters. Both stations lie in areas of Mollisols, but for</u>

<u>Medicine Hat the suborder is Borolls.</u>

The ustic regime also applies to certain low-latitude climates, particularly the wet-dry tropical climate (3) and adjacent areas of the semiarid subtype of the tropical dry climate (4s). These strongly seasonal climates have a large shortage (D), and either a very small surplus (R) or no surplus at all. As an example, we use text Figure 8.12, Raipur, India. Another is Kaduna, Nigeria, Ex. 8-D(b). These were compared in that exercise and found to be very similar.

(10) In what important ways does the budget of Raipur differ from that of Dodge City?

<u>Seasons of shortage and recharge are out of phase in these two budgets. In</u>

<u>Raipur, the large shortage is mostly generated in the dry season of low sun</u>

<u>(Nov-Mar), which at Dodge City is the season of recharge (+G). Raipur has a</u>

<u>substantial surplus (R); Dodge City has none. Raipur's surplus is generated in</u>

<u>the high-sun season because two months show a very large precipitation, far</u>

<u>exceeding water need (Ep), whereas at Dodge City, Ep remains far in excess of</u>

<u>P throughout the summer.</u>

Xeric Regime

The xeric regime is associated with the dry climate subtypes (7s, 7sd) of the Mediterranean climate, which in the US are confined to the coastal ranges and valleys of California. However, the "Mediterranean effect" of a sharp reduction in summer precipitation persists in the precipitation cycle of all parts the northwestern states that lie west of the Rocky Mountains. For that reason, soil scientists extend the xeric regime into eastern Washington and Oregon, southern Idaho, and western Utah. Here the dry suborder of the Mollisols is classed as Xerolls (M5 on the US soils map). For our example, however, we choose to stick with the California xeric regime, which has a long, severe summer drought. We select San Bernardino, a city due east of Los Angeles. It is located on a low plain at the foot of the San Bernardo Mountains.

(11) Locate and label San Bernardino on both US maps. What soil order and suborder is indicated as likely to be present? In what climate is it located? (Note: San Bernardino lies outide the map area designated as E2S2, Entisols.)

Soil order: Alfisols, Xeralfs

Climate: Mediterranean, semiarid subtype (7s)

(12) What is the most important difference in regimes between San Bernardino and those of El Paso and Dodge City? Why is it an important difference?

In the Mediterranean regime of San Bernardino, the cycles of Ep and P are exactly 180° out of phase, meaning that the peak in Ep of the one matches the minimum of P in the other, and vice versa. In the aridic and ustic regimes, Ep and P are matched in phase; i.e., peak Ep coincides with peak P.

(13) What effect on soil development can you predict from the unique cycles of Ep and P in the xeric (Mediterranean) regime?

Maximum P, as rain, falls in winter, when soil temperatures are low. Since this is the recharge period, leaching (illuviation) is intensified. In summer, with very low or no P, evaporation of soil water is intensified, keeping the soil dry for several months, during which chemical reactions are minimized in the soil.

(14) Find in text Chapter 9 a water budget similar to that of San Bernardino. Compare it with that of San Bernardino.

Los Angeles (Figure 9.11) has a budget almost identical in all respects. The two stations are very close together.

(15) Find in Exercise 9-D a similar soil-water budget. What soil order and suborder is found at that location?

Sevilla, Spain, has a similar xeric soil-water budget. Xeralfs are shown on the world soils map (Figure 19.11)

Udic Regime

The udic soil-water regime is characterized by a moderate to substantial water surplus (R), despite a short season of shortage (D) during which soil water is withdrawn. This description fits climates ranging from the equatorial zone to the arctic zone. It is best to recognize two quite different regions of the udic regime. One is the low-latitude belt represented by two climates: wet equatorial (1) and monsoon and tradewind littoral (2). The other is found in midlatitude and high-latitude climates and includes the following climates: moist subtropical (6h), marine west-coast (7h), moist continental (10h), and boreal forest (11h). All in the second group are strongly seasonal, with large annual ranges in Ep, peaking strongly in the summer and falling to low or zero values in winter. The list of soil orders and suborders is also a long one, with five orders represented (see Table A). Our udic regime example is Urbana, IL. in the heartland of the Middle West.

(16) Find and label Urbana on the US maps. With what soil order, suborder, and climate is this station associated?

      Soil order: __Mollisols__   Soil suborder: __Udolls__

      Climate type: __Moist continental (12h)__

(17) Describe the budget of Urbana with special emphasis on its deficit and surplus.

Because of abundant rainfall in the summer, the shortage is very small. The surplus is quite large (considering that the location is in the western part of the climate zone). Much of the surplus is held in the frozen state until March, when spring thaw sets in.

(18) Find a similar budget in text Chapter 9. Compare it with Urbana's budget. Pittsburgh, PA, Figure 9.27. These budgets are almost identical in every way. One difference is that the precipitation for Urbana shows a much more marked winter minimum than Pittsburgh.

(19) Do both Urbana and Pittsburgh have soils of the same soil order? If not, explain why not. (Hint: Examine text Figures 18.15 and 18.17.)

The US soils map shows the Pittsburgh area as having Ultisols, whereas for Urbana they are Mollisols (Udolls). Both have a udic regime. Presence of thick loess and hotter summer temperatures may help explain the development of the Udolls in eastern Illinois. Also, the plateau surrounding Pittsburgh was never glaciated, whereas eastern Illinois was covered by the continental ice sheet in the Wisconsinan Epoch and subsequently acquired a very young loess layer. The Ultisols are much older, giving time for development of the horizons of eluviation and illuviation.

(20) The udic soil-water regime of the equatorial and tropical zones looks very different on the graphs from those of midlatitudes. Aparri, Luzon, text Figure 8.5, is a good illustration. Point out the important differences between the soil-water regime of Aparri and those of Urbana and Pittsburgh.

Aparri has no winter and is warm all year, therefore low values of Ep are lacking. Aparri's precipitation cycle is much more strongly developed and is not in the same phase as the US examples. Therefore, Aparri's shortage (D) occurs just after spring equinox (Mar-Apr-May), rather than in midsummer. Aparri's surplus, R, is much greater, but this is not a crucial difference.

(21) What soil orders are associated with the udic regime in low latitudes?

Ultisols (of the suborder of Udults). Some Oxisols are in a climate with a short dry season. (Most Oxisols are in the perudic regime.)

Perudic Regime

The perudic soil-water regime is characterized by a very large water surplus (R), often exceeding 100 cm. There is a very small shortage (D), or none. This regime is important in both low latitudes and the latitudes of the midlatitude zone. Obviously, the perudic regime is largely concident with the wet equatorial climate (1), which occupies vast areas of equatorial South America, Africa, and the East Indies. Very different is the perudic regime of coastal zones and islands both north and south of latitude 40°, where it coincides with the perhumid climate subtypes, 8h and 10h. In the western Pacific, the perudic regime extends to lower latitudes in Japan, China, and Taiwan, where the moist subtropical climate, perhumid subtype (6p), is present.

Our American example is Aberdeen, Washington, a coastal city on the 47th parallel. The graph of Aberdeen is rendered oversize and rather grotesque by the very high monthly precipitation values in winter months.

(22) Locate and label Aberdeen on the US soils and climate maps. Identify the climate type and subtype.

Climate type and subtype: Marine west-coast, perhumid (8p).

(23) No perudic example of the marine-west coast climate is found in your textbook, so compare Aberdeen with Cork, Ireland, text Figure 9.17, representing the humid subtype of the same climate. How are they similar and how do they differ?

The budgets are very similar in form, differing mostly in only one important

respect: the extremely great monthly precipitation amounts in winter in

Aberdeen. This extreme precipitation drives up the surplus (R) to 148 cm,

nearly four times greater than at Cork.

(24) Where else in North America is the perudic regime found? What is the climate type and subtype associated with these locations?

Southeastern Quebec and eastern Labrador, the Gaspé, Nova Scotia, Prince

Edward Island, Newfoundland. Perhumid subtypes of two climates are present:

moist continental (10p) and boreal forest (11p).

(25) What soil-forming processes and soil orders would you expect to find in the perhumid regimes of North America?

<u>Eluviation and illuviation would be strongly in evidence, producing spodosols in well-drained sandy parent materials. Histosols would also be present in bog environments.</u>

We turn next to examples of the perudic regime in the wet equatorial climate.

(26) Describe the soil-water budget of Singapore, text Figure 8.2. In what respect does it qualify for the perudic regime? How does it differ from the budget of Aberdeen?

<u>The large surplus, 69 cm, and lack of any shortage (D) qualify it as perudic. The main difference between this and Aberdeen is that water need (Ep) of Singapore has uniform monthly values throughout the year, along with uniformly warm temperatures.</u>

(27) What soil order or orders would you expect to find near Singapore? Comment on the fertility of these soils.

<u>Either Oxisols or Ultisols (udic suborder, Udults). They would be of very low fertility in terms of food crops, lacking in important nutrient bases.</u>

Exercise 19-C   Soil Orders and Climate
    [Text p. 434-51, Figures 19.10, 19.11, 19.12.]*

Physical geography brings together its several subject areas in meaningful ways. Your study of world climates didn't just end with the last page of text Chapter 9. Climates were referred to many times thereafter in the chapters on landforms, because climate influences the processes that shape landforms. Now we reach back again to world climates, this time for relationships between soil orders and climate types.

We begin by relating the soil profiles of text Figure 19.12 to the climate type of the place from which each was taken. For this purpose, we offer three special maps, none of which is in your textbook. Maps A and B show the United states and the adjacent part of Canada lying south of the 51st parallel. Map A shows soil orders and some of the most important suborders. Map B shows climates of the same area, using the same boundaries and climate types as in text Chapter 7 and Figure 7.9. These two maps will enable you to locate places much more closely in terms of state and province boundaries, and other reference features such as rivers and lakes. Map C is a detailed world soils map showing suborders and their various combinations.

NOTE: Retain maps A, B, and C for use in exercises 20-A and 20-B.

---

*Modern Physical Geography, 3rd Ed., p. 394-401; Pl. L.1, L.2, L.4.

(1) Taking the soil profile photographs text Figure 19.12 in order from A through K, find the location of each on the US-Canada soils map, Map A, and enter the identifying profile letter on the map. Be sure the letter falls within the map area of the correct soil order or suborder. For foreign countries, use the world soils map, Map C. Find each locality on the US-Canada climate map, Map B, or the world climate map, text Figure 7.9. Enter the climate numbers in the blank spaces below.

Profile	Soil Order	Soil Suborder	Climate Type & Subtype Number Name
A	Oxisols	Torrox*	1 Wet equatorial
B	Ultisols	Udult*	10h Moist continental (humid)
C	Vertisols	Ustert*	5 Dry subtropical
D	Alfisols	Udalf	10h Moist continental (humid)
E	Alfisols	Ustalf	9sd Dry midlatitude (semidesert)
F	Spodosols	Orthod*	8h Marine west-coast (humid)
G	Mollisols	Boroll	11s Boreal forest (semiarid)
H	Mollisols	Udoll	6sh Moist subtropical (humid)
I	Mollisols	Ustoll	9s Dry midlatitude (semiarid)
J	Mollisols	Rendoll*	9s Dry midlatitude (semiarid)
K	Aridisols	Argid*	5d Dry subtropical (desert)

*You are not required to know these suborders, as they are not shown on your soils maps and are not explained in your textbook.

Oxisols

(2) Carefully compare the Oxisol profile of text Figure 19.12A with that in Figure 19.13. In what way are the two soil profiles similar?

These are photographs of the same soil exposure. Every detail in the close-up profile is matched in the more distant view, including the large boulder on the surface and the boulder on which the man is standing.

_____

(3) On the world soils map, text Figure 19.11, determine the latitudinal range of the Oxisols. What is their northern limit; their southern limit? Name the locality in which each limit lies.

Northern limit: 22°N   Place: Hawaiian Islands

Southern limit: 25°S   Place: Madagascar (Malagasy)

Ultisols

(4) Compare the Oxisol profile (A) with that of the Ultisol (B). What is the most important difference between the two? Before answering, examine also text Figure 19.14 and read the figure legend.

The Ultisols have a distinct horizon of eluviation (E horizon) below the A

horizon. [Note that the E horizon is called the A2 horizon in earlier versions.] It

is pale in color. Below it is a dense B horizon of illuviation, called the argillic

horizon. Both of these horizons are lacking in the Oxisols.

(5) Find and describe the plinthite zone in text Figures 19.12B and 19.14. At what depth below the surface does the plinthite first appear?

The plinthite zone is mottled in appearance. It begins at a depth of about 120 cm

(48 in.) in both profiles. [Note different scales in the two photographs.]

(6) On the world soils map, text Figure 19.11, determine the northern and southern latitudinal limits of the Ultisols. Compare your findings with those for the latitude limits of the Oxisols.

    Northern limit: 41°N   Place: New Jersey, USA

    Southern limit: 44°S   Place: Chile

The Oxisols have a much smaller range, limited to the equatorial and tropical

latitude zones. The Ultisols extend into the lower part of the midlatitude

zone.

(7) Following up on Question 6, it is obvious from the world soils map that by far the largest area of Ultisols is in low latitudes, i.e., in the equatorial, tropical, and subtropical zones. Using the world climate map, text Figure 7.9, list below the low-latitude climates in which important areas of Ultisols are found.

Wet equatorial (1), Monsoon and trade-wind littoral (2), Tropical wet-dry

(3).

Vertisols

(8) Study the surface details of the Vertisol profile in text Figures 19.12C and 19.15. In your opinion, are horizons clearly visible as horizontal layers? What minor features of soil structure are clearly shown in these profiles? What is the origin of these structures?

Horizons are not identifiable by color or structure. However, slick (shiny) surfaces that are nearly vertical are clearly shown. They resemble shingles on a very steep roof. These slickensides are the result of vertical slippage movements in the soil.

(9) The accompanying schematic drawing, Figure A, shows how Vertisols evolve by the alternate widening and closing of deep soil cracks. During this annual cycle of change, the soil "swallows itself." In the space below the drawing, label the following: Dry season; Wet season. Three stages are represented by three cracks, each undergoing a different process. Below each of the three, enter a description of what is taking place. (Make three columns.)

Alfisols

As explained in your textbook (p. 446), the Alfisols are subdivided into four suborders, each representing a different soil-water regime: Boralfs, Udalfs, Ustalfs, and Xeralfs. The world soils map, text Figure 19.11, shows the distributions of these suborders. Map C also shows the suborders. (Exercise 19-B developed the soil-water regimes for these suborders.)

(10) Boralfs are not shown in photos in your textbook, but they are a major suborder in terms of extent. Find the areas of Boralfs on the world soils map, text Figure 19.11. Describe the areas in terms of latitude and continental location. What soil order lies adjacent to the Boralfs on the low-latitude side?

All are in the northern hemisphere; mostly north of 50°. In western Canada they occupy a large part of Saskatchewan and Alberta. In Eurasia, they form an almost continuous belt between 50° and 65° from the Baltic shores to eastern Siberia. Mollisols lie along the southern boundary in all these areas.

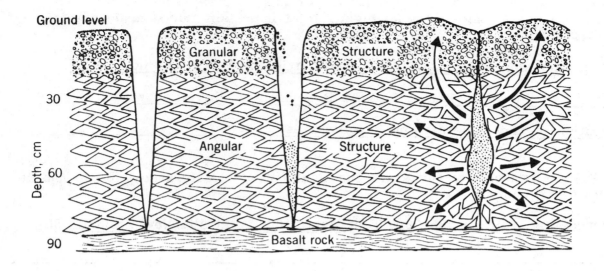

Ground level

Depth, cm

30

60

90

Granular   Structure

Angular   Structure

Basalt rock

Stage A	Stage B	Stage C
Dry season		Wet Season
Soil crack opens	Soil particles	Rains soak soil.
wide, to depth	fall into crack,	Soil swells, expands,
of 80 cm.	filling it.	pushes soil outward
		and upward.

Figure A Schematic diagram of the effects of seasonal shrinking and swelling of the soil to produce the structure of the Vertisols. (From H.D. Foth, Fundamentals of Soil Science, p. 281, Figure 10-23. Copyright © 1984 by John Wiley & Sons, New York. Used by permission.)

(11) Compare the areas of Boralfs with the climate map, Figure 7.9. With what climate types are they closely associated? (Give number/letter codes.) Are these rated as moist or dry climates?

Canada: 11h and 11s. Russia: 10sh. Siberia: 11s, 11sh. Most are continental

climates of semiarid or subhumid subtypes.

(12) Udalfs, the moist climate suborder of Alfisols, are represented by text Figure 19.12D, an example from Michigan. What important horizons does this profile display?

The ochric epipedon is marked "AP." Below it is the E (or A2) horizon of eluviation, pale in color. Below that the B horizon of illuviation with concentration of clay minerals and sesquioxides.

(13) With what two climate types are the Udalfs most closely associated in middle latitudes? (Give codes.)

N. America: 10h. Europe: 8h, 8p. East Asia: 10h. Australia/New Zealand: 8h, 8sh, 8p.

(14) Ustalfs are shown in text Figures 19.12E and 19.19. Attach a piece of tracing paper over each of these photos. Draw the rectangular outline. On each, draw lines to separate the soil horizons. Label the A, E (A2), and B horizons. (Draw a line to show the soil surface.) Construct and label depth scales on both profiles, starting with zero at the surface. In Figure 19.12E, take the length of the spade handle as 30 cm. In Figure 19.19, the marks on the measuring tape are set at intervals of 15 cm). Transfer the completed tracing sheet to the space provided below.

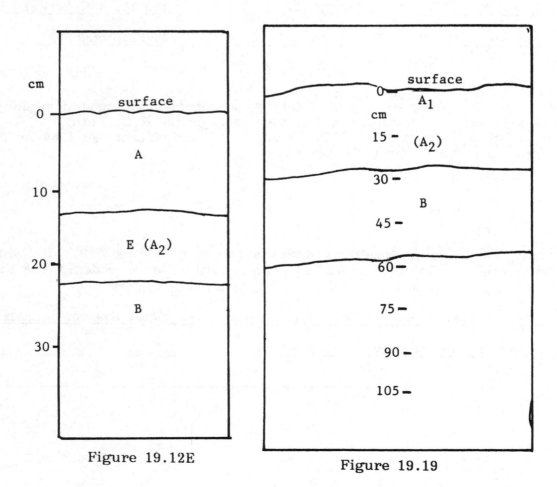

Figure 19.12E

Figure 19.19

(15) Study the distribution of the Ustalfs as shown on the world soils map, text Figure 19.11. (a) In what latitude zones do the large areas lie? (b) Name the countries or continental regions in which these large areas occur (from east to west, from northern to southern hemisphere). (c) Name the climate types found in these regions.

(a) Latitude zones: equatorial, tropical, subtropical

(b) Countries or continental regions: India/Pakistan/Burma, Sahel of Africa, Tanzania/Kenya, Brazilian eastern highlands, Angola/Zambia/Mozambique, northern Australia.

(c) Climate types: Wet-dry tropical (3), Dry tropical, semiarid and semidesert (4s, 4sd).

(16) Xeralfs are Alfisols of the Xeric soil-water regime, found in the regions of Mediterranean climate (7). Find these regions on the world soils map. List them by name as you did for the Ustalfs.

California and southern Oregon, Spain, Italy, Yugoslavia, Chile, South Africa, southern Western Australia.

Spodosols and Histosols

Spodosols are of wide extent in the northern hemisphere. Here, Histosols are found as isolated patches of bog or muck within the Spodosol areas. The two orders are commonly combined as one unit on the soils map. Spodosols and Boralfs also occur as a mixture of patches within the same area.

(17) Identify by geographical names the major areas of Spodosols shown on the world soils map, text Figure 19.11. Give the approximate latitudinal range of each.

Northern Manitoba/Saskatchewan (55-60°), Ontario/Quebec/Newfoundland (45-55°), Netherlands/Germany/Denmark (50-55°), Sweden/Finland/USSR (60-70°).

(18) With what climate type or types are these Spodosol regions associated?

Boreal forest climate (11s, 11sh, 11h), northern parts of moist continental climate (10h).

Mollisols

As explained in your textbook, p. 448-50, the Mollisols include four important suborders: Borolls, Udolls, Ustolls, and Xerolls. A fifth suborder, the Rendolls, are of much less importance on a world scale, they occur locally on limestones or other forms of carbonate parent matter. The example, Figure 19.12J, is from Argentina. We will concentrate on the four widespread orders. They are shown as separate areas on Maps A and C. We will limit our investigation of the Mollisols to North America.

(19) (a) Where are the Borolls found? Name states and provinces. Give the latitudinal range. (b) With what climate types and subtypes are the Borolls associated? (Use Figure B.)

(a) Southern Alberta, Saskatchewan, and Manitoba; Montana, North Dakota, western Minnesota, northern South Dakota. Latitude range: 43-51°.

(b) Northern part of Dry midlatitude climate, semiarid and subhumid subtypes (9s, 9sh). Southern part of Boreal forest climate, semiarid subtype (11s).

(20) For the Udolls, give same information as for Question 19.

(a) Southern Minnesota, Iowa, northern Missouri, eastern Kansas, northeastern Oklahoma, Illinois, and small areas in NW Indiana and southern Wisconsin; floodplains of Mississippi and Ohio rivers. Latitudinal range: 35-45°.

(b) Moist continental climate, humid and subhumid subtypes (10h, 10sh).

(21) For the Ustolls, give the same information as for Question 19.

(a) South Dakota, Nebraska, Kansas, Oklahoma, northern and central Texas, northwestern Colorado, southeastern New Mexico, southern edge of Colorado Plateau Province in Arizona and New mexico. Latitudinal range: 28-46°.

(b) Dry midlatitude climate, semiarid subtype (9s).

(22) For the Xerolls, give the same information as for Question 19.

(a) Eastern Washington and Oregon, northern California and Nevada, southern Idaho, central and eastern Utah. Latitude range: 36-49°.

(b) Dry midlatitude climate, semiarid subtype (9s).

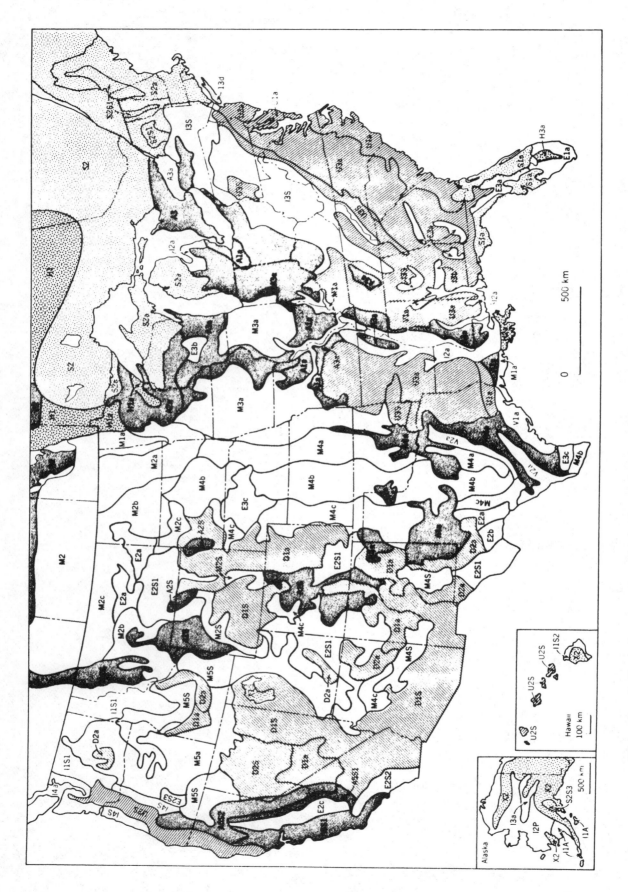

Map A Soil orders and suborders of the United States and southern Canada. Legend on facing page. (Soil Conservation Service, U.S. Department of Agriculture.)

ALFISOLS

AQUALFS
A1a—Aqualfs with Udalfs, Haplaquepts, Udolls: gently sloping.

BORALFS
A2a—Boralfs with Udipsamments and Histosols: gently and moderately sloping.
A2S—Cryoboralfs with Borolls, Cryochrepts, Cryorthods and rock outcrops: steep.

UDALFS
A3a—Udalfs with Aqualfs, Aquolls, Rendolls, Udolls. and Udults; gently or moderately sloping.

USTALFS
A4a—Ustalfs with Ustochrepts, Ustolls, Usterts, Ustipsamments, and Ustorthents; gently or moderately sloping.

XERALFS
A5S1—Xeralfs with Xerolls, Xerorthents, and Xererts: moderately sloping to steep.
A5S2—Ultic and lithic subgroups of Haploxeralfs with Andepts, Xerults, Xerolls, and Xerochrepts; steep.

ARIDISOLS

ARGIDS
D1a—Argids with Orthids, Orthents, Psamments, and Ustolls; gently and moderately sloping.
D1S—Argids with Orthids, gently sloping; and Torriorthents, gently sloping to steep.

ORTHIDS
D2a—Orthids with Argids, Orthents, and Xerolls; gently or moderately sloping.
D2S—Orthids, gently sloping to steep, with Argids, gently sloping; lithic subgroups of Torriorthents and Xerorthents, both steep.

ENTISOLS

AQUENTS
E1a—Aquents with Quartzipsamments, Aquepts, Aquolls, and Aquods; gently sloping.

ORTHENTS
E2a—Torriorthents, steep, with borollic subgroups of Aridisols; Usterts and aridic and vertic subgroups of Borolls; gently or moderately sloping.
E2b—Torriorthents with Torrerts; gently or moderately sloping.
E2c—Xerorthents with Xeralfs, Orthids, and Argids, gently sloping.
E2S1—Torriorthents; steep, and Argids, Torrifluvents, Ustolls, and Borolls; gently sloping.
E2S2—Xerorthents with Xeralfs and Xerolls; steep.
E2S3—Cryorthents with Cryosamments and Cryandepts; gently sloping to steep.

PSAMMENTS
E3a—Quartzipsamments with Aquults and Udults; gently or moderately sloping.
E3b—Udipsamments with Aquolls and Udalfs; gently or moderately sloping.
E3c—Ustipsamments with Ustalfs and Aquolls; gently or moderately sloping.

HISTOSOLS

HISTOSOLS
H1a—Hemists with Psammaquents and Udipsamments; gently sloping.
H2a—Hemists and Saprists with Fluvaquents and Haplaquepts; gently sloping.
H3a—Fibrists, Hemists, and Saprists with Psammaquents; gently sloping.

INCEPTISOLS

ANDEPTS
I1a—Cryandepts with Cryaquepts, Histosols, and rock land; gently or moderately sloping.
I1S1—Cryandepts with Cryochrepts, Cryumbrepts, and Cryorthods; steep.
I1S2—Andepts with Tropepts, Ustolls, and Tropofolists; moderately sloping to steep.

AQUEPTS
I2a—Haplaquepts with Aqualfs, Aquolls, Udalfs, and Fluvaquents; gently sloping.
I2P—Cryaquepts with cryic great groups of Orthents, Histosols, and Ochrepts; gently sloping to steep.

OCHREPTS
I3a—Cryochrepts with cryic great groups of Aquepts, Histosols, and Orthods; gently or moderately sloping.
I3b—Eutrochrepts with Uderts; gently sloping.
I3c—Fragiochrepts with Fragioquepts, gently or moderately sloping; and Dystrochrepts, steep.
I3d—Dystrochrepts with Udipsamments and Haplorthods; gently sloping.
I3S—Dystrochrepts, steep, with Udalfs and Udults; gently or moderately sloping.

UMBREPTS
I4a—Haplumbrepts with Aquepts and Orthods; gently or moderately sloping.
I4S—Haplumbrepts and Orthods; steep, with Xerolls and Andepts; gently sloping.

MOLLISOLS

AQUOLLS
M1a—Aquolls with Udalfs, Fluvents, Udipsamments, Ustipsamments, Aquepts, Eutrochrepts, and Borolls; gently sloping.

BOROLLS
M2a—Udic subgroups of Borolls with Aquolls and Ustorthents; gently sloping.
M2b—Typic subgroups of Borolls with Ustipsamments, Ustorthents, and Boralfs; gently sloping.
M2c—Aridic subgroups of Borolls with Borollic subgroups of Argids and Orthids, and Torriorthents; gently sloping.
M2S—Borolls with Boralfs, Argids, Torriorthents, and Ustolls; moderately sloping or steep.

UDOLLS
M3a—Udolls, with Aquolls, Udalfs, Aqualfs, Fluvents, Psamments, Ustorthents, Aquepts, and Albolls; gently or moderately sloping.

USTOLLS
M4a—Udic subgroups of Ustolls with Orthents, Ustochrepts, Usterts, Aquents, Fluvents, and Udolls; gently or moderately sloping.
M4b—Typic subgroups of Ustolls with Ustipsamments, Ustorthents, Ustochrepts, Aquolls, and Usterts; gently or moderately sloping.
M4c—Aridic subgroups of Ustolls with Ustalfs, Orthids, Ustipsamments, Ustorthents, Ustochrepts, Torriorthents, Borolls, Ustolls, and Usterts; gently or moderately sloping.
M4S—Ustolls with Argids and Torriorthents; moderately sloping or steep.

XEROLLS
M5a—Xerolls with Argids, Orthids, Fluvents, Cryoborolls, Cryoborolls, and Xerorthents; gently or moderately sloping.
M5S—Xerolls with Cryoboralfs, Xeralfs, Xerorthents, and Xererts; moderately sloping or steep.

SPODOSOLS

AQUODS
S1a—Aquods with Psammaquents, Aquolls, Humods, and Aquults; gently sloping.

ORTHODS
S2a—Orthods with Boralfs, Aquents, Orthents, Psamments, Histosols, Aquepts, Fragiochrepts, and Dystrochrepts; gently or moderately sloping.
S2S1—Orthods with Histosols, Aquents, and Aquepts; moderately sloping or steep.
S2S2—Cryorthods with Histosols; moderately sloping or steep.
S2S3—Cryorthods with Histosols, Andepts and Aquepts; gently sloping to steep.

ULTISOLS

AQUULTS
U1a—Aquults with Aquents, Histosols, Quartzipsamments, Aquepts, Dystrochrepts, and Aquults; gently or moderately sloping.

HUMULTS
U2S—Humults with Andepts, Tropepts, Xerolls, Ustolls, Orthox, Torrox, and rock land; gently sloping to steep.

UDULTS
U3a—Udults with Udalfs, Fluvents, Aquents, Quartzipsamments, Aquepts, Dystrochrepts, and Aquults; gently or moderately sloping.
U3S—Udults with Dystrochrepts; moderately sloping or steep.

VERTISOLS

UDERTS
V1a—Uderts with Aqualfs, Eutrochrepts, Aquolls, and Ustolls; gently sloping.

USTERTS
V2a—Usterts with Aquolls, Orthids, Udifluvents, Aquolls, Ustolls, and Torrerts; gently sloping.

Areas with little soil
X1—Salt flats.
X2—Rock land (plus permanent snow fields and glaciers).

Slope classes
Gently sloping—Slopes mainly less than 10 percent, including nearly level.
Moderately sloping—Slopes mainly between 10 and 25 percent.
Steep—Slopes mainly steeper than 25 percent.

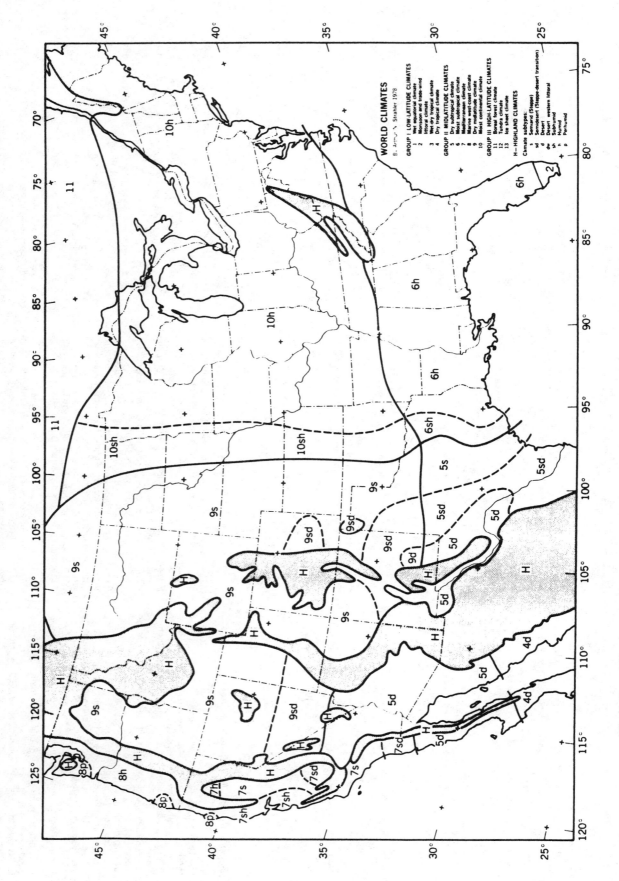

Map B Climate types and subtypes of the contiguous 48 United States and southern Canada. (A.N. Strahler.)

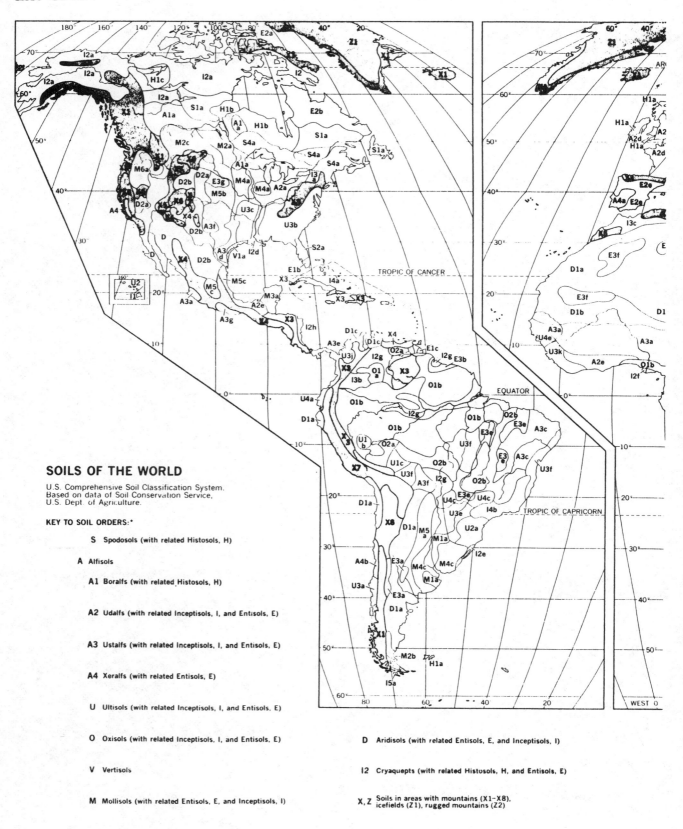

SOILS OF THE WORLD

U.S. Comprehensive Soil Classification System.
Based on data of Soil Conservation Service,
U.S. Dept. of Agriculture.

KEY TO SOIL ORDERS:*

S Spodosols (with related Histosols, H)

A Alfisols

A1 Boralfs (with related Histosols, H)

A2 Udalfs (with related Inceptisols, I, and Entisols, E)

A3 Ustalfs (with related Inceptisols, I, and Entisols, E)

A4 Xeralfs (with related Entisols, E)

U Ultisols (with related Inceptisols, I, and Entisols, E)

O Oxisols (with related Inceptisols, I, and Entisols, E)

V Vertisols

M Mollisols (with related Entisols, E, and Inceptisols, I)

D Aridisols (with related Entisols, E, and Inceptisols, I)

I2 Cryaquepts (with related Histosols, H, and Entisols, E)

X, Z Soils in areas with mountains (X1–X8),
icefields (Z1), rugged mountains (Z2)

Map C Soils of the World.   (Legend on p. 598.)

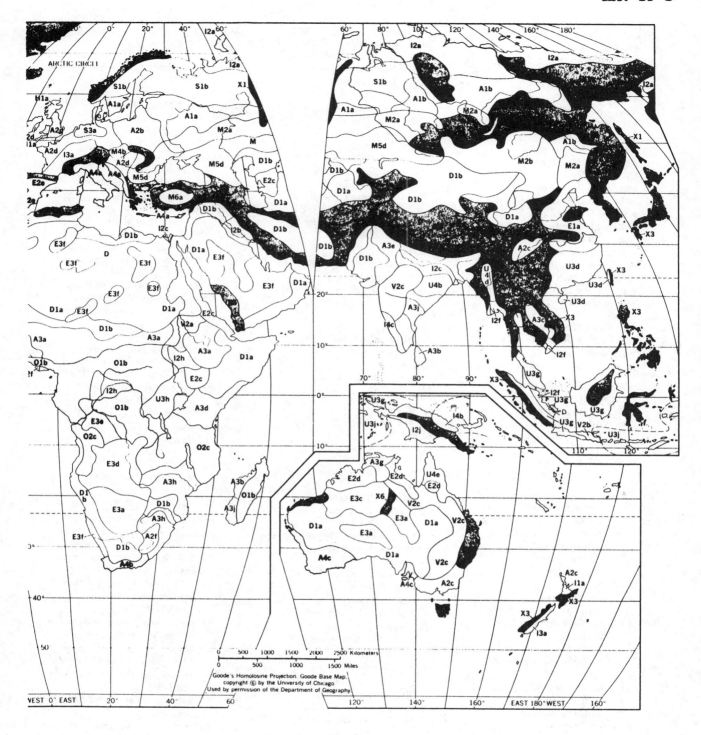

A **Alfisols**
   A1 *Boralfs*
       A1a   with Histosols
       A1b   with Spodosols
   A2 *Udalfs*
       A2a   with Aqualfs
       A2b   with Aquolls
       A2c   with Hapludults
       A2d   with Ochrepts
       A2e   with Troporthents
       A2f   with Udorthents
   A3 *Ustalfs*
       A3a   with Tropepts
       A3b   with Troporthents
       A3c   with Tropustults
       A3d   with Usterts
       A3e   with Ustochrepts
       A3f   with Ustolls
       A3g   with Ustorthents
       A3h   with Ustox
       A3j   Plinthustalfs with
             Ustorthents
   A4 *Xeralfs*
       A4a   with Xerochrepts
       A4b   with Xerorthents
       A4c   with Xerults

D **Aridisols**
   D1 *Aridisols, undifferentiated*
       D1a   with Orthents
       D1b   with Psamments
       D1c   with Ustalfs
   D2 *Argids*
       D2a   with Fluvents
       D2b   with Torriorthents

E **Entisols**
   E1 *Aquents*
       E1a   Haplaquents with
             Udifluvents
       E1b   Psammaquents with
             Haplaquents
       E1c   Tropaquents with
             Hydraquents
   E2 *Orthents*
       E2a   Cryorthents
       E2b   Cryorthents with Orthods
       E2c   Torriorthents with
             Aridisols
       E2d   Torriorthents with Ustalfs
       E2e   Xerorthents with Xeralfs
   E3 *Psamments*
       E3a   with Aridisols
       E3b   with Orthox
       E3c   with Torriorthents
       E3d   with Ustalfs
       E3e   with Ustox
       E3f   of shifting sands
       E3g   Ustipsamments with
             Ustolls

H **Histosols**
   H1 *Histosols, undifferentiated*
       H1a   with Aquods
       H1b   with Boralfs
       H1c   with Cryaquepts

I **Inceptisols**
   I1 *Andepts*
       I1a   Dystrandepts with
             Ochrepts

I2 *Aquepts*
    I2a   Cryaquepts with Orthents
    I2b   Halaquepts with Salorthids
    I2c   Haplaquepts with
          Humaquepts
    I2d   Haplaquepts with
          Ochraqualfs
    I2e   Humaquepts with
          Psamments
    I2f   Tropaquents with
          Hydraquents
    I2g   Tropaquepts with
          Plinthaquults
    I2h   Tropaquepts with
          Tropaquents
    I2j   Tropaquepts with
          Tropudults
I3 *Ochrepts*
    I3a   Dystrochrepts with
          Fragiochrepts
    I3b   Dystrochrepts with Orthox
    I3c   Xerochrepts with Xerolls
I4 *Tropepts*
    I4a   with Ustalfs
    I4b   with Tropudults
    I4c   with Ustox
I5 *Umbrepts*
    I5a   with Aqualfs

M **Mollisols**
   M1 *Albolls*
       M1a   with Aquepts
   M2 *Borolls*
       M2a   with Aquolls
       M2b   with Orthids
       M2c   with Torriorthents
   M3 *Rendolls*
       M3a   with Usterts
   M4 *Udolls*
       M4a   with Aquolls
       M4b   with Eutrochrepts
       M4c   with Humaquepts
   M5 *Ustolls*
       M5a   with Argialbolls
       M5b   with Ustalfs
       M5c   with Usterts
       M5d   with Ustochrepts
   M6 *Xerolls*
       M6a   with Xerorthents

O **Oxisols**
   O1 *Orthox*
       O1a   with Plinthaquults
       O1b   with Tropudults
   O2 *Ustox*
       O2a   with Plinthaquults
       O2b   with Tropustults
       O2c   with Ustalfs

S **Spodosols**
   S1 *Spodosols, undifferentiated*
       S1a   cryic regimes, with Boralfs
       S1b   cryic regimes, with
             Histosols
   S2 *Aquods*
       S2a   Haplaquods with
             Quartzipsamments
   S3 *Humods*
       S3a   with Hapludalfs
   S4 *Orthods*
       S4a   Haplorthods with Boralfs

U **Ultisols**
   U1 *Aquults*
       U1a   Ochraquults with Udults
       U1b   Plinthaquults with Orthox
       U1c   Plinthaquults with
             Plinthaquox
       U1d   Plinthaquults with
             Tropaquepts
   U2 *Humults*
       U2a   with Umbrepts
   U3 *Udults*
       U3a   with Andepts
       U3b   with Dystrochrepts
       U3c   with Udalfs
       U3d   Hapludults with
             Dystrochrepts
       U3e   Rhodudults with Udalfs
       U3f   Tropudults with Aquults
       U3g   Tropudults with
             Hydraquents
       U3h   Tropudults with Orthox
       U3j   Tropudults with Tropepts
       U3k   Tropudults with Tropudalfs
   U4 *Ustults*
       U4a   with Ustochrepts
       U4b   Plinthustults with
             Ustorthents
       U4c   Rhodustults with Ustalfs
       U4d   Tropustults with
             Tropaquepts
       U4e   Tropustults with Ustalfs

V **Vertisols**
   V1 *Uderts*
       V1a   with Usterts
   V2 *Usterts*
       V2a   with Tropaquepts
       V2b   Tropofluvents
       V2c   with Ustalfs

X **Soils in areas with mountains**
   X1   Cryic great groups of Entisols,
        Inceptisols, and Spodosols.
   X2   Boralfs and cryic great groups of
        Entisols and Inceptisols.
   X3   Udic great groups of Alfisols,
        Entisols, and Ultisols;
        Inceptisols.
   X4   Ustic great groups of Alfisols,
        Inceptisols, Mollisols, and
        Ultisols.
   X5   Xeric great groups of Alfisols,
        Entisols, Inceptisols, Mollisols,
        and Ultisols.
   X6   Torric great groups of Entisols;
        Aridisols.
   X7   Histic and cryic great groups of
        Alfisols, Entisols, Inceptisols, and
        Mollisols; ustic great groups of
        Ultisols; cryic great groups of
        Spodosols.
   X8   Aridisols; torric and cryic great
        groups of Entisols, and cryic
        great groups of Spodosols and
        Inceptisols.

Z **Miscellaneous**
   Z1   Ice sheets
   Z2   Rugged mountains, mostly
        devoid of soil (includes glaciers,
        permanent snowfields, and in
        some places, areas of soil.)

Name _____ Date _____

_____ _____

Exercise 20-A   The Forest Biome
          [Text p. 472-79, Figure 20.12.]*

Scarcely a week goes by without news media mention of the rapid destruction of the low-latitude rainforests and their unique ecosystem. Environmental activists among the geographers and ecologists in the highly developed nations of the Anglo-European culture cry out against this devastation, seeing it as an environmental disaster. All well and good, perhaps, but managers of those developing nations where the rainforests stood sometimes see the issue in a different light. They point out that Europeans, whose rapidly expanding population desperately needed food, centuries ago destroyed their own native forests to place the land in cultivation. Under continued pressure, emigrant Europeans then settled North America, destroying the forests and prairies. Have we forgotten the loss of those primeval forest and prairie ecosystems? From Brazil to Indonesia hostility greets the environmentalists who would save the rainforest. Have you a practical solution to this deep-seated conflict?

Physical geography can be thought of as having layers of knowledge, stacked one upon another. One such stack consists of world climates, at the base, world soils upon that, and world vegetation as the top layer. These layers interact, of course, and the idea is more an abstraction than a reality, but it gives order to the way physical geography can be studied. The exercises for Chapter 19 emphasized the relationship of world soils to climate. Now, we turn to the relationship of world vegetation to both soils and climate. Taking on this task is made easier when we concentrate on one biome at a time, and on the formation classes within a biome, one at a time.

Map B (end of exercise) shows details of the world distribution of the formation classes. Refer to it when you need information not available on your textbook map of world vegetation, Figure 20.12. You will also need to use maps A, B, and C of Exercise 19-C.

Low-Latitude Rainforest

(1) Enter the basic requirements of the low-latitude rainforest (including equatorial rainforest and tropical rainforest):

    (a) Climate types and subtypes (number and name): <u>(1) Wet equatorial,</u>

    <u>(2) Monsoon and trade-wind littoral</u> _____

    (b) Soil-water regimes (from Ex. 19-B): <u>udic, perudic</u>

    (c) Soil orders and suborders: <u>Oxisols, Ultisols (Udults)</u>

---

*Modern Physical Geography, 3rd Ed., p. 446-56; Pl. M.2.

(2) From figures in text Chapter 8, select those stations whose soil-water budgets are linked closely to the low-latitude rainforest. Enter below the station name and country. Follow it with corresponding entries selected from the options in (a) (use climate code only), (b), and (c), above. Locate and label each station on the world vegetation map, Map B of this exercise.

Station: __Singapore, Malaysia__   (a) __(1)__

(b) __perudic__   (c) __Oxisols__

Station: __Aparri, P.I.__   (a) __(2)__

(b) __udic__   (c) __Oxisols__

(3) Follow the same instructions as Question 2 for stations shown by climographs of Exercise 8-C and soil-water budgets of Exercise 8-D.

Station: __Cristobal, Panama__   (a) __(1)__

(b) __perudic__   (c) __Oxisols__

Station: __Saigon, Vietnam__   (a) __(2)__

(b) __udic__   (c) __Ultisols__

Station: __Cairns, Australia__   (a) __(2)__

(b) __udic__   (c) __Ultisols__

Station: __Iauarete, Brazil__   (a) __(1)__

(b) __perudic__   (c) __Oxisols__

Monsoon Forest

Monsoon forests of India, Ceylon, Burma, Thailand, and Kampuchea (Cambodia) once held the great teakwood tree, sought after the world over for its fine lumber. Now, most of that resource is gone, and with it the Indian elephant, once widely domesticated in great numbers to handle the teakwood logs. The remaining domesticated elephants now have little to do but appear in fesitvals and public ceremonies. Those remaining in the wild state face extermination at the hands of hunters, despite local laws to curb poaching.

Monsoon forest, also known as rain-green forest, is a tropical-zone formation class developed in certain parts of southeast Asia, Central and South America, Africa, the East Indies, and Australia. On the world vegetation map, text Figure 20.12, it is included within a much larger class called tropical raingreen vegetation, most of which falls in the savanna biome as savanna woodland and thorntree tall-grass savanna. Find monsoon forest on Map B, designated as Fmo.

(4) Enter the basic requirements of the monsoon forest:

(a) Climate types and subtypes: (3) Wet-dry tropical (also bordering

parts of 4s)

(b) Soil-water regimes: ustic

(c) Soil orders and suborders: Ultisols (Ustults) and Alfisols (Ustalfs)

(5) From text Chapter 8 and Exercise 8-C select those stations whose climographs and soil-water budgets are linked closely to the monsoon forest. Enter the station name and country. For each, enter the appropriate options for (a), (b), and (c), above. Locate and label each station on the world vegetation map, Map B of this exercise.

Station: Raipur, India   (a)   (3)

(b)   ustic   (c)   Ultisols (Ustults)

Station: Nagpur, India   (a)   (3)

(b)   ustic   (c)   Ultisols (Ustults)

Subtropical Evergreen Forests

The subtropical evergreen forests vary widely in composition from place to place in different world localities in which they occur. Some are composed broadleaved trees that hold their leaves through the season, others consist largely of needleleaf evergreen trees, particularly the pines, and there are also mixed broadleaf/needleleaf forests. The latitudinal span is mostly within the subtropical zone, but it extends farther poleward in the western Pacific Ocean in both Japan and New Zealand, showing the strong marine influence of that great ocean. Map B shows the distribution of the subtropical evergreen forest, Fbe.

(6) Enter the basic requirements of the subtropical evergreen forest:

(a) Climate types and subtypes: (6h, 6p) moist subtropical, (8h, 8p)

marine west-coast (S. hemisphere only)

(b) Soil-water regimes: udic, perudic

(c) Soil orders and suborders: Ultisols (Udults) and Alfisols (Udalfs)

(7) From text Chapter 9 and Exercise 9-D, select those stations whose climographs and soil-water budgets are linked closely to the subtropical evergreen forest. Enter the station name and country. Follow it with corresponding entries selected from the options in (a), (b), and (c), above. Locate and label each station on the world vegetation map, Map B.

Station: Baton Rouge, LA   (a)   (6h)

(b)   udic   (c)   Ultisols (Udults), Inceptisols

Station: New Plymouth, N.Z.   (a)   (6p)

(b)   udic   (c)   Alfisols (Udalfs) or Ultisols (Udults)

For formation classes in the US and southern Canada, we supplement the world vegetation map with Map A, showing vegetation types of that region. It gives specific information on the composition of forest, grassland, and shrub formation classes.

(8) Examine Map A closely in southeastern region of the United States. On the world vegetation map, text Figure 20.12, this area shows subtropical evergreen forest, with an arrow reading "Southern pine forest." What kinds of forest vegetation are shown here in Map A? With what kind of land surface (upland or marsh land) is each associated?

LLP-longleaf-loblolly-slash pines (southeastern pine forest) covers most of the

area on well-drained upland surfaces. CT--cypress-tupelo-red gum (river bottom

forest) on floodplains of the major rivers.

Note: Text Figure 9.6 shows the Evangeline oak, or "live oak," which is an evergreen species of oak. The accompanying text (p. 194) mentions the magnolia as another broadleaved evergreen characteristic of this Gulf Coast zone. Areas of these species are very small and are not shown on the US-Canada map. Text p. 475 mentions these species, along with trees of the laurel family as comprising a "laurel forest," and states that this forest is also found in southern China and southern Japan. Figure 9.9 illustrates the southeastern pine forest.

Midlatitude Deciduous Forests

The midlatitude decidious forests, often called "hardwood forests," once covered most of North America from the Atlantic coast to the Mississippi valley and from Ontario to Georgia. Much of that area is today cultivated farmland, with isolated timber plots. In contrast, the rugged upland surfaces of the Appalachian mountains and higher plateaus maintain these forests, albeit mostly regrown since settlers cut over the climax forest they found as they spread westward through what was then called "the wilderness."